1987

University of St. Francis
GEN 595.146 L432
Lee, K. E.
Earthworms :

Earthworms

Earthworms
Their Ecology and Relationships with Soils and Land Use

K. E. Lee
CSIRO
Division of Soils
Adelaide

ACADEMIC PRESS
(Harcourt Brace Jovanovich, Publishers)

Sydney Orlando San Diego New York
London Toronto Montreal Tokyo

1985

ACADEMIC PRESS AUSTRALIA
Centrecourt, 25-27 Paul Street North
North Ryde, N.S.W. 2113

United States Edition published by
ACADEMIC PRESS INC.
Orlando, Florida 32887

United Kingdom Edition published by
ACADEMIC PRESS, INC. (LONDON) LTD.
24/28 Oval Road, London NW1 7DX

Copyright © 1985 by
ACADEMIC PRESS AUSTRALIA

All rights reserved. No part of this publication may be
reproduced or transmitted in any form or by any means,
electronic or mechanical, including photocopy, recording,
or any information storage and retrieval system, without
permission in writing from the publisher.

Printed in Australia

National Library of Australia Cataloguing-in-Publication Data

Lee, K. E. (Kenneth Ernest), 1927- .
  Earthworms, their ecology and relationships with
  soils and land use.

  Bibliography.
  Includes index.
  ISBN 0 12 440860 5.

  1. Earthworms - Ecology. I. Title.

595.1'46045

Library of Congress Catalog Card Number: 84-73226

Contents

Foreword ix

Avant-propos xi

Preface xv

Acknowledgements xvii

Part I Ecology

1
Earthworms and Their Environment — 3
   I. The Nature of Earthworms — 3
   II. The Soil Environment — 11
   III. Basic Environmental Requirements of Earthworms — 15
   IV. Food and Digestive Capabilities — 17

2
The Physical Environment — 33
   I. Water Relationships — 33
   II. Temperature — 39
   III. Respiratory Requirements and Adaptations — 47
   IV. Light — 53
   V. Soil Texture — 54

3
The Chemical Environment — 56
   I. Carbon, Nitrogen and C:N Ratio — 56
   II. pH — 58
   III. Electrolyte Concentrations — 60
   IV. Oxidation–Reduction Potentials — 63
   V. Earthworms in Aquatic Habitats — 64

4
Phenology — 67
  I. Life Cycles — 67
  II. Age Structure of Populations — 83
  III. Rhythmic Behaviour Patterns — 84

5
Populations and Associations — 89
  I. Abundance and Biomass — 89
  II. Species Associations — 94
  III. Ecological Strategies — 102
  IV. Models of Earthworm Populations — 131

6
Ecological Energetics — 135
  I. Ecological Energetics of Earthworms — 135

7
Predators, Parasites and Pathogens — 147
  I. Predation on Earthworms — 147
  II. Parasites and Pathogens — 150

8
Dispersal — 154
  I. Dispersal — 154
  II. Peregrine Species — 156

9
Altitudinal Zonation — 166

Part II Relationships with Soils and Land Use

10
Physical Effects on Soils — 173
  I. Casts — 173
  II. Burrows — 182

11

Chemical Effects on Soils	200
I. Contribution to Organic Matter Incorporation and Decomposition	201
II. Nitrogen Transformations Due to Earthworms	211
III. Phosphorus	223
IV. pH and Exchangeable Cations	224
V. Heavy Metals	228
VI. Radioactive Elements	241

12

Earthworms and Pedogenesis	244
I. Effects on Soil Profile Development	244
II. Earthworms and Landscape Processes	251

13

Earthworms and Plant Growth	256
I. Promotion of Plant Growth	256
II. Influence on Soil Fertility and Effects on Plant Production	263
III. Production of Metabolites that Stimulate Plant Growth	270
IV. Production of Other Biologically Active Substances	272
V. Ingestion, Egestion and Viability of Seeds	273
VI. Relationships with Plant Pathogens	274
VII. Earthworms as Plant Pests	276

14

Earthworms and Land Use Practices	278
I. Land Clearance and Revegetation	278
II. Cultivation	282
III. Mulching	286
IV. Irrigation	288
V. Fertilizers	289
VI. Biocides and Earthworms	292

15

Use of Earthworms for Waste Disposal	315
I. Suitable Earthworm Species	316
II. Methods of Waste Disposal	320

16
Earthworms as a Protein Source   328
   I. Nutrient Content and Food Quality of Earthworms   328
   II. Economic Problems   329
   III. Disease Hazards   329
   IV. Heavy Metal and Biocide Toxicity   330

Appendix   Field Sampling Methods   331
   I. Passive Methods   334
   II. Behavioural Methods   339
   III. Mark–Recapture Methods   345
   IV. Counting Casts   346
   V. Summary of Sampling Methods   347
   VI. Biomass Estimation   347

References   351

Author Index   389

Subject Index   399

Foreword

The Foreword was written in French by Dr Marcel B. Bouché; his original text begins on p. xi.

The understanding of natural systems so that they might be better utilized is in *theory* the aim of all basic and applied research in the ecological, agricultural and environmental sciences. In *practice* our work as specialists (because of our individual limitations) aims at a global understanding while concentrating on minute fragments.

And so humanity knows little about its most important commensals: the earthworms. We are unaware of the nocturnal, hidden, subterranean activity of the most important animal biomass that shares with us the earth's land surface. Using increasingly powerful physical and chemical methods, we decide to remodel the landscape, to disturb the soils, to pulverize chemicals, to release fumes and waste water ... ignoring the principal animal that inhabits the environments we alter.

This paradoxical situation, this ecological divorce of humanity from the environment, must undoubtedly have some explanation. Despite the significance of their role, earthworms manage their publicity badly; they are animals with few attractive features, their subterranean habits and labours are unseen, they do no harm to crops.

If we compare, for example, the significance accorded to ornithology and the multitude of birdwatchers studying about one kilogram of birds per hectare, with the extremely limited number of research workers interested in the hundreds of kilograms or tonnes per hectare of earthworms, we must conclude that our knowledge of ecosystems is fundamentally distorted by our above-ground, visual perception of nature and our ignorance of life below-ground.

It would be no more than a regrettable academic oversight that we have not established a major scientific discipline devoted to earthworms if, as a consequence, humanity's powerful influences were not blindly altering a poorly understood environment of limited natural resources, which themselves become dissipated.

x *Foreword*

This is why *the work* that I have the great honour to dedicate *marks an important event* ... an event whose significance will probably become apparent only little by little: it is *the* critical synthesis of knowledge of the role and the uses of earthworms. For about the last quarter of a century, ecological research on earthworms has been developed as the outcome of many specialized projects by a few research workers, from the work of specialists in related fields, or from the work of students who have worked for short periods on the subject. Their results have been published in many languages and also in various reviews, some of which are not readily available. Earthworms occur everywhere, and often the studies have been restricted to local conditions and confined to very specialized aspects of the subject. Some multi-author volumes, some publications of local significance and some compilations of information (often made by authors who are not earthworm specialists) do exist, but no comprehensive overview of high quality has been available.

Now, thanks to the extensive work that has been done by K. E. Lee over a period of seven years, a homogeneous and masterly work, interpreted and synthesized by one author, provides access to contemporary knowledge of the ecology and usefulness of earthworms. In a presentation that is clear and suited to the various aspects of knowledge of the subject, the author puts at our disposal a book that permits a clear appraisal of an important part of our biological heritage. Henceforth, pedologists, agronomists, ecologists, environmental technologists, biologists ... may no longer ignore a factor that plays an important part in their professional fields.

From this point of view the work marks an important step, because technologists, research workers, decision makers or students, taking advantage of it, will be able to make progress from a new level of knowledge. The problems that arose from our former ignorance were too many to allow the study of earthworms to remain peripheral indefinitely and this book will greatly assist the emergence of a major scientific discipline at the level of the problems that await resolution.

This substantial work could only come from the labour of a research worker with a profound knowledge of his subject, on which he has worked since 1948, and from an authority who is familiar with the many relevant agronomic and environmental disciplines. Moreover, it has necessitated long personal effort to achieve clarification of the essentials, to organize the information critically, and finally to produce an instructive and remarkable work. I believe that only Dr Lee could do it, and the work bears the stamp of his individuality.

In former centuries, beginning in 1789, Gilbert White recognized the role of earthworms as "intestines of the earth" and J. C. Savigny (1826) the diversity of earthworms, but it required all the prestige of Charles Darwin to draw attention to the importance of these animals in the economy of nature. His volume, in 1881*, marks an important step forward, but it has remained isolated as scientific disciplines have evolved towards ever more narrow specialization. Analytical approaches can only poorly integrate knowledge of this animal group, with its pedological and microbiological influence on agronomy: earthworms have the misfortune to be "transdisciplinary". One has only to consider the paucity of ecological knowledge in the general books of Stephenson (1930), Pickford (1937), Avel (1959) and in the work compiled by Stolte (1962-1969) to appreciate how feeble are the echoes of Darwin's message in this field.

The work of K. E. Lee, which for the first time places earthworms on a world-wide scale in the economy of nature and of humanity, takes up again a century later, in modern terms, the message of the great naturalist. By its critical analysis, its synthetic approach and its opening up of all relevant subjects that are accessible to rigorous understanding, this volume of K. E. Lee takes its place as the direct descendant of that of Charles Darwin.

MARCEL B. BOUCHÉ
Secretary, Soil Zoology Committee
International Society of Soil Science
Secretary, Soil Zoology Section
International Union of Biological Sciences

Avant-propos

Comprendre la nature pour mieux l'utiliser, c'est *en théorie* l'objectif de toute recherche scientifique et technique dans les sciences écologiques, agronomiques et de l'environnement. Percevoir cette nature en n'en étudiant qu'une parcelle infime (en raison de nos limites), c'est *notre pratique* quotidienne de spécialiste.

Ainsi l'homme cultivé ignore son premier commensal: les lombriciens. Il ne connaît pas le rôle souterrain, nocturne, discret de la première masse animale qui cohabite avec lui sur les terres émergées. Il décide, avec des moyens physiques et chimiques de plus en plus puissants de modeler le paysage, de retourner les sols, de pulvériser des produits, d'émettre des

---

* Darwin, C.R. (1881). "The Formation of Vegetable Mould through the Action of Worms, with Observations on their Habits". Murray, London.

fumées ou des eaux usées, ... en ignorant le principal animal qu'hébergent les milieux sur lesquels il agit.

Cette situation paradoxale, ce divorce écologique de l'homme avec son milieu, a certes des causes. Même si leur rôle est considérable, les lombriciens soignent mal leur publicité; ce sont des animaux peu attrayants, dont on ne peut observer les moeurs et le travail sous terre, qui ne font pas de dégâts aux cultures.

Que l'on compare par exemple l'importance de l'Ornithologie et sa cohorte d'amateurs étudiant environ un kilogramme d'oiseaux à l'hectare avec l'extrême rareté des chercheurs s'intéressant aux centaines de kilogrammes ou aux tonnes/hectare de lombriciens, et l'on constate que notre connaissance des écosystèmes est nécessairement fort déséquilibrée par notre perception aérienne et visuelle de la nature et notre ignorance du domaine souterrain.

Ce ne serait qu'une carence académique regrettable de ne pas avoir créé une discipline majeure dévolue aux lombriciens, si de ce fait les puissants outils de l'homme n'agissaient pas en aveugle sur un milieu qui nous échappe et si des richesses naturelles limitées n'étaient pas dilapidées.

C'est pourquoi *l'ouvrage* que j'ai le grand honneur de dédicacer *constitue un évènement* ... dont la portée n'apparaîtra probablement que peu à peu: il est *la* synthèse critique sur les rôles et usages des lombriciens. Depuis un quart de siècle environ, la recherche écologique sur les lombriciens s'est développée à partir de quelques chercheurs et de nombreuses études ponctuelles, de spécialistes de disciplines connexes, ou d'étudiants travaillant temporairement sur ce sujet. Ces travaux sont alors publiés dans de nombreuses langues, et dans des revues variées dont certaines sont difficiles à consulter. Les lombriciens étant omni-présents, ces études ont souvent trait à des approches locales, dans des domains très spécialisés. Certes des ouvrages à auteurs multiples, des publications locales importantes, des compilations (faites souvent par des non spécialistes des lombriciens) existent mais aucune vue d'ensemble de haute qualité n'était disponible.

Aujourd'hui grâce au travail considérable conduit pendant sept années par Kenneth E. Lee, un ouvrage homogène, maîtrisé, interprété et synthétisé par an auteur donne accès aux acquis modernes sur l'écologie et l'usage des lombriciens. Dans un langage clair et adapté aux diverses facettes de la connaissance, l'auteur met à notre disposition un livre qui permettra une prise en compte effective d'une part importante de notre patrimoine biologique. Dorénavant, le pédologue, l'agronome, l'écologue, le technicien de l'environnement, le biologiste, ... ne pourra plus ignorer un élément qui joue un rôle important dans son domaine professionnel.

De ce point de vue l'ouvrage marque une étape importante, car il permet aux techniciens, chercheurs, décideurs ou étudiants, en s'appuyant sur lui, de franchir de nouvelles étapes. Les problèmes soulevés par notre ignorance antérieure sont trop nombreux pour que l'étude des lombriciens reste indéfiniment marginale et ce livre permet de faciliter grandement l'éclosion d'une discipline majeure à la hauteur des problèmes à résoudre.

Ce travail considérable ne pouvait être que l'oeuvre d'un chercheur ayant une connaissance approfondie d'un sujet qu'il travaille depuis 1948 et d'un responsable ayant une familiarité avec les nombreuses disciplines argonomiques, ou environnementales en cause. Enfin, il fallait par un long travail personnel arriver à dégager l'essentiel, à classer de façon critique l'information, pour finalement aboutir à une ouvre didactique remarquable. Je crois que seul K. E. Lee pouvait le faire et l'ouvrage est la marque de sa personnalité.

Si aux siècles antérieurs, dès 1789, Gilbert White avait reconnu le rôle d'"intestine of the earth" et J. C. Savigny (1826) la diversité des lombriciens, il a fallu tout le prestige de Charles Darwin pour attirer l'attention sur l'importance de ces animaux dans l'économie de la nature. Son ouvrage de 1881* marque une étape importante qui est restée isolée, les disciplines scientifiques évoluant vers des spécialisations de plus en plus étroites. L'analyse ne pouvait que mal intégrer cet élément zoologique jouant un rôle pédologique et microbiologique en agronomie: les lombriciens ont le tort d'être "transdisciplinaires". Il suffit de voir la minceur des connaissances écologiques dans les ouvrages généraux de Stephenson (1930), Pickford (1937), Avel (1959) et dans la compilation de Stolte (1962-69) pour mesurer le faible échos du message de Darwin en ce domaine.

L'oeuvre de K. E. Lee en replaçant pour la première fois les lombriciens dans l'économie de la nature et l'économie humaine au niveau planétaire, reprend un siècle plus tard, en le modernisant, le message du grand naturaliste. Par son analyse critique, par son esprit synthétique, par son ouverture sur tous les thèmes actuellement accessibles à la connaissance rigoureuse, l'oeuvre de Kenneth E. Lee s'inscrit en filiation directe de celle de Charles Darwin.

MARCEL BOUCHÉ
Docteur ès-Science
Secrétaire du Comité de zoologie de
l'Association international des sciences du sol
Secrétaire de la section de zoologie du sol de
l'Union internationale des sciences biologiques

---

* Darwin, C.R. (1881). "The Formation of Vegetable Mould through the Action of Worms, with Observations on their Habits". Murray, London

Preface

> *It may be doubted whether there are many other animals which have played so important a part in the history of the world, as have these lowly organized creatures.*
> C. R. Darwin (1881).

I began writing this book seven years ago and had hoped to have it completed and published in 1981, to coincide with the centenary of the publication of Charles Darwin's book on the significance of earthworms as agents of soil formation and soil fertility. The book has taken longer than I expected, but I believe that the time has been well spent in attempting to produce a synthesis and an evaluation of contemporary knowledge of a group of animals whose significance is little appreciated.

The book comprises two principal parts. Part I describes the more basic and theoretical aspects of the ecology of earthworms, their limitations and environmental requirements, demography, ecological strategies, energy relationships, and some special aspects of their relationships with predators, parasites and pathogens. Part II deals with more applied aspects of the effects of earthworms on soils and plant growth, their relationships to patterns of land use, their use as agents of waste disposal and for protein production.

Every taxonomic group of animals (or plants), such as earthworms, is defined in terms of a shared set of morphological characters supposed to be genetically determined and unique to the group; subgroups are defined in terms of a range of variations in the shared character set and a range of variations in characters defined as subordinate to those unique to the group. The unique set of characters implies genetically determined limits of mechanical "design" that can be accepted as coming within the group, and these and the subordinate characters imply genetically determined limits of behavioural and physiological adaptation and compromise within which populations may survive in a range of environments. In this book I have interpreted the word "ecology" broadly and have put considerable emphasis on relationships between the behaviour and physiology of earthworms and what might be more strictly regarded as their ecology. Much can be learned in this way, and there is a need for more work integrating behavioural, physiological and ecological studies. I hope that the book will encourage others to contribute to the study of these relationships and to other aspects of the ecology of earthworms and their significance in soils, land use and environmental problems.

Generic and specific names and the names of higher taxa follow the comprehensive schemes of Easton (1983) for Lumbricidae, Sims and Easton (1972) for the *"Pheretima"* s.l. group of genera, and Jamieson (1978) for other taxa.

Charles Darwin (1837, 1881) first made scientific studies of the relationships between earthworms and soils and was the founder, at the same time, of the wider field of soil biology. His work and that, particularly, of Michaelsen (1900), Stephenson (1930), Pickford (1937) and Benham (many papers from 1886 to 1950) encouraged me to begin work on earthworms in 1948. Many friends and colleagues throughout the world have contributed to the understanding of these "lowly organized creatures", and I thank them for their cooperation and interest in the preparation of this book.

I do not pretend to have covered all the published work that might be considered relevant to the subject, but have selected, I believe, most of the appropriate and significant contributions up to about the middle of 1983, and some that have appeared later, up to the beginning of 1984. A comprehensive bibliography of research papers on earthworms, containing about 2000 references and covering the period 1930–1980, was prepared by Satchell and Martin (1981) for the Darwin Centenary Symposium on Earthworm Ecology, which commemorated the publication of Darwin's book on earthworms in 1881. Papers presented at this Symposium were published (Satchell 1983a) and represent a major contribution to the literature of earthworm ecology.

This book presents a personal viewpoint that owes much to the contributions of others but does not always agree with their conclusions.

Acknowledgements

I am particularly grateful to Dr Marcel Bouché, of Montpellier, who contributed a foreword to this book and made many valuable comments that have helped me to improve the manuscript, and to Dr J. E. Satchell, of Grange-over-Sands, who also criticized the manuscript and made many valuable suggestions for its improvement. Dr K. G. Tiller, of CSIRO Division of Soils, read and commented on the section on earthworms and heavy metals and Mr P. G. Allen, of the South Australian Department of Agriculture, made comments on the section on earthworms and biocides; their suggestions have helped me to improve the quality of these sections.

My wife, Norma, and my daughter, Christina, have contributed greatly to the tedious task of reading proofs. My wife has cheerfully tolerated my preoccupation through the years I have worked on the manuscript and I could not have written the book without her cooperation and support.

The line drawings were prepared by Denise Truscott in the Publications Section of CSIRO Division of Soils, and the manuscript was typed by Cathy Skinner, Kathy Simpson and Debbie Smith, also of CSIRO Division of Soils. Their skilled assistance and their cooperation are much appreciated.

Part I
Ecology

1
Earthworms and Their Environment

I. The Nature of Earthworms

Earthworms are found in all but the driest and the coldest land areas of the world. They include about 3000 species of the suborders Alluroidina, Moniligastrina and Lumbricina of the subclass Oligochaeta (Annelida: Clitellata), as defined by Jamieson (1978). Among these suborders are some aquatic forms, not usually regarded as earthworms, and others that are aquatic, but are closely related to terrestrial forms, and are generally regarded as earthworms that have secondarily adopted aquatic habits. Most earthworms are inhabitants of soils, including litter layers and above-ground habitats such as animal dung, rubbish heaps, rotting logs on the ground surface, moss-covered and fern-covered tree trunks, under the bark of standing trees and in organic material accumulated at the bases of epiphytes or in leaf bases of some subcanopy forest trees. A few species inhabit the intertidal zone.

It seems likely that earthworms are among the most ancient of terrestrial animal groups. Species of the closely allied, largely marine, polychaete worms are known as fossils found in South Australia in pre-Cambrian sediments 650–570 million years old (Glaessner *et al.* 1969). The oligochaete worms may have been derived from polychaetes or the two groups may have had common ancestral forms. The limits of evolutionary and ecological diversity that can be achieved by an animal group, given such a vast length of time, must be defined by inherent limitations of their mechanical "design" and behavioural and physiological adaptability.

4   *Ecology*

A. Basic "Design"

The basic "design" of earthworms is simple and varies little (Figs 1-3). They are cylindrical animals that consist essentially of two concentric tubes, the body wall and the gut, separated by a fluid-filled cavity, the coelom, divided into segments by septa.

Structural elements of the body wall (Richards 1978, Seymour 1978, Trueman 1978) include:

(a) a tough, thin, laminated cuticle that commonly contains several layers of nearly unstretchable collagen fibres, with alternate layers aligned in left-handed and right-handed helices around the body, forming a "fibrous skeleton" (Seymour 1978);

(b) the epidermis, which includes supporting columnar cells that are the main outer covering of the body, and to which the cuticle is attached;

(c) a layer of circular muscle fibres, and internal to these a layer of longitudinal muscles, the two layers mutually antagonistic and thus able to extend or contract the segments by pressure generated on the coelomic fluid;

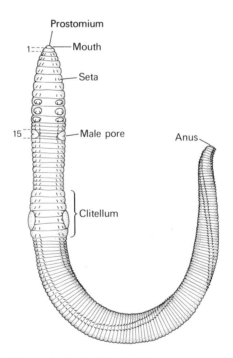

Fig. 1   External morphology of a "typical" lumbricid earthworm. (From Handreck 1978).

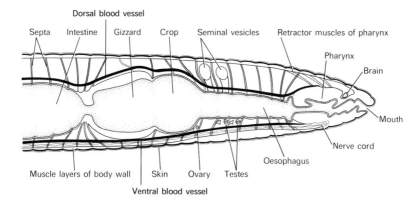

Fig. 2 Schematic longitudinal section of the anterior 24 segments of a "typical" lumbricid earthworm, illustrating the disposition of the principal organs of ingestion, digestion, reproduction, blood circulation, locomotion and coordination. (From Handreck 1978).

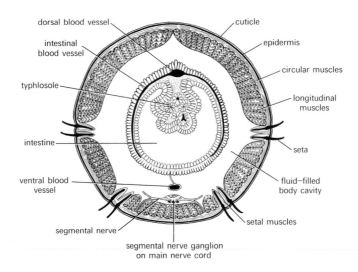

Fig. 3 Schematic cross section of the intestinal region of a "typical" lumbricid earthworm, illustrating the structure of the body wall, the intestine and associated organs. (From Handreck 1978).

(d) a thin, membranous peritoneum that defines the inner boundary of the body wall and confines the coelomic fluid.

The gut also has two muscle layers, an outer layer of longitudinal fibres and an inner layer of circular fibres, bounded internally and externally by complex epithelial layers.

The septa that divide the coelom are perforated by small pores, but these are surrounded by sphincters which may be opened or closed (Stephenson, 1930), so that the septa may behave as watertight bulkheads or may allow transfer of pressure between adjacent segments. Septa are mainly thin, with a layer that contains muscle fibres, connective tissue and blood vessels sandwiched between two layers of peritoneal cells (Stephenson 1930). In many species some anterior septa, and occasionally some of the most posterior septa, are greatly thickened and strongly muscular; when the body is relaxed the anterior muscular septa are conical in form, with their attachment to the oesophagus posterior to their attachment to the body wall.

Setae may be either (a) arranged symmetrically in pairs (usually four pairs, sometimes more), or (b) may be more or less evenly spaced, with eight to c. 100 on each segment. Each seta is enclosed in a follicle and has opposing sets of protracting and retracting muscles, which are independent of the circular and longitudinal muscles of the body wall, and allow the seta to be protruded or retracted as required.

Nephridiopores and dorsal pores, opening through the body wall to the exterior, have sphincter muscles to control loss of coelomic fluid.

B. Locomotion

Each segment acts as a separate hydraulic system, which can be more or less sealed, and can be shortened and broadened by contraction of the longitudinal muscles, or alternatively lengthened and narrowed by contraction of the circular muscles; contraction of one muscle layer is accompanied by relaxation of the opposing set. Movement is achieved through waves of alternate contraction and relaxation of the opposing muscle layers that pass along the body wall from segment to segment, in a peristaltic manner. The waves are retrograde, i.e., they move in the direction opposite to that of the animal, and the cycles of muscle contraction and relaxation are coordinated. For example, if a wave involves ten segments the state of contraction or relaxation of the corresponding muscles of adjacent segments will be one-tenth of a cycle different from each other. As each segment contracts longitudinally and expands laterally the setae are protruded to provide a grip on the substrate or burrow walls and then retracted as the cycle continues and the segment elongates. Waves can be propagated in either direction, so that the worm can move forward or backward. All the longitudinal muscles can be contracted simultaneously, with the setae taking hold only at the anterior or posterior region, so that the opposite end is withdrawn very rapidly. This is done in response to sudden stimulation and the movement is mediated by three giant nerve fibres that run along the dorsal region of the nerve cord (Gardner 1976, Dorsett 1978).

Yapp (1956) noted that earthworms were capable of moving vertically up a sheet of polished metal or clean glass, where the setae cannot provide anchoring points, and examined the mode of locomotion. When earthworms climb the sides of a glass beaker the anterior region of the body is extended, the peristomium is flattened against the glass, so that the mouth is used as a sucker; the anterior region behind the peristomium is lifted free of the glass and then contracts, pulling the body upward. I have observed a similar, but more striking form of very rapid movement across the soil surface by *Neodrilus polycystis*, a New Zealand acanthodrilid, in which the anterior end is raised and extended, the worm attaches its peristomium to the surface, the whole body is pulled forward, arching upward so that the posterior end is close to the peristomium, then the anterior end is moved forward to take another grip, so that the worm moves across the surface like a leech or a geometrid caterpillar.

In making a burrow, the prostomium is inserted between soil particles and then pushed forward by elongation of the anterior portion of the body, with the setae gripping further back. Anterior setae are then protruded, the more posterior setae are retracted, and the body is dragged forward by contraction of the longitudinal muscles, simultaneously pushing soil particles sideways and widening the burrow by lateral expansion of the body. Septal muscles provide a restraint that limits the lateral expansion of the body and so maintains its cylindrical form. If no space is available for insertion of the prostomium, earthworms will evert the pharynx, engulfing soil particles, and then retract the pharynx, dragging soil back into the gut and eventually excreting it as casts. The strongly muscular cone-shaped septa of anterior segments of some species are probably used to provide additional thrust in burrowing and rapid withdrawal of the body.

Seymour (1978) examined the deformation of the cuticle during the cyclic changes in diameter and length of segments as an earthworm moves. The layers of collagen fibres in the cuticle deform like a cylindrical trellis as the segments elongate and shorten. The shape of the segments varies between cylindrical when elongated and barrel-shaped when contracted. For the cylindrical shape maximum volume is attained when the angle between the fibres and the long axis of the body is 55°; and for the barrel shape, when the body wall is spherically curved and the angle between the fibres and the body axis is 45°. Some latitude in segmental volume is permitted by (a) rhythmic backward and forward bulging of the septa in response to pressure differences as the peristaltic waves move along the body, (b) induced pressure waves in the gut contents with corresponding decreases and increases in gut volume across segmental limits and (c) controlled flow of coelomic fluid backwards and forwards through the pores in intersegmental septa. But fluctuations in segmental volume are confined within rather narrow limits by the fibrous framework of the cuticle.

Pressures inside the coelom of *Lumbricus terrestris*, measured when at rest, were < 200 Pa with little fluctuation (Trueman 1978), and reached a maximum of *c.* 7500 Pa when the earthworm was moving (Seymour 1978). The lateral force that can be exerted to push soil particles aside during burrow construction must relate to the force generated by the internal pressure and to the surface area to which the force is applied, and assuming a non-rigid body wall these forces should be equal. Seymour (1978) measured the force exerted by *L. terrestris* on a "bridge", under which the worms were forced to crawl, and found that it exceeded, by *c.* 145%, the amount calculated from measurement of internal pressure and area of contact with the "bridge". He attributed the additional force to the intrinsic rigidity developed by contracting and stiffening the muscles of the body wall.

C. Size

The dimensions of earthworms, especially their lengths, are not easily compared since measurements are sometimes based on living specimens, which may be in any state between maximum and minimum extension, and sometimes on preserved specimens, which are usually more or less contracted. The smallest species (adult) are mainly forest litter-inhabiting or arboreal megascolecids from the southern hemisphere, with species in a number of genera varying from 10 mm to 20 mm in length and from 1.0 mm to 1.5 mm in diameter (see, e.g., Lee 1959, 1981). A few peregrine lumbricids are also in or near this size range (see, e.g., Graff 1957), and their small size has probably facilitated their accidental dispersal by man. Very large earthworms, *c.* 1.0–1.5 m in length and 20–30 mm in diameter, are known in several families of earthworms and are found in temperate and tropical regions of the northern and southern hemispheres, usually inhabiting deep soil layers in forest or sometimes in grasslands. The "giant Australian earthworm", *Megascolides australis*, is perhaps the most widely known of these; specimens up to 1.4 m in length and 20 mm in diameter, i.e., weighing 400–450 g, were recorded by Spencer (1888). *Glossoscolex giganteus*, from

Brazil, is somewhat larger; specimens up to 1.2 m in length and 30 mm in diameter, weighing 500–600 g, were recorded by Johansen and Martin (1965). Bouché (1972) described a lumbricid *Scherotheca (Opothedrilus) occidentalis thibauti*, 1.05 m in length, 18–25 mm in diameter, weighing 106 g, and this is probably the largest of all lumbricid earthworms. Truly gigantic earthworms have been reported from South Africa, e.g., a newspaper reports a specimen (presumed to be *Microchaetus* sp.) of *c.* 7 m in length and 75 mm in diameter (Ljungström and Reinecke 1969). Excluding the South African reports, the smallest accurately measured earthworms must weigh *c.* 10 mg and the largest *c.* 500–600 g, a ratio of 50 000–60 000 between largest and smallest. If the South African reports are accepted, individual earthworms may be up to rather more than 30 kg in weight, and the ratio of largest to smallest would then be 3 000 000. No other terrestrial invertebrate would come within an order of magnitude of the weight of such earthworms. It seems unlikely that the internal hydraulic pressures necessary to move a 30 kg body could be sustained without rupturing the thin cuticle necessary for respiration, and doubtful that the circulatory system could survive the blood pressure necessary to maintain blood circulation (p. 14) in so large an earthworm.

D. Surface : Volume Ratio

The ratio of surface area to body volume ($cm^2 : cm^3$) is generally *c.* 2–4 over the full size range of earthworms. The critical consequences of surface:volume ratio relate to control of rates of water loss from the body during desiccation, which are minimized by low surface:volume ratios, and to maintenance of acceptable rates of respiratory gas exchange, which are maximized by high surface:volume ratios.

E. Vascular System and Blood Circulation

Unlike most invertebrates, earthworms have a closed vascular system. The basic structure of the circulatory system is as follows.

The main vessels are the dorsal vessel, which runs longitudinally along the top of the gut, two to five pairs of large circumoesophageal vessels that arise from the dorsal vessel to connect around the oesophageal region of the intestine to the ventral vessel, which runs longitudinally along the bottom of the gut. Blood is collected by the dorsal vessel from somatic vessels, including oxygenated blood from the subcuticular capillaries, and is pumped forward by a rhythmic peristalsis of the vessel's walls to flow into the paired circumoesophageal vessels. The circumoesophageal vessels are equipped with valves, are strongly muscular, and function as "hearts", providing pressure to pump the blood into the ventral vessel, from which it is distributed by segmentally arranged lateral branches to the somatic vessels.

Few comprehensive studies have been made of the functioning of the circulatory system. The best documented studies are those of Johansen and Martin (1965, 1966), who described the circulatory system and its performance in the giant Brazilian glossoscolecid *Glossoscolex giganteus*, which has a body volume up to *c.* 600 cm^3, surface area up to *c.* 950m^2, and surface : volume ratio *c.* 1.6. Johansen and Martin measured the blood pressure in resting and active *G. giganteus*. The systolic blood pressure in the dorsal vessel, i.e., the maximum pressure attained during contraction in the peristaltic cycle, was *c.* 2500 Pa (*c.* 20 mm Hg) in resting individuals and *c.* 6500 Pa (*c.* 50 mm Hg) in active individuals. In the ventral vessel systolic pressures were *c.* 4000-5000 Pa (*c.* 30-40 mm Hg) at rest and *c.* 10 000 Pa (*c.* 75 mm Hg), sometimes rising as high as 12 250 Pa (*c.* 92 mm Hg), in active individuals. The only other recorded blood pressures are for *Lumbricus terrestris*, which is no larger than *c.* 30 cm in length, *c.* 9 mm in diameter, and *c.* 20 g in weight; maximum blood pressures for *L. terrestris* are *c.* 600-750 Pa (4-5 mm Hg) at rest and *c.* 1250 Pa (9-10 mm Hg) when active. Maximum pressures recorded in *G. giganteus* are *c.* eight times those for *L. terrestris*. The systolic pressures recorded for *G. giganteus* approach those of the human vascular system (normal systolic pressures *c.* 15 000-20 000 Pa [110-150 mm Hg] at rest), and must be presumed to be necessary to maintain circulation through the extremities of the body against the peripheral resistance resulting from the capillaries. If the results of Johansen and Martin for *G. giganteus* were extrapolated to earthworms of the gigantic dimensions reported from South Africa (p. 9), pressure in the ventral vessel would be so high as to be unsupportable; it might, however, be reduced to more reasonable levels if branches of the main system closer to the peripheral vessels were themselves contractile and could function as auxiliary "hearts".

F. Sense Organs

A wide variety of specialized individual epidermal and subepidermal cells, free nerve endings in the epidermis, and some aggregations of cells into more complex structures are associated with the reception of tactile, positional, chemical and light stimuli, and are scattered over the body surface. In *Lumbricus terrestris* most of them, on each segment except the first, are arranged in two distinct circumferential bands, one band near the anterior margin and one near the setae; they are most numerous towards the extremities. Langdon (1895) counted 1900 sense organs on segment one and the prostomium, 1200 on segment ten and 700 on segment 56 of *L. terrestris*.

Details of the structure and function of the sense organs were reviewed by Mill (1978), who distinguished:
(a) receptors for tactile stimuli, probably free nerve endings in the epidermis;
(b) proprioceptors, that register deformations and stress in the body arising from the animal's own movement, its weight, or from external forces, most likely a type of multiciliate epidermal cell whose cilia are bent over and lie horizontally beneath the cuticle;
(c) chemoreceptors, which probably include some multiciliate sensory cells, and the nuchal organs, which are pits or folded lobes on the body surface;
(d) photoreceptors, single celled and widely distributed over the body surface, each containing an optic organelle composed of a transparent hyaline vacuole surrounded by a network of neurofibrils (the retinella). Hess (1925) showed that the structure of the optic organelles was such that light, irrespective of its direction, would be focussed onto the retinella.

II. The Soil Environment

Soil is the upper weathering layer of the earth's crust, in which life exists, in contrast with the lifeless lithosphere, which is the outer mineral layer of the earth, beneath the soil. The soil includes surface layers of plant litter, living organisms, roots and other underground parts of plants (Taylor and Pohlen 1962). Its lower boundary is sometimes sharply defined, but more commonly there is no clear boundary but a gradation through progressively less weathered materials to the underlying unweathered lithosphere. The soil is usually, but not always, differentiated into horizons that are more or less parallel to the surface. An idealized soil profile, a vertical section through the soil, with horizon designations and notes on the diagnostic characters of the horizons, is illustrated in Figure 4. Soils are open systems that vary continuously in space and time; inorganic and organic materials are added and lost, but cyclic processes of energy accumulation and release (C-cycle) and of plant and animal nutrients maintain a "dynamic equilibrium" that regulates soil fertility, and in which earthworms frequently play an important part.

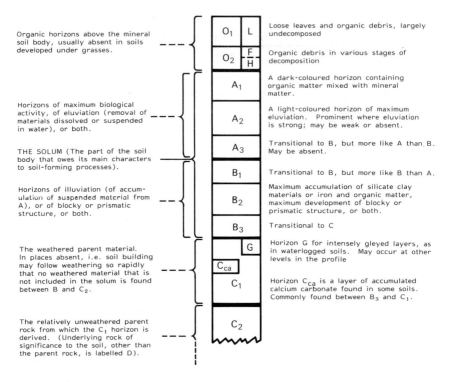

Fig. 4 An idealized soil profile, with horizon designations and notes on the diagnostic characters of the horizons. (From Taylor and Pohlen 1962).

The soil is a three-dimensional framework that admits little light and is made up of mineral particles that range in size from clay (< 2 μm) particles to large masses of rock, organic materials that include plant litter and logs at the soil surface, dead and living plant roots, organic compounds and dead tissue exuded or sloughed from root surfaces, microbial and fungal tissue, dead and living bodies of soil invertebrates, and water-filled and air-filled spaces between and on the surfaces of soil aggregates. Organic matter, which is the food of earthworms, ranges from undecomposed plant tissue to the resistant residues (humic materials) that remain after microbial and faunal decomposition of plant tissue. Most of the soil organic matter is at or near the surface, with the concentration falling rapidly with increasing depth. Some organic matter is bonded electrostatically to the surfaces of clay particles or incorporated with mineral particles in aggregates. The soil water includes ionic and non-ionic solutes derived from the weathering of mineral particles and from decomposing plant material, some of which are essential to the earthworm's physiology, some of no significance to earthworms, and some that are toxic. The soil's physical texture and aggregate size, availability of water and oxygen, pH, concentration of electrolytes and redox potentials (which depend largely on the state of aeration and saturation of the soil) vary, but must be within limits that are tolerable for earthworms to survive. Ambient temperatures exercise an overall control, since most, if not all earthworms cannot survive for long at temperatures outside the range $c.$ 0°C–35°C.

The favourableness of the environment varies seasonally and also spatially on a small scale that is determined by variability in litter and soil depth, plant cover, temperature, the proportion of air-filled to water-filled pore spaces, drainage, chemical composition of the inorganic and organic components of the soil, and the nature of the underlying material, whether solid rock, fractured rock or weathered material.

For soil-inhabiting organisms in general, Vannier (1973) proposed the term *porosphere* to describe the environment provided by porous media, i.e., more or less solid media with an internal surface, such as soils. Vannier (1983) distinguished a succession of changing environments in the porosphere as a soil is progressively dried, ranging from *aquatic*, when the soil is saturated, pF $\leqslant$ 2.5 and the relative humidity (R.H.) in any air-filled spaces is 100%, through *edaphic*, when water is confined to films on the surfaces of soil aggregates or is held in pore spaces by capillary forces, pF = 2.5–4.7 and R.H. in air-filled spaces is 100% or slightly less, to *aerial*, when free water is no longer present, pF > 4.7 and R.H. in air-filled spaces $\leqslant$ 95.4%. Earthworms can maintain an active life only in the edaphic range of Vannier's scheme, in the pF range that corresponds approximately to that between field capacity and the wilting point of plants. At pF 2.5 water retention forces are $c.$ 350 g/cm^2, while at pF 4.7 they are $c.$ 50 kg/cm^2 and at least some earthworms can apparently retain water against a negative pressure of this magnitude.

A. Ecological Significance of Burrows

Some earthworms, mostly in forests but also in some grasslands, are confined to surface litter layers (O horizons of Fig. 4) or live in aboveground habitats, but most species live in subsurface horizons (A, B and C of Fig. 4) in burrows that they excavate themselves. By living in burrows they are to some extent protected from diurnal and seasonal variability in the physico-chemical environment and from predatory pressure.

Burrows are of three principal kinds:

(a) the more or less permanent refuges in underlying soil horizons of species that feed on surface litter. These are usually vertical for most of their length, sometimes branching near the top to several entrances. They may extend deep into the soil, e.g., *Lumbricus terrestris* burrows go down 3 m or more in deep soils of the northeast United States (G. E. Gates, pers. comm.). They have smooth linings built up by compression of the soil and mucus secreted by the earthworms, which probably help to maintain high humidity and reduce water loss. Burrows enable earthworms to select the conditions that suit them best from the range of microenvironments available in one or more soil horizons, while retaining access to forage for food at the surface at times when conditions (especially temperature and humidity) are tolerable. Earthworms that live in such burrows are the *anéciques* of Bouché's (1972) ecological classification (p. 108);

(b) the more extensive burrows of geophagous species that forage for food in subsurface soil horizons. These are predominantly horizontal, but have some vertical components and some openings to the soil surface. They may extend deep into the soil, e.g., I have recorded *Octochaetus multiporus* at depths of 3–5 m from road cuttings in deeply weathered soils near Wellington, New Zealand. Earthworms that make such burrows are the *endogées* of Bouché's (1972) ecological classification (p. 108). Some endogées live in the organic matter-rich A horizon, some in the B or C horizon feeding on dead roots or organic matter that has been washed down cracks from the surface. Their burrows are smooth walled, with a surface layer usually thinner than that found in burrows of anéciques. Many burrows are partly or wholly packed with casts or with soil from overlying horizons, carried down by water;

(c) more or less vertical burrows made by earthworms that live near the surface as they retreat to enter a resting state (diapause, quiescence—p. 85) in deep soil horizons during dry or cold seasons, or return to surface horizons when conditions permit them to resume an active life. These burrows generally lack distinct linings; they are apparently made quickly, are used only once, and are ephemeral. They often terminate in roughly spherical mucus-lined chambers where individual earthworms have taken refuge, and the chambers are frequently filled with humus-stained earthworm casts derived from soil in overlying horizons, presumably deposited by the former inhabitants as they returned to an active life in surface soil horizons.

The intensity of burrowing activity in geophagous species is related to the amount and quality of food available. Martin (1982a) kept large juvenile *Lumbricus rubellus* and *Aporrectodea caliginosa* in thin layers of soil, held between glass sheets, with powdered grass for food mixed with the soil in the proportions 1 : 16, 1 : 64 and 1 : 256, and recorded the length of burrows made by the earthworms over a period of eight days. The extent of burrowing, particularly of *L. rubellus*, was directly proportional to the concentration of food provided. Similarly, Lavelle *et al.* (1983b) showed an inverse relationship between soil ingestion rate and concentration of water-soluble plant extract added to soil in the Ivory Coast megascolecid *Millsonia anomala*; McColl (1982) showed an inverse relationship between soil macroporosity, apparently due to more extensive burrowing and increased casting of *A. caliginosa*, and density of ryegrass roots (a favoured food of *A. caliginosa*) in pot experiments. Most of the energy used by geophagous species is allocated to burrowing and feeding, with little available for tissue production; quantitative and qualitative differences in available food must have important implications for growth rates and reproductive potential.

III. Basic Environmental Requirements of Earthworms

The basic requirements for life are summarized below. In subsequent Sections the nutritional, physiological, behavioural and ecological requirements and capabilities of earthworms are related to the physico-chemical environment provided by soils.

A. Adequate and Suitable Food Supplies

Earthworms feed on dead and decaying plant (and animal) remains and on free-living soil microflora and fauna (pp. 17–29). The primary source of food for all the soil biota is dead plant material, especially plant litter, which varies greatly in physical and chemical composition, palatability, and nutritional quality. Because of their limited capacity for moving about, earthworms are obliged to live very close to their sources of food.

B. Adequate Moisture

Given adequate and suitable food, perhaps the most important requirement of earthworms is adequate moisture. Water conservation mechanisms are poorly developed in earthworms; respiration depends upon diffusion of gases through the body wall, which must be kept moist, and much water is lost by most species in urine.

C. Suitable Temperatures

The temperature limits for survival of earthworms vary between species and some intra-specific individual adaptation is possible, but in general they can survive long exposure to temperatures only within the range of $c.$ 0°C–35°C. High soil tempratures are commonly associated with desiccation and moisture stress, and it is not always easy to distinguish the effects of these factors.

D. Respiratory Exchange

Oxygen for respiration can be obtained from air or from oxygenated water. Absorption from air requires the maintenance of a moist surface layer and so involves water loss except when the relative humidity of the air is 100% or close to 100%, and survival is threatened by prolonged exposure to dry air. Respiration of oxygen from solution in water requires that water surfaces in contact with air are sufficient to maintain the concentration of dissolved oxygen at a level adequate for the earthworm's demands. Earthworms are able to tolerate carbon dioxide (CO_2) concentrations much higher than those characteristic of soil air.

E. Protection from Light

Earthworms are injured and may be killed by exposure to light, and are particularly affected by ultra-violet wavelengths. They vary in their tolerance of light, with more pigmented species or individuals being less sensitive than the less pigmented, but in general they retreat from light.

F. Suitable Soil Texture

Earthworms are generally absent or rare in soils with very coarse texture, probably due to physical abrasion of their body surface by coarse mineral materials and to the susceptibility to drought of such soils. They are also generally absent from soils with high clay content in regions of high rainfall, because of the susceptibility of such soils to seasonal or permanent oxygen deficit.

G. Suitable pH and Electrolyte Concentrations

Earthworms are rare in soils with pH < 4, but within the pH range generally encountered in soils and associated habitats (slightly acid to slightly alkaline) pH is not a limiting factor. Much attention has been given to pH as a factor controlling earthworm populations, but its significance should be seen as only one aspect of the electrolytic status of the soil solution. Chloride concentration is particularly significant, and so too is calcium ion concentration, especially for species that have active calciferous glands. Redox potentials, which are a measure of the balance between oxidizing and reducing conditions in soils, seem also to be important in determining earthworm survival and abundance. Some ions and non-ionic substances are toxic and can be tolerated only at low concentrations.

IV. Food and Digestive Capabilities

Earthworms are saprophagous animals, i.e., their diet comprises mainly organic detritus in various stages of decay and incorporation into the soil. Although the bulk of food ingested is dead plant tissue, living microorganisms, fungi, microfauna and mesofauna and their dead tissues are also ingested, and there is evidence that they are an important part of the diet. Earthworms may be fairly clearly divided into *detritivores*, that feed at or near the soil surface, mainly on plant litter or dead roots and other plant debris in the organic matter-rich surface soil horizons or on mammalian dung, and *geophages*, that feed deeper beneath the surface, ingesting large quantities of soil, usually selecting portions with higher than normal organic matter content. Some species that feed in deep soil horizons are mostly dead root-feeders and so are detritivores rather than geophages. The group that comprises detritivorous earthworms are more or less equivalent to the humus formers, and geophagous species to humus feeders, the two principal morpho-ecological categories of earthworms recognized by Perel' (1977) (p. 114). Detritivores include épigées and anéciques, while geophages include most endogées in the ecological groupings of Bouché (p. 108).

A. Mull and Mor

The organic matter-rich litter and underlying surface layers of soils (the 0 and A_1 horizons of Figure 4) vary greatly in structure and form. Two basic forms of organic matter-rich soil layers or humus types were first recognized in Danish woodlands by Müller (1878) who named them *mull* and *mor*. Many intergrading forms between the two have been recognized and classified; I refer the reader to Satchell (1974) and Swift *et al.* (1979) for concise reviews.

Mull was described by Müller (1878) as a soil surface layer composed of plant detritus intimately mixed with mineral soil material, and with its upper portion consisting almost entirely of earthworm casts. Mor was described as a peaty surface layer, composed almost entirely of organic matter, with a sharp lower boundary where it overlies the largely mineral material of the A horizon.

Mull is characterized by low carbon to nitrogen (C:N) ratio (often < 15), organic matter with a low proportion of cellular components (i.e., well humified), pH slightly acid to about neutral (c. 5–7), high bacterial counts, absence of matted fungus mycelia, low soil arthropod (mesofauna) populations, and high earthworm populations, usually including several species that are stratified vertically. It is the classic earthworm–habitat association of lumbricids in the soils of European deciduous forests.

Mor is characterized by high C : N ratio (usually > 20), organic matter poorly decomposed, with a high proportion of recognizable cellular components (i.e., raw humus), pH acid (usually c. 3.5–5.0), low bacterial counts, litter in surface layers often matted with fungus mycelia, high mesofauna populations, and earthworms rare or absent, often only one or two species living close to the surface.

It should be appreciated that mull and mor represent extremes and are rarely to be found; more commonly humus types are intermediate between the two. The terms can be applied to many tropical humus types, but not to all, particularly not when litter decomposition processes are dominated by termites (see Lee and Wood 1971b).

B. Detritivorous and Geophagous Species

Much information on the food preferences of earthworms has been derived from detailed examination of the contents of the alimentary canal. Some of it is contradictory, and indicates that the same species does not necessarily have the same diet in different localities.

Direct observation of burrows made by *Aporrectodea rosea* by Bolton and Phillipson (1976) indicated that feeding consisted of a "grazing" procedure, in which soil materials were removed from the burrow wall, frequently producing a non-linear fan-shaped depression. Examination of the gut contents of large immature and adult *A. rosea* showed that the gizzard contained mainly fragments of plant tissue and inorganic soil particles, with some finely divided organic matter, while at the posterior end of the intestine there was only finely divided organic matter and mineral material. Analyses of material from the posterior gut showed an organic matter content of 14.17%, 15.25% and 15.40%, in small immature, large immature and adult specimens respectively, compared with 11.25% in well-mixed surrounding soil. It was concluded that *A. rosea* is a geophagous species, but that it actively selects organic in preference to inorganic materials from the soil.

Piearce (1972a) removed material separately from the oesophagus, crop/gizzard, anterior intestine and posterior intestine of *Aporrectodea caliginosa* and *Lumbricus rubellus*. Examination of smears of this material showed that there were three principal components, which he named; (a) raw humus, in which regularly arranged intact plant cells were recognizable, (b) amorphous humus, consisting of brown or black organic material with no recognizable intact cells and (c) mineral material. A high proportion of raw humus would indicate a predominantly detritivorous diet and of amorphous humus and mineral material a predominantly geophagous diet. Both species examined by Piearce had ingested much amorphous humus, but there was a significant preponderance ($P < 0.01$) of raw humus in the gut contents of *L. rubellus*, indicating that it is a detritivore, and a significant preponderance ($P < 0.01$) of mineral material in those of *A. caliginosa*, indicating that it is a geophage.

In further studies of six lumbricid species from the same pasture, Piearce (1978) found recognizable fragments of grass leaves and roots in crop/gizzard contents. Among other recognizable organic materials algal cells were most common and were found in most specimens examined. Minor recognizable components included leaf fragments from dicotyledonous plants, earthworm setae, roots, seeds, fungi, protozoans and cuticle of arthropods. Piearce concluded that *Lumbricus castaneus* and *L. rubellus* consume relatively little-decomposed plant remains (raw humus), i.e., they are detritivores; *Aporrectodea caliginosa* and *Allolobophora chlorotica* consume relatively much-decomposed plant remains (amorphous humus), i.e., they are geophages; while *Aporrectodea longa* and *Satchellius mammalis* ingest mixtures of the two, i.e., they are intermediate detritivorous/geophagous species.

Bouché and Kretzschmar (1974) examined crop/gizzard contents of lumbricids from a pasture and a forest habitat in France. They showed that in pastures *Aporrectodea rosea* and *A. chlorotica* were root feeders (cf. these species as geophages, previous paragraph), and *A. longa* a leaf feeder, while in forest *A. chlorotica* and *Dendrodrilus rubidus* were leaf feeders, *A. caliginosa* (adults and juveniles) a root feeder, while *Lumbricus terrestris* (adults and juveniles) fed mainly on leaves, with roots as a minor component of the diet.

Ferrière (1980) was able to distinguish the species of plant remains eaten by comparing epithelial cells of plant fragments from the crop/gizzard contents of ten lumbricid species with those from living plants in a pasture. Adult épigées and épi-anéciques were shown to feed mainly on little-decomposed and readily identifiable tissue from leaves or roots (only in *Lumbricus terrestris*) of the plants in the pasture, adult anéciques mainly on partially decomposed but identifiable fragments of the above-ground parts of the pasture plants, while adult endogées fed mainly on much-fragmented roots and leaves at an advanced stage of decomposition, mixed with much unidentifiable organic matter. Of the juveniles studied, *L. terrestris* (épi-anécique) apparently fed on the same materials as do adults, three species of *Allolobophora* (anéciques) fed on fine plant debris in an advanced stage of decomposition, while three other species of *Allolobophora* (endogées) fed mainly on much-decomposed unidentifiable organic matter, with a minor component of small fragments of leaves.

Earthworm communities in tropical grasslands are dominated by geophagous species. Lavelle *et al.* (1980) have found that the earthworm community of a pasture in Mexico comprised 100% geophagous species, of two north Guinean savannas in the Ivory Coast 88.1% and 77.8% geophagous species, and of three south Guinean savannas in the Ivory Coast 69.7%, 89.2% and 97.0% geophagous species. The small proportion of detritivorous species in these tropical grasslands contrasts with the opposite situation reported by Bouché (1975c) for lumbricids in a permanent pasture in France, where there were 86.1% detritivorous species and 13.1% geophagous species.

Lavelle *et al.* (1980) fed individuals of three common megascolecid species from the south Guinean savannas, *Millsonia anomala* (young and adults), and young *Millsonia ghanensis* and *Dichogaster terrae-nigrae*, with soil taken from several depths under shrub savanna. Some characteristics of the organic fraction of the soil layers used are set out in Table 1.

Mean specific daily rates of soil ingestion, i.e., grams of dry weight soil consumed per gram of fresh weight of earthworm per day, and daily percentage change in fresh weight of earthworms are shown in Figure 5. Young *M. anomala* have their highest soil ingestion rate and increase in weight most rapidly in soil from the 5–10 cm layer, but are not unduly stressed over the depth range of 0–25 cm; adult *M. anomala* have a narrow optimum range centred on the 2–5 cm layer. Young *D. terrae-nigrae* and young *M. ghanensis* show similar patterns, with maximum growth rate in soil from the 10–25 cm layer, but *M. ghanensis* maintains reasonably high growth rates over a wider range of soil depths than does *D. terrae-nigrae*.

Table 1. Some characteristics of organic matter in the different layers of the shrub savanna soil used for earthworm cultures. (From Lavelle et al. 1980).

	Depth (cm)	0–1	2–5	5–10	10–25	30–40
Whole soil	Organic carbon %	11.8	9.8	8.3	7.0	4.1
	Total nitrogen %	0.58	0.50	0.40	0.30	0.30
	C:N ratio	20.4	19.5	20.8	23.3	13.6
	Total organic matter %	20.4	16.8	14.3	12.0	7.0
Light organic fraction	%	0.29	0.22	0.24	0.10	0.15
	C %	26.7	36.7	42.2	41.6	29.6
	N %	7.4	5.6	4.6	3.6	3.8
	C:N ratio	3.6	6.6	9.3	11.7	7.9
Humic acids	Total fulvic acids (in C %)	2.53	1.91	1.57	1.54	1.36
	Total humic acids (in C %)	0.61	0.74	0.74	0.43	0.16
	FA:HA ratio	4.18	2.58	2.12	3.69	8.25
	Total humic matter (in C %)	3.14	2.65	2.31	2.02	1.53
	Humification coefficients	26.5	27.1	27.8	29.0	37.6

Lavelle et al. (1980) concluded from the growth data that *M. anomala* is truly adapted to feeding on the organic matter of the upper soil horizons and *M. ghanensis* and *D. terrae-nigrae* to feeding on the organic matter of deeper soil horizons. This is further confirmed by the general fall in rate of soil ingestion with increasing depth in *M. anomala* and the opposite, rising rate in *D. terrae-nigrae* and *M. ghanensis*. *Millsonia anomala* is found in the field mainly at depths of 0–10 cm, *D. terrae-nigrae* at 10–30 cm, and *M. ghanensis* at 10–40 cm.

The energy content (calorific value) of the soil layers in which these earthworms live and which they ingest is low. For the 0–10 cm layer it is 376 kJ/g, for the 10–30 cm layer 210 kJ/g, and for the 10–40 cm layer 167 kJ/g; this compares with c. 17 000 kJ/g for litter, which is the food source of detritivores. Geophagous species must ingest much larger quantities of soil than detritivores to satisfy their energy requirements. For example, Lavelle (1983a) found that young individuals of *Millsonia anomala*, at Lamto, ingest 20–36 times their own body weight of dry soil per day, the ratio falling as the worms grow older; an average *M. anomala* may ingest (dry weight) 3–5 kg of soil per year, and a field population of 215 000/ha c. 500 t ha^{-1}y^{-1}. Two other geophagous species at Lamto (*M. ghanensis* and *D. terrae-nigrae*) have lower ingestion rates, but the three species together ingest 850–1200 t ha^{-1}y^{-1}. Litter-dwelling detritivorous species at Lamto (*M. lamtoiana* and *D. agilis*) ingest (dry weight) 180–980 kg of litter per year, the quantities varying between different savanna associations.

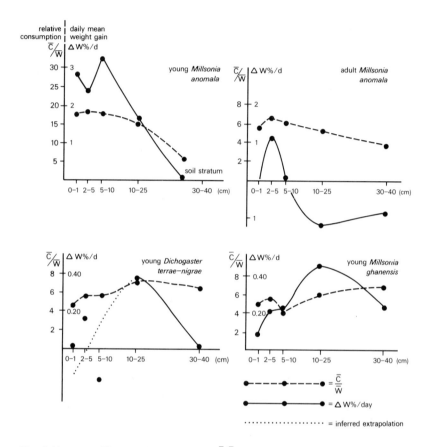

Fig. 5 Mean specific rate of soil ingestion ($\bar{c}/\bar{w}$ = consumption in g dry wt soil per g fresh wt earthworm per day) and daily % change in earthworm weight ($\Delta W\%/d$) of three species of geophagous earthworms fed with soil taken from different depths in the soil profile. Data for earthworms and soils from a shrub savanna at Lamto, Ivory Coast. (After Lavelle et al. 1980).

It is of interest to compare soil ingestion rates of these tropical savanna earthworms with rates for a forest-inhabiting European lumbricid. Zicsi (1983) estimated that *Lumbricus polyphemus*, with a mean fresh weight of 7.64 g, consumed soil and produced casts (dry weight per gram of earthworm fresh weight) at a rate of 100.8 mg g^{-1} d^{-1}, i.e., 770 mg/d per individual, or c. 0.1 times their own body weight. A field population of c. 40 000/ha would ingest c. 8.1 t $ha^{-1}y^{-1}$ dry weight of soil.

C. Dung-feeding

Numerous species of earthworms are known to feed on mammalian dung, and it is apparently a highly nutritious food for earthworms. In Europe *Aporrectodea caliginosa, A. longa,* and *L. rubellus* are reported to feed on dung in grazed pastures; in Australia *A. caliginosa* is recorded; in New Zealand *A. caliginosa, L. castaneus, L. rubellus* and *L. terrestris* are recorded; in Japan *Eisenia japonica, Amynthas hilgendorfi* and *Pheretima* spp. were recorded by Nakamura (1975).

Barley (1959b) fed *A. caliginosa* from South Australia for 40 days on various diets and found that body weight decreased by 53% on unamended soil, 26% when roots or leaves of *Phalaris* were added, and 2% when clover roots were supplied, but increased by 71% when sheep dung was supplied at the soil surface and by 111% when the dung was mixed into the soil.

Kale *et al.* (1982) studied the growth rate, cocoon production and survival over a one year period of pairs of *Perionyx excavatus* fed in laboratory cultures on sheep, cow, horse and poultry dung. *Perionyx excavatus* is an Indian species that commonly inhabits compost heaps and similar accumulations of organic matter and might be regarded as an Indian equivalent of *Eisenia fetida*. Fed on sheep dung (means of 17 experiments) *P. excavatus* grew to maturity in 124 days, produced 65 cocoons in the following 41 days, and in 35% of the experiments failed to survive to the end of the year's observations. Fed on cow dung (means of 11 experiments) they grew to maturity in 183 days, produced 25 cocoons in the following 102 days, and all survived to the end of the year. Fed on horse dung (means of 12 experiments) they grew to maturity in 177 days, produced 42 cocoons in 55 days, and all survived to the end of the year. Poultry dung was toxic to *P. excavatus*, and the earthworms did not reproduce. The experiments of Kale *et al.* (1982) aimed to assess the capability of *P. excavatus* to dispose hygienically of organic wastes, and for this purpose it is apparent that the species is particularly suited to disposal of sheep dung. The experiments also demonstrated that production is closely related to food quality, and have implications for protein production from earthworm tissue (Chapter 16).

D. Fungi and Microorganisms as Food

It is well known that earthworms will not generally ingest plant litter until it is partly decomposed, and it has frequently been suggested that fungi and microorganisms ingested with decomposing litter are important components of the diet. There is also much evidence of a resident gut microflora and of many microrganisms passing through the gut unharmed (p. 26). Much of the experimental and field data on the significance of microorganisms as food are inferential, and sometimes contradictory, but there is good evidence that at least some fungi and microorganisms are important in the diet of some earthworms, and a strong possibility that they have a significant role more generally among earthworms.

Aichberger (1914) found that nearly all living organisms from earthworm gut contents were forms with a strong outer coat, apparently protecting them from digestion, and in contrast with Parle (1963a) he found very few live yeasts. Day (1950) recorded heavy mortality of *Bacillus cereus* var. *mycoides* when soil inoculated with this bacterium passed through the gut of *Lumbricus terrestris* and concluded that spores probably survived while vegetative cells were digested; *Serratia marcescens* also suffered high mortality in transit through the gut of *L. terrestris*, as did *Eschericia coli* when ingested by *L. terrestris* (Brusewitz 1959) or by *Pheretima* sp. (Khambata and Bhat 1957), and it may reasonably be presumed that bacterial cells were digested.

Some fungal spores survive passage through the gut, while others apparently do not. Viable spores of 17 species of fungi were isolated by Hutchinson and Kamel (1956) from the hind-gut of *L. terrestris*, while Keogh and Christensen (1976) in New Zealand recorded zero viability of spores of the fungus *Pithomyces chartarum* after transit through the gut of *L. rubellus*. *Pithomyces chartarum* spores occasionally occur in vast numbers in pasture litter in New Zealand, and their ingestion by grazing mammals, especially sheep, causes a serious liver disease (facial eczema). Keogh and Christensen concluded that the presence of large populations of the litter-feeding earthworm *L. rubellus* in soils of grazed pastures might significantly reduce the risk of facial eczema in livestock.

Wright (1972) noted that *L. terrestris* fed preferentially in the field on apple leaves that showed clear indications of microbial growth (predominantly *Pseudomonas* sp.) on their surfaces compared with leaves that were apparently clear of microorganisms. Apple leaves were softened in a moist chamber, autoclaved, then spread with a pure culture of *P. aeruginosa*. Discs punched from these inoculated leaves and from similar leaves without bacteria were fed to laboratory cultures of *L. terrestris* and it was found that the presence of the bacteria resulted in an increase of 35% in rate of consumption of the discs compared with consumption of uninoculated discs. When filter paper discs were substituted for leaf discs the increase in feeding rate on inoculated compared with uninoculated discs was 62%. Wright concluded that *L. terrestris* finds bacteria attractive as food and proposed that they could be an important source of dietary protein.

Cooke and Luxton (1980) isolated two fungi, *Mucor hiemalis* and *Penicillium* sp., and a bacterium, *Pseudomonas fluorescens*, from soil and cultured them on discs of sterilized filter paper placed on the surface of nutrient agar plates. Inoculated discs and similar uninoculated discs were placed on the soil surface in laboratory cultures of *Lumbricus terrestris*. The earthworms fed preferentially on the fungus-inoculated discs but not on the *Pseudomonas*-inoculated discs relative to the uninoculated controls. Analyses of the filter paper discs showed a higher nitrogen content in fungus-infected discs than in the bacterium-infected or control discs, indicating that the earthworms may be attracted by the high protein content of the fungi.

In further experiments, Cooke (1983) confirmed the preference of *L. terrestris* for discs inoculated with *M. hiemalis*, and for discs with *Fusarium oxysporum, Alternaria solani* and *Trichoderma viride*, and their lack of preference for or rejection of discs inoculated with *Cladosporium cladosporioides, Poronia piliformis, Chaetomium globosum* and *Pencillium digitatum*. The preferences were shown not to be influenced by the polyphenol, nitrogen or carbon content of the fungi, but the quantity of fungus on the discs and calcium and moisture concentrations were shown to be strongly correlated with selection. Analysis of the data using stepwise multiple regressions indicated that fungus weight and moisture content together accounted for 75% of the variance, while addition of calcium did not significantly alter the coefficient of determination. Cooke concluded that his data did not unequivocally establish the characteristics that determined palatability of the discs. He considered that a number of factors might be involved, including attraction to moist substrates to maintain the body water content or because moisture content provides an index of the level of organic matter decomposition and hence of the presence of a high-energy microbial substrate, a connection between high calcium content and some unmeasured pH effect, or an unmeasured capability of *L. terrestris* to taste differences between fungi and to select those that are most palatable.

Atlavinyte and Pociene (1973) demonstrated a close correlation between numbers of lumbricid earthworms and algal cell numbers, including green, blue-green algae and diatoms, in pot and plot experiments, and showed by direct examination of earthworm gut contents that algae are an important component of the earthworms' diet. Algae and, to a lesser extent fungi, were found by Piearce (1978) to be a significant component of the diet of six lumbricid species.

Eisenia fetida, hatched from microbiologically sterile cocoons, were introduced by Miles (1963a) to soil that had been sterilized and then re-inoculated with bacteria and fungi. Growth of the earthworms was retarded until mixed cultures of protozoans, mostly ciliates and flagellates were also introduced, and Miles concluded that protozoans were an essential component of the diet of *E. fetida*. Piearce and Phillips (1980) introduced living *Colpidium campylum*, a ciliate, to fluid from the mid-gut of

Lumbricus terrestris and found that they were immediately immobilized and frequently disintegrated, indicating that some ciliates may be digested. Some other ciliates, in contrast, appeared to be resident and unharmed in the midgut of *L. terrestris*.

Rouelle (1983) took cocoons of *Eisenia fetida* and *Lumbricus terrestris*, surface sterilized them by immersion for 20 minutes in 2.5% Javel solution, rinsed them in distilled water, then allowed them to hatch in soil extract agar medium with combinations of antibiotics, to obtain juvenile earthworms that were demonstrably aseptic. The juveniles were then cultured on autoclaved soil and injected (into the gut) with cultures of the common soil amoebae *Saccamoeba stagnicola*, *Thecamoeba* sp. and *Acanthomoeba triangularis*. Subsequent culture of whole sections of the gut and its contents of both species of earthworms revealed no active amoebae of *S. stagnicola* or *Thecamoeba* sp., but cysts of *A. triangularis* were found. Rouelle concluded that amoebae were a component of the diet of earthworms; three hours after injection of the active amoebae none could be found in the gut, indicating that they were quickly killed and digested.

Perhaps the most convincing evidence that microorganisms are significant in the diet is from studies of the fatty acid content of earthworm lipids (Hansen and Czochanska 1974, 1975). The investigation was based on mixed samples of *Lumbricus rubellus* and *Aporrectodea caliginosa*, collected from New Zealand pastures. Of the fatty acids identified a substantial proportion, quantitatively and qualitatively, were those characteristically synthesized by microorganisms especially *iso-*, *anteiso-* and isoprenoid branched-chain fatty acids, and not by animals. Presumably they are assimilated in the digestive tract from bacteria or other microorganisms.

E. Symbiotic and Synergistic Relationships with Microorganisms

Satchell (1967) considered the evidence that the earthworm gut may have an autochthonous microflora. Bassalik (1913) isolated more than 50 species of bacteria from the gut of *Lumbricus terrestris*, while Parle (1963a) cultured organisms from three lumbricid species; both of these authors found no microorganisms that were not common in surrounding soil and plant debris, and Satchell (1967) concluded that there was probably no special gut microflora. There may be exceptions to this generalization. For example, Márialigeti (1979) cultured gut contents from *Eisenia lucens*, a wood-digesting earthworm found in decaying pieces of beech wood in Hungarian forests. By using a suitable medium he found that 73% of all bacteria isolated were of a large number of strains of a gram negative, facultatively anaerobic *Vibrio* sp. which are able to ferment glucose, arabinose and xylose, and some strains could also ferment sucrose. He concluded that this *Vibrio* sp. probably plays an important part in the degradation of ingested food and that it is a true autochthonous gut microorganism. Digestion of wood is not common in earthworms, and Márialigeti's findings may indicate that some species that live in unusual conditions are able to do so because they have symbiotic relationships with microorganisms.

Contreras (1980) extended Márialigeti's work with a study of the actinomycete flora of the gut of *E. lucens*. Of 145 isolates of *Streptomyces* spp., 122 were identified as *S. lipmanii*, which is not a common species. Contreras considered it likely that, in earthworms in general, one or a few actinomycetes may grow, multiply, and become dominant in the gut microflora.

Selective proliferation of a variety of microorganisms in the earthworm gut is further supported by some earlier work. Stöckli (1928) recorded an increase in populations of actinomycetes, pigmented bacteria and bacteria of the *Bacillus cereus* group in soil as it passed through the lumbricid gut. Parle (1963a) showed that actinomycetes and bacteria multiply rapidly as they pass through the gut of *Lumbricus terrestris, Aporrectodea caliginosa* and *A. longa*, while yeasts and fungi do not proliferate; samples of intestinal contents of *L. terrestris* taken from the fore-gut, the mid-gut and the hind-gut showed a logarithmic increase in bacterial and actinomycete numbers, with the populations of the hind-gut about 1000 times those of the fore-gut (means of counts from 25 specimens).

Some explanation of the contradictory results of investigations of the gut microflora may be found in experimental results of Lavelle *et al.* (1983b). The geophagous species *Millsonia anomala* was fed soil with varying substrate quality, established by addition of a series of concentrations of a water-soluble extract of plant leaves. When the earthworms fed on soil with low concentrations of added extract, their casts had higher water-soluble contents and higher respiratory (microbial) activity than the initial soil levels, while in soil with high concentrations of added extract the casts had lower water-soluble contents and lower respiratory activity than the initial soil. Lavelle *et al.* (1983b) suggested, on the basis of these experiments, that at low levels of substrate quality microbial activity in the gut may be stimulated, resulting in release of simple water-soluble compounds from more complex organic materials, while at high levels of substrate quality the gut microflora may not be stimulated.

Satchell (1983b) has investigated the origin of increased phosphatase, apparently produced in the gut of earthworms, and considered to be important in increasing the lability of organically bound phosphorus in soils. Preliminary experiments were conducted with *Eisenia fetida* and

Dendrobaena veneta, fed on cellulose waste from a paper mill, which is relatively free of phosphatase, with 2.5% added phytin (calcium inosotol hexaphosphate) as a phosphate source to simulate the organic phosphorus fraction of plant litter. After three to four weeks' culture at 20°C with *E. fetida* and *D. veneta* phosphatase was determined on faecal material and on control samples of the culture medium, using a modified version of the disodium phenyl phosphate method, in which phosphatase activity is measured as micrograms of phenol liberated per gram of dry weight of substrate. The results showed a value of 631 for *E. fetida* and 756 for *D. veneta* cultures, compared with 50 for control cultures without earthworms, indicating a clear increase of phosphatase activity in earthworm faeces. Further experiments included *E. fetida, D. veneta, Lumbricus rubellus* and *Aporrectodea caliginosa*, with phosphatase activity determined in a series of buffered solutions with pH steps of one pH unit in the range pH 2-10. These experiments showed an increase in phosphatase activity for all four species, with clear increases at pH 3-5 and at pH 9-10. It was concluded that the peak at pH 3-5 was probably due to enhanced microbial activity, while that at pH 9-10 was probably due to alkaline phosphatases, which are generally regarded as typical of earthworms; the possibilty that alkaline phosphatases were also produced by gut microorganisms could not be excluded, but was considered unlikely.

Lavelle *et al.* (1983a) have proposed a symbiotic relationship between earthworms and their gut microflora. They propose that mucus, produced by glands in the anterior region of the gut, passes back through the intestine with ingested food and provides a favourable substrate for symbiotic microorganisms that decompose complex organic compounds in the ingesta, to the mutual benefit of the earthworms and the symbionts. Mucus-secreting cells that produce mucopolysaccharides are known to be present in the anterior portion of the gut in *Eisenia fetida* and *Pontoscolex corethrurus*, and the general presence of these cells in earthworms can be inferred from direct observation of copious amounts of mucus in the intestinal contents of many earthworms. It has previously been considered that the mucus functions as a lubricant but Lavelle *et al.* (1983a) consider it unlikely that such an energy-rich material would function only in this way. They measured the amount of mucus mixed with the intestinal contents of *Millsonia lamtoiana* and *Pontoscolex corethrurus* and calculated that in a field population of *M. lamtoiana* in the Ivory Coast intestinal mucus production would be c. 239 kg $ha^{-1}y^{-1}$, while in a field population of *P. corethrurus* in Mexico intestinal mucus production would be c. 50 t $ha^{-1}y^{-1}$. For *P. corethrurus* this represents 2.2 times the total organic matter content of soil ingested annually by the field population. Little mucus is excreted in the casts of either species studied, and it is most likely that the mucus is resorbed in the posterior intestine. There is no microbiological evidence for the proposed symbiosis.

F. Nematodes and Other Soil Fauna as Food

Piearce and Phillips (1980) found that *Lumbricus terrestris*, feeding on cattle dung with a high nematode population, had living nematodes in the pharynx and oesophagus but none could be found in the crop or gizzard, and considered that they were digested. Nematodes were also recorded as a food item of *Lampito mauritii* by Dash *et al.* (1980). Yeates (1981) compared soil nematode populations in field and glasshouse trials with and without lumbricid earthworms. He found that the presence of earthworms reduced total soil nematode populations by 37%-66% relative to controls without earthworms, and suggested that if soil nematodes were ingested and digested by earthworms it is likely that other living soil microfauna are also taken and may be a significant part of their diet. Lavelle (1981b) observed an individual of the octochaetine species *Agastrodrilus dominicae* as it ingested a living eudrilid earthworm (*Stuhlmannia porifera*), and suggested (Lavelle 1983a) that two other species of this genus (*A. multivesiculatus* and *A. opisthogynus*) are also probably predators of other earthworms.

G. Integumentary Uptake of Nutrients

Uptake of dissolved organic nutrients by absorption through the integument, in quantities sufficient to be of nutritional significance, is well known in many groups of invertebrates, including polychaetes and the enchytraeid *Enchytraeus albidus*. Richards and Arme (1982) have investigated the capabilities of *Eisenia fetida* and *Lumbricus rubellus* to absorb amino acids, monosaccharides and fatty acids from solution. They demonstrated that transintegumentary absorption of all three classes of compounds does occur in both species. However, rates of uptake for all three classes were found to be low. Uptake was shown to have a diffusion and a mediated component, but concentration ratios across the integument did not exceed equilibrium, and the contribution of transintegumentary uptake to the nutritional budget was not considered to be significant. It was suggested that the phenomenon may represent a vestigial nutrient transport system.

H. Digestive Capabilities

The energy consumption rates of earthworms and their significance in the dynamics of energy flow in ecosystems are discussed in Chapter 6. The end result of their feeding, digestion and incorporation of excreta with inorganic soil materials is to greatly affect the decomposition of soil organic matter, the cycling of plant nutrients, soil fertility, plant growth, and soil structure; these effects form the subject of much of Part II of this book. Here it is intended to discuss the physiological capabilities of earthworms to digest the plant, microbial and animal tissues that are ingested at the surface or with soil.

Laverack (1963, pp. 18–23) reported the presence of lichenase, protease, cellulase, chitinase, amylase and lipase in extracts of the tissue of the alimentary tract of earthworms. The principal structural component of plant tissue is cellulose, and the presence of cellulase, apparently produced by cells in the intestinal wall of *Lumbricus terrestris* (Lesser and Taschenberg 1908) was considered to be of particular interest. The enzymes mentioned have distinctive pH optima, but Laverack (1963) reported little variation in pH at various points within the gut (pH 6.4–6.6 in *Lumbricus terrestris*, pH 6.5–7.7 in *Allolobophora* sp., and pH 7.0–7.5 in *Pheretima* sp.), so some of the enzymes, though present, may not be very effective. Nielsen (1962) examined the carbohydrases present in the gut of a variety of soil animals, including some earthworms; he found cellulase in the litter-feeders *L. terrestris* and *Dendrobaena octaedra*, but not in the soil-inhabiting species *Aporrectodea rosea*.

The biochemical properties of enzymes known from the gut of earthworms were reviewed by Jeuniaux (1969) and by Michel and de Villez (1978); some of the enzymes identified from tissue extracts have not been identified from the gut contents, so it cannot be said for certain that they play any part in extracellular digestion.

Neuhauser and Hartenstein (1978) investigated the ability of homogenates of earthworm tissue (*Octolasion tyrtaeum, L. terrestris, Eisenia fetida, Allolobophora chlorotica, Amynthas hupeiensis*) to digest lignin and several phenolic compounds. They found no evidence of lignin decomposition, but found that they were able to convert the ring-carbon atoms of phenolics (e.g. vanillin) to respired carbon dioxide. They pointed out that peroxidase-catalysed reactions with phenolics proceed by an orderly sequence of one-electron steps, through a "ping-pong" mechanism (Cleland 1973) of reaction, followed usually by polymerization (Musso 1967; Pew and Connors 1969; Hartenstein 1973). In view of the observed reactions and the probability of polymerization it was suggested that earthworms may enhance the polymerization of aromatic compounds, perhaps producing humic compounds as an end product. Neuhauser *et al.* (1978) further investigated the possibility of lignin degradation by *E. fetida*, feeding them for 10 days with ^{14}C-labelled synthetic lignin; no ^{14}C-labelled respiratory carbon dioxide was detected from *E. fetida* fed on ring ^{14}C-labelled lignin, methoxyl ^{14}C-labelled lignin, or side chain ^{14}C-labelled lignin.

The evidence of enzymology confirms that of ecological energetics (Chapter 6), that earthworms are able to digest very little of the plant tissue they ingest. Some may digest a little cellulose, others apparently can not. Probably no lignin is digested, but some more simple phenolic materials may be partially decomposed, and polymerized to contribute to humus formation; oligosaccharides and simple sugars are probably digested. The presence of chitinase, lipase and proteases strengthens the probability that fungal and microbial tissue, protozoa and other soil fauna are significant items in the diet (pp. 23–29), but it is not known for certain that the enzymes are all produced by earthworms, and some may be derived from gut microflora.

The digestive capabilities of earthworms and the implications for organic matter decomposition in soils remain an area that would repay further investigation.

I. Palatability and Toxicity

Satchell and Lowe (1967) attempted to identify characteristics of litter correlated with palatability to *L. terrestris*, which feeds primarily on leaf litter. Palatability was shown to correlate positively with nitrogen and soluble carbohydrate content, negatively with polyphenolic content, especially the tannin content of leaves. Heath and King (1964) had shown that concentrations of particular phenolic compounds in leaves were more important in determining palatability to soil animals than total polyphenol content; in particular, high concentrations of gallic and protocatechuic acids were shown to be unpalatable. Satchell and Lowe (1967) tested the effects on palatability to *L. terrestris* of eight phenolic compounds found in leaf litter by offering them filter paper discs impregnated with the compounds. Consumption of discs impregnated with gallic acid, D-catechin, phloroglucinol, protocatechuic acid and tannic acid was respectively 50%, 27%, 26%, 14% and 0% of those taken when discs with no phenolic materials were offered. Swift *et al.* (1979) pointed out that consumption of paper discs may not be strictly comparable with that of leaves and that further experiments, using leaves with differing concentrations of phenolic materials, are necessary to establish a causal relationship.

Satchell and Lowe (1967) noted that microbial degradation of tannins in leaves appeared to increase their palatability to *L. terrestris*. They found no correlation of palatability with the physical toughness or thickness of leaves, but Wright (1972) showed that *L. terrestris* fed little on fresh apple leaves but fed readily on them after leaching with water, when they were softened and consequently more easily fragmented. It might be concluded that leaching with water had removed, or at least reduced the content of, unpalatable leaf constituents, but Wright (1972) also found that *L. terrestris* would feed at similar rates on leaves that had different concentrations of polyphenols but similar texture, and that unsoftened leaf material was readily eaten if it was fragmented so that it could be swallowed whole. Darwin (1881) considered that earthworms have a well-developed "sense of taste", and this was to some extent confirmed by Wright (1972) who found that *L. terrestris*, previously starved for five days, would eat small pieces of aromatic plants (lavender, sage, mint) while those that had been fed previously on softened apple leaves would not feed on the pieces of aromatic plants until they had been starved for two to four days. However, these results can not be interpreted without information on the nature of the aromatic substances contained in the aromatic plants used in the experiments.

A variety of aromatic substances is produced by living and decomposing plant material and by microorganisms associated with decomposition processes, and many of them must be ingested by earthworms. High concentrations of water-soluble polyphenols and of tannins in plant detritus are associated with slow rates of decomposition (Handley 1954, King and Heath 1967, Edwards and Heath 1975). The toxicity to earthworms (*Eisenia fetida*) of 38 low molecular weight aromatic compounds (benzene and hydroxybenzene derivatives), two samples of humic acid and five samples of lignins, was investigated by Hartenstein (1982). The data indicated that the presence of aldehyde, methoxyl, methyl or hydroxyl groups on low molecular weight compounds is potentially toxic to earthworms and that they are repelled by such compounds; high molecular weight compounds with these groups, e.g., humic acids and lignins, are non-toxic and are tolerated. Hartenstein concluded that there may be a delicate balance in the timing of release of some low molecular weight decomposition products from plant materials, their polymerization and their ingestion by earthworms.

2
The Physical Environment

I. Water Relationships

If soil-inhabiting earthworms are kept in aerated fresh water, so that they become fully hydrated, their water content is c. 80%-90% of body weight. Earthworms collected from the field or maintained in laboratory cultures are partially dehydrated, with water contents of c. 65%-75% (Oglesby 1969, 1978). Though they will live continuously in water for long periods in the laboratory, if given a choice they move out of the water into moist soil, where they lose some water.

Because of their cutaneous respiratory system, which requires the maintenance of a moist body surface, and their excretion of nitrogen as ammonia and urea, which are toxic and thus require copious losses of hypotonic urine, earthworms are obliged to lose body water in large quantities. Wolf (1940) showed that *Lumbricus terrestris*, kept in water, produces urine at a rate of c. 60% of body weight per day. This rate of water loss would be intolerable in most terrestrial situations, but it is likely that losses of 10%-20% of body weight per day are common, and this water must be replaced if earthworms are to survive. In terms of their respiratory and excretory physiology earthworms more closely resemble aquatic than truly terrestrial animals, and though some species are able to survive seasonal or intermittent desiccation by entering a resting stage, a population can not persist unless it can find moisture conditions that satisfy the individuals' requirements for active life at a frequency and for a duration sufficient to ensure their survival and reproductive success.

It will be seen in the discussion that follows that earthworms require water in large quantities, and that they can acquire it only from their food or by direct imbibition of water from surface films or water-filled pores in the soil or other habitats. They must obtain oxygen by absorption in solution through the cuticle, and though a few species have special morphological adaptations that enable them to live in saturated soils and obtain oxygen from the soil surface (p. 50), most species can not survive for long in saturated soils, except where oxygenated groundwater flows downslope on hillsides. Most earthworms have limited physiological and behavioural defences against desiccation, but are able to survive loss of much of their body water. In general, active life is restricted to periods when soils contain moisture sufficient to maintain plant growth (pF 2.0–4.2), except that some species have burrows that extend from the surface deep into the soil, and so are able to spend short periods in desiccating conditions but return to spend longer periods at favourable soil moistures.

A. Water Uptake and Loss

Experiments of Ghabbour (1975) have shown that *Aporrectodea caliginosa* and *Metaphire californica* are unable to absorb water from saturated air. The earthworms were suspended on fine-mesh nylon gauze in sealed vessels, containing at the bottom various dilutions of sulphuric acid to provide constant levels of relative humidity. Experiments were conducted at room temperature (19°C–22°C). Both species survived for three days at 100% R.H.; they survived only two days at 97% R.H.; *A. caliginosa* also survived for two days at 96% R.H. and at 90% R.H., but *M. californica* survived only one day; *A. caliginosa* survived for one day at 82% R.H., while *M. californica* did not survive; neither species survived a day at 72% R.H. Water is lost even at 100% R.H., and Ghabbour (1975) showed in some experiments with *A. caliginosa* that the weight loss was equal to the weight of urine produced, but in other experiments weight loss greatly exceeded urine production, indicating loss from the body surface. *Metaphire californica* consistently lost water more rapidly than *A. caliginosa*. Ghabbour concluded that water can be taken up by earthworms only by ingestion or absorption of free water from the soil. The ability of *A. caliginosa* to survive at lower humidities than *M. californica* was related to previous work (Khalaf El-Duweini and Ghabbour 1968) that showed their soil moisture preferenda in a clay soil to be 20%–45% and 35%–55% respectively.

Carley (1978) similarly found that *Lumbricus terrestris*, kept at R.H. 70%–80%, lost water rapidly; after 24 hours only two of ten earthworms survived at this relative humidity, and the two survivors had lost 55% and 63% respectively of their body water. At 100% R.H. Carley found that body weight of *L. terrestris* declined to *c.* 80% in 24 hours, probably due to mucus and urine production, but thereafter body weight declined only slowly for several days. Differences in rates of water loss at 100% R.H. and at 70%–80% R.H., and changes in rates were observed by Carley, who kept ten *L. terrestris* in sealed culture vessels in which the water vapour content of the air initially contained 5% D_2O (heavy water), and measured the rate of dilution of the D_2O. Only two of ten earthworms survived 24 hours at 70%–80% R.H., and for these the D_2O saturation of the air was < 50% of the initial value, while all ten earthworms survived at 100% R.H. and the D_2O concentration fell only to *c.* 85% of the initial value. Measurement of the osmotic pressure of coelomic fluid at 100% R.H. and at 70%–80% R.H. showed that *L. terrestris* can not control the osmotic pressure as dehydration progresses; there was an increase of 9 kPa for each 1% of body weight lost as water. However, large increases in osmotic pressure, from an initial 450 kPa to *c.* 900 kPa were tolerated. The rate of water loss was shown to reduce as the humidity decreased, indicating that there is some physiological control of water loss, probably through regulation of urine excretion and integumental water permeability. Carley (1978) drew attention to the similar capabilities of some polychaetes, brachyuran crabs and amphibia when subjected to desiccation and to the ability of trout, *Salmo gairdneri*, to reduce its integumental permeability when it moves from fresh water to the sea, and concluded that control of permeability may be a widespread phenomenon among aquatic and semi-aquatic animals.

B. Permeability of Cuticle

It has been demonstrated (Zimmermann 1971, 1973; Carley 1975) that *Lumbricus terrestris* can reduce the permeability of its body surface in response to desiccation, and that the regulatory mechanism may be hormonal.

The cuticle of earthworms consists of layers of unbanded collagen fibres, with successive layers lying at right angles to one another, embedded in a fine fibrillar matrix containing mucopolysaccharides, and it has small epicuticular projections. Gland cells in the underlying epidermis secrete mucus, which acts as a lubricant for the body during burrowing and maintains the moist surface layer necessary for respiration. The mucus-producing cells are of three types, one of which, described by Richards (1978) as reticulate monochromatic cells, produces a carboxylated acid mucus with low viscosity that is trapped by the epicuticular projections and provides the respiratory surface film. Interspecific differences in the nature of the collagen have been recorded, e.g., its thermal transition point differs between *Aporrectodea caliginosa* and *Eisenia fetida* and the difference is correlated with the species' temperature preferenda (Richards 1978).

It is known that the permeability of the skin of amphibia is inversely related to its lipid content, and Ghabbour (1976) examined the possibility that this may also be true of earthworms. He determined the total body lipid content of *A. caliginosa*, which has an optimum soil moisture content of 35%, *Metaphire californica*, with an optimum level of 45%, and *Alma stuhlmanni*, which is aquatic and does not survive for long in soils with < 55% moisture content. The lipid contents of the three species were respectively 5.77%, 3.46% and 10.65% on a dry weight basis. *Aporrectodea caliginosa* and *M. californica* both had a low lipid content, but if this were the only factor influencing their ability to absorb or lose water through the cuticle, it would be expected that *A. caliginosa* would have a lipid content lower than that of *M. californica*. Tolerance of water loss and of moisture availability from the environment would also be expected to relate to the osmotic pressure of body fluids. Examination of the ash (inorganic salt) content of the two species (Ghabbour 1976) showed that *A. caliginosa* contained 5.31% and *M. californica* 19.18% of dry weight. Ghabbour concluded that this difference enabled *A. caliginosa* to lose more water than *M. californica* before suffering acute physiological stress. These two species have much lower lipid contents than the aquatic *A. stuhlmanni* and it may be inferred that this gives them an advantage in tolerating desiccation; the additional advantage of a lower osmotic pressure in the body fluids of *A. caliginosa* may compensate for its higher lipid content compared with *M. californica*.

C. Reaction to Desiccation

Some earthworms survive drought by entering resting stages, in which their rate of water loss is reduced (p. 85). Other species can tolerate the loss of much of their body water, going into a torpid state during periods of desiccation, followed by rehydration and return to an active life when soil water again becomes more plentiful. For example, Schmidt (1918) recorded that a specimen of *Eisenia fetida* lost 61.6% of its body weight (c. 70% of its water content) in a dry atmosphere, but recovered when placed on a moist surface; Hall (1922) similarly revived a specimen of *Allolobophora chlorotica* that had lost water equivalent to 69.6% of its body weight (c. 80% of its water content).

Lethal limits of water loss for *Amynthas hupeiensis*, *Aporrectodea caliginosa* and *Eisenia fetida* were determined by Grant (1955a) by keeping the earthworms (after evacuation of the gut in a moist environment) over calcium chloride, at 20.5°C, and measuring weight losses. *Amynthas hupeiensis* died after losing 48.6% of body weight (59.2% body water), *A. caliginosa* after losing 63.4% of body weight (74.7% of body water), and *E. fetida* after losing 58.8% of body weight (70% of body water). Similar experiments of Madge (1969) with *Hyperiodrilus africanus* and *Eudrilus eugeniae* showed 50% mortality after loss of 45% of body weight (51% of body water) for *H. africanus* and after loss of 43.3% of body weight (52% of body water) for *E. eugeniae*. One hundred per cent mortality resulted after loss of 69.4% of body weight (79% of body water) for *H. africanus* and after loss of 62.3% of body weight (74% of body water) for *E. eugeniae*. For *Microchaetus modestus*, a South African microchaetine, Reinecke and Ryke (1970a) recorded a lethal limit of 53% loss of body weight (61% loss of body water). Comparing *M. modestus* with *A. caliginosa* (from Grant 1955a) Reinecke and Ryke found 26% of all specimens of *M. modestus* in soils with moisture content < 18%, 12% at moisture content < 15% and only 2% at moisture content > 25.2%, while Grant found that *A. caliginosa* preferred 29.9% moisture content and was absent at moisture contents < 18%.

Madge (1969) introduced *H. africanus* and *E. eugeniae* to soils with a graded series of moisture conditions, to test their preferences. Surface soil was dried at 25°C, crumbled and mixed, then subsamples were wet to 0.9%, 3.8%, 7.8%, 12.5%, 17.2% and 22.4% water content. The subsamples were put in boxes, with the soil of each moisture content separated from its neighbours by a vertical band of air-dried soil. A hundred specimens of mixed sizes of each species were then evenly distributed over the soil surface. After 48 hours the number of surface casts on each moisture-sector were counted, the soil was removed and the earthworms from each sector were counted. About 65% of the earthworms were found in the sectors with 12.5% and 17.2% moisture. No casts were found in sectors with moisture levels up to 7.8%, and > 80% of casts were found in the sectors with 12.5% and 17.2% moisture content. Further confirming the moisture preferences of these species, Madge (1969) distributed 90 earthworms evenly over the surface of a box of soil evenly wet to 35% moisture content, and after 48 hours found that they were evenly distributed in the soil.

Several physiological mechanisms for water conservation are known in earthworms, but they are relatively inefficient, and each is known to operate only in a limited number of species. They include limited control over water loss through the cuticle (p. 35), regulation of urine output (p. 34), excretion of urine into the gut with subsequent resorption of much of its water content (p. 212) and some evidence that in some species calciferous glands may have a role in reducing water loss, although in other species they may be responsible for continuous loss of water (p. 224).

D. Significance of Soil Water Content in Field Conditions

Variation in soil moisture is a critical factor in determining earthworm populations, as well as in controlling levels of activity of individuals, and this is borne out by many population studies (see, for example, data of Phillipson et al. [1976], p. 126 and Table 13).

Optimum soil moisture contents for earthworm activity are not the same for all species and it is apparent that even within species there is considerable scope for adaptation. Ljungström and Emiliani (1971) and Ljungström et al. (1973) found that in seasonally arid regions of Santa Fe Province, Argentina, the peregrine species *Aporrectodea caliginosa, A. rosea* and *Microscolex dubius* were abundant in soils with c. 25% moisture content and were present, though in decreasing numbers, down to c. 15% moisture content. In contrast, Baltzer (1956) and Zicsi (1958c) reported that lumbricids in Europe go into diapause below c. 25%–30% soil moisture content and practically all die below c. 20% soil moisture. Experimental determination of moisture preferences of earthworms from semi-arid South African soils showed that *A. caliginosa* preferred soil moisture levels of c. 30% and avoided soil with < 18% moisture, while for the endemic *Microchaetus modestus* < 2% of individuals were found at soil moisture levels > 25.2%, 26% were found at < 18% and 12% at < 15% moisture levels (Reinecke and Ryke 1970a). Lumbricids in Argentina were better able to tolerate drought than introduced and endemic megascolecoid and glossoscolecoid species, while in South Africa the endemic microchaetines have the advantage over lumbricids; in both countries lumbricids apparently tolerate soil moisture levels lower than they do in Europe.

Moisture tolerances of earthworms might better be expressed in relation to matric potential, or pF, which is a measure of the free energy of water held in a soil and relates to the energy required to extract water from soil, rather than as per cent moisture on a dry-weight basis, which gives no precise indication of the energy required to extract water.

Lavelle (1974) related rates of food consumption and growth to soil moisture in *Millsonia anomala* (Megascolecidae) from savanna lands of the Ivory Coast. At pF 4.2, which is about the permanent wilting point of plants, c. 50% of field populations were inactive and some were inactive at pF 3. Maximum activity was at c. pF 2.5–2.0, which is near field capacity; as pF became lower the soils became saturated, with increasing mortality. In experimental conditions, using soil ground and sieved to 1 mm mesh size, juvenile *M. anomala* died after a few days at pF 4.2 (4.9% soil moisture content), fed but lost weight at moisture levels up to c. pF 3 (8.9% soil moisture content), attained maximum growth rate (6.5% increase in weight per day) at pF 2.5 (11% soil moisture), continued to feed and grow at high but declining rates up to pF 2 (17% moisture), then declined rapidly until growth rate became nil or negative, with high mortality, at pF 0 (flooded soil, moisture > 30%). Adult *M. anomala* reacted similarly, but required slightly more moisture at the lower limit to avoid death.

Lumbricidae from various localities in Sweden (*Allolobophora* s.l. spp.) were active in surface soil layers in the range pF *c.* 2.0–3.2, and activity declined at pF > 3.2–3.5. *Lumbricus terrestris* remained active when pF exceeded 4.3 in the 0–5 cm soil layer (Nordström 1975, Nordström and Rundgren 1972), but *L. terrestris* lives in deep burrows and needs to be exposed to surface-soil conditions only for short periods when it emerges to feed.

Bouché (1983), on the basis of non-parametric analyses of Mazaud (1979), related earthworm activity (proportion of active earthworms, captured by behavioural techniques, to total population, estimated by passive techniques) to pF and ambient temperature. He defined limits of pF within which earthworm activity exceeded the mean of measurements throughout the year. For all species present, mean activity was exceeded between pF 2.31 and 2.82, over the temperature range 5.1°C to 17.9°C, with optimum levels between pF 2.31 and 2.57, within the temperature range 12.8°C to 17.9°C. Some differences were evident between the ecological categories of earthworms defined by Bouché (1971; p. 107). The pF limits for anéciques were low (pF 2.06 to 3.08), for endogées higher (pF 2.31 to 3.33), while those of épigées were not clearly defined.

The effects of excess soil water result from respiratory stress, and they are discussed in Section III of this chapter.

II. Temperature

It is not easy to define soil temperature for particular earthworms in field conditions, as soil temperature varies with depth and with diurnal and seasonal changes in ambient temperature. Bouché (1983) used a miniaturized sucrose inversion method developed by Ranc (1980) to measure the internal temperature of the earthworms themselves. This method offers possibilities for a better understanding of the significance of ambient soil temperatures, but has as yet been little used.

It is, however, relatively easy to determine upper and lower temperature limits, optimum temperatures, the effects on temperature tolerances of previous acclimatization, on individual species in the laboratory and, to a limited extent, field populations, and to interpret the findings in relation to other ecological, behavioural and physiological data.

A number of vital requirements of earthworms make it difficult to isolate the effects of ambient temperature on earthworm behaviour or on population parameters. For example:
(a) high temperatures are often associated with drought, and therefore with moisture stress;
(b) in the absence of moisture stress, earthworms can more easily extract water from soil as temperatures increase, because pF decreases with increasing temperature;
(c) metabolic rate varies with body temperature.

For earthworms Q_{10} is about 2, i.e., metabolic rate increases c. 2 times for an increase in body temperature of 10°C (p. 139). Some evaporative cooling is possible (Hogben and Kirk 1944), so that body temperature is not necessarily the same as ambient temperature, but evaporative cooling necessitates high rates of water loss and can be tolerated for only a short time unless the water lost can be replaced from the surroundings.

A. Temperature Limits

Upper lethal temperature limits have been determined for a variety of earthworms, lower limits for only a few species (Table 2). The conditions of exposure to experimental temperatures and the level of mortality regarded as significant vary, so it is not possible to directly compare some of the data. The temperature that causes 50% mortality (LT_{50}) has been used most commonly, but exposure times are varied.

It is apparent from the data of Reinecke (1974) for *Aporrectodea rosea* and *Microchaetus modestus* and of Madge (1969) for *Eudrilus eugeniae* and *Hyperiodrilus africanus* that earthworms can survive for short periods at temperatures much higher than can be tolerated for longer periods. The high lethal temperatures (LT_{50}) recorded by Khalaf El-Duweini and Ghabbour (1965b) were for only 30 minutes' exposure; LT_{50} for periods of 2880 minutes (48 hours) are probably more realistic. It seems most likely that in field conditions all earthworms have upper lethal temperature limits within the range c. 25°C–35°C, if exposed for long enough to come to equilibrium with ambient temperature. For species that are native to temperate climates limits are generally in the lower half of the range, while for species native to hotter climates limits are near the top of the range.

The upper lethal temperature limits are remarkably low. Death may be due to inability of respiratory exchange to maintain adequate oxygen supply to the tissues as metabolic rate increases with increasing temperature (Q_{10} = c. 2, p. 139). Rigby (1967, 1968) showed that collagen fibres from the body wall of *Digaster longmani*, *Amynthas megascolidioides* and *Aporrectodea caliginosa* melted when incubated at 22°C in hydrochloric acid at pH 1.0,

Table 2. Lethal temperature limits and optimum temperatures of earthworms.

Species	Location	Lethal temperature limits (°C)				Optimum (°C)	
		Conditions of exposure	Lower limit	Upper limit	Conditions of of exposure	Temperature	Reference
Lumbricus terrestris	England	400 min exposure	—	27.5–28.5			Wolf 1938
	England	720 min exposure	—	29			Hogben and Kirk 1944
	England				Field observations	10.5	Satchell 1967
Lumbricus rubellus	Germany				Field observations	15–18	Graff 1953b
Aporrectodea caliginosa	USA	LT$_{50}$ 2880 min exposure	0	26.3	Laboratory exposure	10–23.2	Grant 1955b
	Egypt	LT$_{50}$ 30 min exposure in water bath	—	39.6			Khalaf El-Duweini and Ghabbour 1965b
		in soil	—	40.8			
	Germany				Field observations	12	Graff 1953b
Allolobophora chlorotica	Germany				Field observations	15	Graff 1953b
Aporrectodea longa	England	LT$_{20}$	—	25.7	Laboratory exposure	10–12	Miles 1963b
	Sweden						Nordstrom 1975
Bimastos samerigera	Israel		−1.3	32.4	Laboratory	14.7–27.8	Bodenheimer 1935
Eisenia fetida	USA		—	35–40			Smith 1902
	USA	LT$_{50}$ 2880 min exposure	0	24.7	Laboratory	15.7–23.2	Grant 1955b

Table 2—continued

Species	Location	Lethal temperature limits (°C)				Optimum (°C)	
		Conditions of exposure	Lower limit	Upper limit	Conditions of exposure	Temperature	Reference
	USA	LT_{70} and LT_{30} in long-term cultures	$5(LT_{30})$	$33(LT_{70})$	Laboratory	20–29	Kaplan et al. 1980a
	England		−(1.6–2.0)	28			Schmidt 1918
	England	LT_{50}	—	33.3			Miles 1963b
	Germany				Field observations	25	Graff 1953b
Aporrectodea rosea	South Africa	LT_{50}: 4.8 min exposure	—	34.3			
		15.5 min exposure	—	33.9			
		31.8 min exposure	—	32.9			Reinecke 1974
		83.0 min exposure	—	32.0			
		LT_{50} calculated for 2880 minutes exposure					
	South Africa		—	29.7	Laboratory	25.0–26.9	Reinecke 1975
	Germany				Field observations	12	Graff 1953b
Dendrodrilus rubidus	Germany				Field observations	18–20	Graff 1953b
Octolasion cyaneum	Germany				Field observations	15	Graff 1953b
Lumbricidae	USA				Field observations	2.2	Hopp and Linder 1947

2. The Physical Environment

Species	Country	Exposure	Temp	Range	Condition	Reference
Amynthas hupeiensis	USA	LT_{50} 2880 min exposure	—	24.9	Laboratory	15–23	Grant 1955b
Metaphire californica	Egypt	LT_{50} 30 min exposure in water bath in soil	— —	37.1 37.8	Laboratory	26–35	Khalaf El-Duweini and Ghabbour 1965b
Eudrilus eugeniae	Nigeria	LT_{100}: 60 min exposure 720 min exposure 180 min exposure	— — 7.5	38.5 34.5 —	Laboratory	23.0–31.5	Madge 1969
	USA				Laboratory culture in activated sludge in horse manure	20–28 24–29	Neuhauser et al. 1979
Hyperiodrilus africanus	Nigeria	LT_{100}: 60 min exposure 720 min exposure 180 min exposure	— — 7.5	37.0 34.5 —	Laboratory	24–26	Madge 1969
Microchaetus modestus	South Africa	LT_{50}: 19.8 min exposure 26.6 min exposure 109.6 min exposure 392.6 min exposure LT_{50} calculated for 2880 min exposure	— — — — —	34.8 34.2 32.4 31.0 29.0	Laboratory	21.0–26.9	Reinecke and Ryke 1972, 1974
Alma nilotica	Egypt	LT_{50} 30 min exposure in water bath in soil	— —	38.8 37.8	Laboratory	24–26	Khalaf El-Duweini and Ghabbour 1965b

and at 34°C–40°C in physiological saline, and suggested that this might account for death at the upper limits of tolerance.

Lumbricidae are known to survive in soils where surface layers are frozen in winter, but the few results in Table 2 show that they die at temperatures not much below 0°C. It is, however, claimed (M.S. Ghilarov, pers. comm.) that *Eisenia nordenskioldi* in Russia will revive after long periods of being entirely frozen. The physiological mechanisms that make this possible are unknown. The only data for lower temperature limits of tropical species (Madge 1969) show LT_{50} after three hours at the remarkably high temperature of 7.5°C for *Eudrilus eugeniae* and *Hyperiodrilus africanus*. Prolonged exposure to temperatures much below 0°C probably results in death due to freezing of coelomic fluid or cell contents, but it is difficult to imagine what might cause death at 7.5°C for the two Nigerian species. Byzova (1974) concluded that an observed decrease in blood volume, implying an increase in ionic content, was important in lowering the freezing point of the blood of lumbricid earthworms that hibernate during winter in the frozen upper soil layers around Moscow.

When given the opportunity, earthworms move quickly away from temperature extremes, seeking optimum conditions. In field conditions most species move deeper into the soil to escape extremes, which usually only affect superficial horizons; even surface-inhabiting species, that have no burrows, may find refuge in cracks in the soil, and under logs or stones.

B. Acclimatization to High Temperatures

There is no evidence that acclimatization at increasing temperatures can increase the absolute maximum temperature that can be tolerated, but there is evidence that it can increase the duration of survival at temperatures approaching the lethal limit.

Grant (1955b) kept cultures of *Amynthas hupeiensis* at 4°C, 9°C and 15°C for one month and then recorded LT_{50} for 2880 minutes (48 hours) exposure. The LT_{50} increased 0.3°C for each 1°C rise in conditioning temperature. As conditioning temperatures were increased to approach 25°C, the effects of conditioning rapidly reduced to zero. *Amynthas hupeiensis* were kept for a month at 15°C, then removed to 22°C, and then after 1, 2, 4, 6, 10, 12, 14 and 16 days removed to a chamber at 24.1°C and tested for survival up to 2880 minutes (48 hours). Conditioning at 22°C made little difference for up to six days, then there was a rapid increase in mean survival time, from c. 10–23 hours to 100% survival after 48 hours after conditioning at 22°C for 10–12 days. Reinecke (1975) showed that the preferred temperature for *Aporrectodea rosea* acclimatized for one week at 20°C was 24.1°C–25.6°C, and for specimens acclimatized at 25°C it was

24.8°C–26.0°C; there was no difference in the lethal limit of earthworms subjected to the two treatments.

Mangum (1978) suggests that thermal acclimatization in earthworms results from variations in oxygen consumption, and is regulated by neuroendocrine secretions that resemble several vertebrate endocrine hormones. The secretions trigger a series of reactions that result in qualitative changes in glucose metabolism and therefore in quantitative changes in oxygen consumption (Rao 1966). Mangum (1978) found little conclusive evidence of increased blood flow with increasing temperature (but see p. 49); he concluded that there may be such increases, and perhaps increases in the oxygen-carrying capacity of the respiratory pigments, paralleling increasing body temperature.

C. Optimum Temperatures

Temperature optima of earthworms have been estimated in two ways, as follows:
(a) field observations of conditions under which feeding, weight gain and reproduction are apparently at a maximum;
(b) laboratory determinations of the preferred range of temperature when earthworms are introduced to soil, usually in a long narrow trough, in which a continuous temperature gradient is established by heating at one end and cooling at the other.

The first method probably provides more realistic data since reproductive success is the critical requirement for survival and maintenance of populations. The second method provides information on the range of temperatures under which survival of the individual is possible. Optimum temperatures, based on either of the two methods, are listed for 16 species in Table 2.

On the basis of field observations Lumbricidae in Europe generally have optima in the range 10°C–15°C, which would correspond with expected soil temperatures in near-surface soil horizons during the spring and autumn months, when earthworms are most active. *Dendrodrilus rubidus*, which is a surface or litter inhabitant, and *Lumbricus rubellus*, which lives close to the surface, have optima in the range 15°C–20°C, while *Eisenia fetida*, which is most common in compost and dung heaps, where temperatures are high due to fermentation, has an optimum of 25°C. Laboratory determinations of temperature optima for Lumbricidae extend over a rather wider range than is apparent from field observations.

It is apparent that some European species that have become established in other parts of the world have very different temperature optima. Reinecke (1975) established the optimum range in laboratory conditions for

Aporrectodea rosea in South Africa as 25.0°C–26.9°C, while Graff (1953b), in Germany, concluded that field populations of *A. rosea* had an optimum of 12°C. Although the methods of determination were different and it is possible that the data indicate that *A. rosea*, as defined, includes more than one species, it is likely that they reflect a real difference resulting from selection for tolerance of high ambient temperatures in field populations, because for other lumbricid species where both methods have been used the optimum for field observations is within the range determined by laboratory experiments. The low optimum (2.2°C) determined by Hopp and Linder (1947) for Lumbricidae in maize fields in Maryland must be regarded as doubtful, and perhaps mistaken, but they may reflect a real difference resulting from selection for tolerance of low ambient temperatures.

The middle eastern lumbricid *Bimastos samerigera* and all species of other families listed in Table 2 have temperature optima in the range 15°C–30°C, all but two (*B. samerigera* and *Amynthas hupeiensis*) in the range 20°C–30°C. All are native to tropical or subtropical regions, where soil temperatures are higher than in northern Europe.

D. Relationships between Temperature Optima and Soil Moisture Content

Reinecke (1975) determined the preferred temperature range of *Aporrectodea rosea* in soil maintained at 28%–31% moisture content and at 13%–17% moisture content. For earthworms acclimatized for seven days at 20°C the preferred temperature range was 24.1°C–25.6°C at the higher soil moisture content and 17.6°C–21.8°C at the lower soil moisture content. Field observations in Swedish forests showed that *Allolobophora* spp. became inactive at pF 4.0 and 20°C, while at pF 4.3 they became inactive at 17°C (Nordström and Rundgren 1972), while Nordström (1975) reported that activity declined in *Allolobophora* spp. at pF > 3.2–3.5 and temperatures above 14°C–16°C.

E. Significance of Temperature Tolerances in Field Conditions

Earthworms that live at or above the soil surface may find temporary shelter from high or low ambient temperatures under logs or stones or in crevices in the soil, in the bark of trees or in deep litter layers. They generally have short generation times (see p. 103 and Fig. 13). Adult mortality is high during seasonal extremes of high or low temperatures but embryos survive in cocoons and populations recover rapidly with the restoration of conditions suitable for active life.

Earthworms that burrow in the soil can escape temperature extremes by retreating to deeper soil horizons, and some are capable of burrowing to depths of 5 m or more. High mortality is related more closely to seasonal variations in soil moisture than in soil temperature (see Table 13, p. 128).

III. Respiratory Requirements and Adaptations

The respiratory system of earthworms is simple, oxygen and carbon dioxide exchange taking place through the cuticle, which is underlain by a network of capillary blood vessels. Within the limitations imposed by this simple system, morphological and behavioural adaptations, the capacity for respiration in air or in oxygenated water, and some capacity for anaerobic metabolism enable them to survive in varied and sometimes apparently hostile environments.

A. Respiratory Function

Capillary blood vessels ramify within the body wall of earthworms, in the circular muscle layer and in some species penetrate between the epidermal cells, so that they lie very close to the inner surface of the cuticle (Stephenson 1930). Exchange of oxygen and carbon dioxide takes place through the cuticle, which must be kept moist so that gases can pass in solution across the surface layer. The degree of vascularization of the body wall varies with its thickness (Weber 1978a). In some species special areas of the body wall are modified for respiratory exchange, e.g., in *Aporrectodea caliginosa* the lateral regions of segments 9-13 have a thinner than normal cuticle, the epithelial cells are cubical or flattened, instead of six to seven times as high as broad, and there is a proliferation of subcutaneous capillaries (Stephenson 1930), while in others that live in habitats where oxygen is in short supply the caudal region is highly modified as a respiratory organ (p. 50).

The respiratory pigment of earthworms is erythrocruorin, a high molecular weight form of haemoglobin (M.W. range $2.7-3.9 \times 10^6$ u; Weber 1978b), which is extracellular, i.e., it is carried in solution in the blood, not in blood corpuscles or similar cells. Erythrocruorin has a very high affinity for oxygen, with loading-unloading tensions considerably lower than those of mammalian haemoglobin. There is some conflict of evidence on the significance of the erythrocruorin in oxygen transport (see review of Weber, 1978b). Much oxygen may be simply dissolved in the blood plasma and the pigment may be significant only when ambient partial pressure of oxygen is low. Weber (1978b) quoted Cosgrove and Schwartz (1965) who found that in *Lumbricus terrestris* erythrocruorin transports oxygen at all environmental tensions including atmospheric, but that its role appeared to be maximal when the partial pressure of oxygen was between 4 kPa and 10 kPa (30–80 mm Hg) decreasing at higher tensions which apparently hinder oxygen unloading in tissues, and at lower tensions, when loading at the respiratory surface is reduced. Cosgrove and Schwartz (1965) found that at the oxygen concentrations characteristic of the air in loamy soils (16%–19%) erythrocruorin is responsible for the transport of only 15%–20% of the oxygen consumed, and concluded that its increasing role with decreasing oxygen concentrations was an important factor in the ability of earthworms to "oxy-regulate", i.e., to maintain the level of oxidative metabolism. Most of the data on the oxygen transporting capabilities of erythrocruorin derive from experiments that involve poisoning with carbon monoxide. Weber (1978b) points out that high concentrations of carbon monoxide not only poison the respiratory pigment but also cytochromes, thus inhibiting cellular respiration, so the conclusions are valid only if control experiments show that the carbon monoxide concentrations used do not inhibit respiration in isolated tissue.

The erythrocruorin content of earthworm blood was shown by Byzova (1975) to increase through a series of habitats with progressive deterioration of oxygen supply, as follows. Blood erythrocruorin concentration was measured in *Lumbricus rubellus*, which lives near the soil surface, *Aporrectodea caliginosa*, from deeper soil layers, *Eisenia fetida*, from compost, *Octolasion lacteum*, from wet soils, and *Eiseniella tetraedra*, from wet soils and flooded ground. The mean erythrocruorin content of the blood of these five species (mg/g) was 3.1, 3.6, 4.2, 5.0 and 7.8 respectively.

The respiration rate of earthworms apparently differs between species, between size classes within species, and with seasonal temperature differences and other environmental variables. There is much contradictory evidence of the significance of observed differences, and there is a need for critical experiments to clarify the conflicting results. The present state of knowledge of factors that affect respiration may be summarized as follows:
(a) body size: Satchell (1967, 1970) pointed out that respiratory rates, at constant temperatures, were higher for small than for large individuals of the same species, a relationship that holds throughout the animal kingdom. Byzova (1965a) found a clear relationship between size and respiratory rate for species that are active for at least part of the time at the ground surface (*Dendrobaena octaedra, Eiseniella tetraedra,*

Lumbricus castaneus, L. terrestris, Octolasion lacteum), but found little size-related difference in subsurface dwelling species (*Aporrectodea caliginosa, A. rosea*) from sampling sites near Moscow. Maldague (1970) showed that oxygen consumption (per gram of fresh weight) of *L. terrestris* was highest (*c.* 90 μL $g^{-1}h^{-1}$) for small specimens (*c.* 1 g fresh weight) and fell, following a hyperbolic curve, to a fairly steady rate (*c.* 50 μL $g^{-1}h^{-1}$) for specimens of *c.* 4 g fresh weight and larger. The mean oxygen consumption rate of *L. terrestris* was *c.* 70 μL $g^{-1}h^{-1}$, while that of small forest-litter dwelling species was *c.* 114 μL g^{-1} h^{-1}. Phillipson and Bolton (1976) calculated mean annual respiratory rates (per gram of fresh weight; μL g^{-1} h^{-1}) at 10°C of adult, large immature and small immature *A. rosea* as 64.17, 72.66 and 78.56 respectively. Phillipson and Bolton (1976) also found that mixed size groups of the litter-dwelling species *L. castaneus* and *Dendrodrilus rubidus* had mean annual oxygen consumption rates (μL $g^{-1}h^{-1}$) at 10°C of 155.83 and 112.02 while the subsurface species *Octolasion cyaneum* consumed 69.35 μL g^{-1} h^{-1}. However, when the data were corrected by subtracting the weight of gut contents the relationships to surface or subsurface habit were no longer apparent; oxygen consumption rates (μL $g^{-1}h^{-1}$) for the litter-dwelling species *L. castaneus* and *D. rubidus* were respectively 194.79 and 142.22, while those for the subsurface species *A. rosea* and *O. cyaneum* were respectively 95.70 and 139.28. No size specific relationships were apparent either within or between species.

(b) activity: Most respiratory measurements are made on resting animals, but there are diurnal and seasonal differences in levels of activity that are reflected in rates of oxygen consumption. Raffy (1930) measured a maximum rate of oxygen consumption by *L. terrestris* 13% higher than the minimum over a 48-hour period. The differences are probably related to temperature changes, and are further discussed below.

(c) oxygen tension: Respiration rates are not much affected by oxygen tensions down to about half that of normal atmospheric concentrations. Oxygen consumption rates of *L. terrestris* at 10% and 5% oxygen concentrations were shown by Johnson (1942) to decrease respectively to 93% and 47% of that at atmospheric concentration (20%). Morphological, behavioural and physiological adaptations that enable some earthworms to obtain oxygen when they live in near-anaerobic environments are discussed below (pp. 50–53).

(d) ambient temperature: Respiration rates increase with increasing temperature, and the increase is associated with increasing rates of pulsation of the dorsal blood vessel, resulting in increasing rates of blood flow (Rogers and Lewis 1914), and with temperature-induced changes in the dissociation curve of erythrocruorin. The effect is modified by seasonal acclimatization. Saroja (1961) found the oxygen consumption rate of *Lampito mauritii*, at 20°C, was 33% higher for large individuals and 300% higher for small individuals collected during winter than rates for individuals collected during summer. In experimental conditions Phillipson and Bolton (1976) demonstrated that oxygen consumption per unit weight of *A. rosea* increased steadily with increasing temperatures of 6°C-15°C.

(e) exposure to light: The respiratory rate of *L. terrestris* exposed to bright light was shown by Davis and Slater (1928) to be *c.* 100% higher than the rate in darkness and to increase by *c.* 30% when the animals were transferred from darkness to diffuse daylight.

(f) contact with soil: The oxygen content of the blood of the giant Brazilian earthworm *Glossoscolex giganteus* was shown by Johansen and Martin (1966) to increase when the worms were placed in contact with soil. Laboratory determination of oxygen content in the anterior portion of the dorsal blood vessel showed concentrations under a variety of conditions of 0.7-9.8 vol.%, corresponding to saturations of 5%-73% (mean 41%). Contact with soil increased oxygen saturation from 46% to 77%.

Carbon dioxide concentrations of soil air appear to have little effect on respiration. Satchell (1967) quotes Shiraishi (1954) who found no behavioural response in *Eisenia fetida* at carbon dioxide concentrations in air up to 25% and Stephenson (1930) who noted only a slight and reversible effect at concentrations up to 50%; these concentrations are much higher than are normally found in soils, where carbon dioxide content rarely exceeds 10% and is usually < 2%. Little is known about the carbon dioxide transport functions of the blood; much carbon dioxide may be carried in solution as bicarbonate, while some is probably carried by the respiratory pigment. The significance of calciferous glands in taking up carbonate ions from the blood and excreting calcium carbonate is discussed on pp. 224-228.

B. Caudal Respiration

Carter and Beadle (1931) described a respiratory adaptation of a species of glossoscolecoid earthworm, probably of the family Almidae, which lives in swamps of the Paraguayan Chaco of South America, in which the caudal portion (1 cm or more) of the body is protruded from the burrow, above the ground surface, and is highly modified for respiratory exchange. The same adaptation is also found in *Alma emini* (Glossoscolecidae) in east African swamps, and was described in more detail by Beadle (1933, 1957). The caudal region is protruded from the burrow, and is first dorsoventrally flattened, then formed into a tube by folding the edges upwards so that they meet, making a funnel-shape. The earthworm then withdraws into its burrow, leaving the open end of the funnel exposed to the air at the burrow entrance (Fig. 6). The dorsal surface of the affected segments is much more heavily vascularized than that of the remainder of the body; the posterior segments are smaller than anterior segments, so that there is a relative concentration of segmental blood vessels, and there are many more subcutaneous capillaries per segment than in more anterior segments. The effect is to provide a "lung"-like organ that enables the earthworms to absorb oxygen from the above-ground atmosphere while most of the body is in a nearly anaerobic environment. The erythrocruorin of *A. emini* was

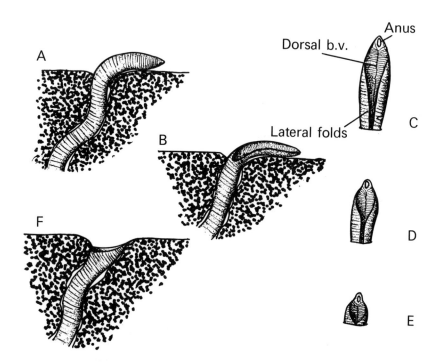

Fig. 6 Exposure to air and formation of "lung" of the posterior dorsal surface in *Alma emini*, from papyrus swamps in eastern Africa. A, Emergence from burrow of posterior region. B, Hollowing of dorsal surface. C–E, Stages in the formation and closure of lateral folds, with progressive withdrawal into burrow; the dorsal blood vessel and numerous lateral branches are clearly visible. F, Final position after retreat into burrow, with tubular "lung" open to the air. (After Beadle 1957).

shown to be saturated with oxygen at a partial pressure of oxygen of 300 Pa in the absence of carbon dioxide, and the dissociation curve of the erythrocruorin to be "phenomenally" insensitive to carbon dioxide tension, with only slight displacement of the curve at a partial pressure of carbon dioxide of c. 27 kPa (200 mm Hg) (c. 30% carbon dioxide by volume). When breathing air through the "lung" *A. emini* is in peculiar circumstances, since if all the body surface were permeable to oxygen the anterior end, in the anaerobic mud, would lose oxygen to the mud. Beadle (1975) showed experimentally that such oxygen loss is minimal, apparently because of the scarcity of subcutaneous capillaries anteriorly and of adaptation for respiration at very low oxygen tension.

Wasawo and Visser (1959) reported that these morphological, physiological and behavioural adaptations were common to the glossoscolecids *A. emini*, *A. stuhlmanni* and *Glyphidrilus* sp. in swamps near Lake Kyoga, Uganda.

Similar phenomena were recorded by Bouché (1970) for two European lumbricid species and the criodrilid, *Criodrilus lacuum*. *Eophila savignyi*, living in reducing conditions in a polluted pond in southern France, were found to lie with the anterior half of the body in the burrow at the bottom of the pond while the posterior half projected into the overlying water. The tail was flattened to form a spatulate sheet, about 2.5 times the width of the anterior segments, and greyish red instead of greyish white in colour, apparently due to augmentation of the subcutaneous blood supply. When individuals were removed to aerobic conditions the form and colour of the caudal region returned to normal. *Allolobophora leoni*, living in saturated soil beside a temporary pond in Romania was observed to project its caudal region into the water on the soil surface, but had no apparent morphological or behavioural modifications like those of *E. savignyi*.

C. Anaerobic Respiration

Short periods of anaerobiosis can be tolerated; stored glycogen is used as an energy source. Gruner and Zebe (1978) showed that glycogen is first broken down to produce succinate, which is subsequently fermented, yielding propionate. Restoration of aerobic conditions results in resynthesis of glycogen from accumulated fatty acids. Weber (1978a) quoted Coles (1970), who found that glycogen decomposition by *Alma emini* in anaerobic conditions resulted in production of propionic, acetic and methylbutyric acids.

The ability to survive short periods of anaerobiosis has obvious advantages for earthworms, since their burrows must occasionally be flooded in nearly all soils during heavy rain, and the oxygen contained in the water would be rapidly exhausted in such circumstances.

IV. Light

Earthworms respond to strong light by the "withdrawal reflex", a rapid shortening of the longitudinal muscles, mediated by the giant nerve fibres, resulting in rapid retreat or retraction of exposed sections of the body from the source of light. The neurological basis of the withdrawal reflex is described by Dorsett (1978). Single-celled photoreceptors are widely distributed over the body surface and are particularly numerous on the prostomium, a few anterior and a few posterior segments. Their morphology and function are discussed by Mill (1978).

Laverack (1963) quoted experiments of Hess (1924), who found that *Lumbricus terrestris*, kept in darkness for some hours and then exposed to a direct beam of light of various intensities moved away from the light except when its intensity was *c.* 1.7 cd/m^2 or less, which provoked movement towards the light source. There appear to be no data confirming this observation during the last 60 years. For example, Laverack quoted the results of Howell (1939), who found that *Amynthas agrestis* moves away from light at all intensities. Various experiments have shown (see Laverack 1963) that earthworms are particularly repelled by the longer wavelengths in the visible spectrum; at least some earthworms are not disturbed by red light, but withdraw rapidly if exposed to blue light. Low intensities of ultra-violet (UV) light are tolerated and the level of tolerance increases with increasing body pigmentation, but the UV intensity in direct sunlight is lethal to earthworms. The threshold of avoidance varies; Stolte (1962) showed that *L. terrestris* seeks shelter when exposed to UV intensities of 64%-69% of its lethal limit, while *Aporrectodea caliginosa* first reacts at 78%-80% of its lethal limit.

The habit of surface-feeding species of coming to the surface only at night or in very low light intensities during the day is probably mainly an avoidance reaction to strong light. However, in most environments relative humidity is higher and ambient temperature lower when light intensity is low than they are when light intensity is high, and as discussed previously these factors also greatly influence earthworm behaviour. Positive phototropism at low light intensities, shown by *L. terrestris*, may be a general character of surface-feeding species, which would gain some advantage in the consequent extension of time available for foraging on the soil surface. The strongly negative phototropism of subsurface-feeding species is correlated with their general lack of pigmentation.

After heavy rain many earthworms are commonly found dead on the ground surface; it is suggested that they are forced to abandon their flooded burrows because of anoxia, or perhaps resulting from acidification of the water in the burrows due to accumulation of dissolved carbon dioxide, and are then killed by the UV of sunlight (Merker and Braunig 1927).

V. Soil Texture

The abrasiveness of coarse textured soils and their susceptibility to drought, as a result of free drainage, influence the species composition and abundance of earthworm populations. The clay content of soils is also significant, and in regions of high rainfall or poor drainage soils with high clay content may have no earthworms, because of their susceptibility to becoming anaerobic when soil water content is high.

In New Zealand a large area in the central North Island is mantled by deep layers of coarse textured volcanic ash, erupted from the vicinity of Lake Taupo during the last 1500–4500 years. Much of the ash was hot when it fell and the previous vegetation was destroyed. The coarse gravelly and sandy soils derived from the ash layers have acted as a selective barrier to the dispersal of earthworms and they are now populated by a small group of megascolecoid species that are able to tolerate the coarse texture of most of the present soils, and whose origins are still apparent from their distribution in areas adjacent to the affected region (Lee 1959). Mean annual rainfall in the region covered by the ash showers ranges between *c.* 1000 and 2500 mm, but in the hot dry summer period soil moisture is low and earthworms are forced to retreat into deep soil horizons. Similarly on the yellow-brown sand soils formed on coastal dunes in New Zealand Lee (1959) found few endemic earthworm species, and those that were found were common to adjacent areas with less sandy soils. The effect of coarse textured soils is to produce an environment for earthworms that is typical of lower rainfall than is actually experienced.

In Egypt Khalaf El-Duweini and Ghabbour (1965a) showed a clear relationship between decreasing populations of *Aporrectodea caliginosa* and increasing proportions of gravel and sand in soils. The effect was modified by soil moisture content and they showed that when the ratio %(water content)/%(gravel and sand) exceeded 1.13 there was no apparent effect on *A. caliginosa* populations. Their data indicate that the effects of coarse soil texture may be primarily due to the moisture relationships of the soil and not directly to soil texture. However, Madge (1969) introduced *Hyperiodrilus africanus* and *Eudrilus eugeniae* to boxes containing moistened sand, which was divided longitudinally into vertical bands of 1 mm, 0.5 mm, 0.25 mm and 0.1 mm particle size, and after 48 hours was able to show a marked preference (P < 0.001) for the 0.25 mm particle size fraction, avoidance of the 1 mm fraction and little difference in preference for the 0.5 mm and 0.1 mm size fractions. Surface casts were produced on the 0.25 mm and the 0.1 mm size fraction but not on the 1 mm and 0.5 mm fractions.

For 15 forest, pasture and heath land sites in Sweden, Nordström and Rundgren (1974) distinguished soils with 0%–5%, 5%–15%, 15%–25% clay content and found a positive correlation (r^2 = 0.34, P < 0.1) between clay content and earthworm abundance. There was no significant correlation between clay content at 0–20 cm depth and abundance of the near-surface dwelling *Lumbricus rubellus*, but correlations were positive and significant between clay content at 0–60 cm and the abundance of *Aporrectodea caliginosa, A. longa, A. rosea* and *L. terrestris*. Nordström and Rundgren conceded that the correlations may not indicate cause and effect, since clay content affects many soil properties, including soil water retention, cation exchange capacity, and other environmental factors that are important for earthworms.

In soils with high clay content, especially when poorly drained, earthworms may be excluded or may be forced to abandon the soil temporarily after rain.

3
The Chemical Environment

I. Carbon, Nitrogen and C : N Ratio

The availability of adequate organic carbon and nitrogen in assimilable forms is essential for the survival and growth of all animals. Absolute deficiency of one or both elements sometimes limits earthworm populations, but more commonly it is the C : N ratio, which varies from *c.* 5 in animal and microbial tissue to >> 100 in some plant litter, that limits populations. The problem is to extract enough nitrogen for tissue production as the C : N ratio of the food source increases.

In the semi-arid agricultural soils of Egypt Khalaf El-Duweini and Ghabbour (1965a) and Ghabbour and Shakir (1982) showed a strong positive correlation ($P < 0.01$) between earthworm abundance and biomass and soil organic matter content. In these soils organic carbon content is always low, i.e., food for soil animals is scarce; earthworms were recorded in soils where the organic carbon content (0-5 cm depth) was *c.* 0.11%-0.83%.

Satchell (1980a, b, c, d) analysed data on earthworm populations, accumulated over a 27 year period at a site in northeast England, where plots of *Calluna* heath on a podzol were initially cleared and planted with birch (*Betula* sp.). It was thought that the change in vegetation might stop the podzolizing process, provide a surface litter richer in plant nutrients, derived from deep soil layers, and lead to the development of a mull, with earthworms that would mix the soil horizons, improve structure, aeration and drainage, the whole process resulting finally in a change of soil from podzol to brown earth. In nearby areas of deciduous forest on brown earths there was a minimum of *c.* 100 g/m^2 (live weight) of earthworms, *c.* 80% *Lumbricus* spp., mainly *L. terrestris* (Satchell 1980a). On the basis of previous estimates of nitrogen requirements of *L. terrestris* for tissue production, efficiency of utilization of ingested nitrogen, the nitrogen content of litter and soil and inputs from rainfall, and assuming that to maintain a mull on a brown earth under forest *L. terrestris* must have a minimum biomass of c. 100 g/m^2, Satchell concluded that the ecosystem could not supply nitrogen at a sufficient rate. The planned duration of the vegetation and soil experiments was 100 years; after 27 years it seemed unlikely that, because of an absolute deficiency of nitrogen (and other factors), earthworms necessary to the change from podzol to brown earth would become established.

Availability of adequate nitrogen appears to be one of the most important factors limiting earthworm populations and their distribution, especially in tropical regions, where the nitrogen content of soils is low compared with that of soils of temperate regions (Lee 1983).

The C : N ratio is a measure of the quality of soil organic matter as an energy source; it is also an important determinant of humus types, is related to soil moisture, pH, and to many other soil properties.

For 67 earthworm taxa in France Bouché (1972) determined optimum C : N ratios from field collections and chemical analyses of soils. He distinguished 49 taxa ("eubiotic" forms) whose C : N optima were < 13 and 18 taxa ("mesobiotic" forms) whose C : N optima were ≥ 13, including only two species with C : N optima > 17. Comparing these two groups with his ecological classification, Bouché found that his anéciques were almost exclusively eubiotic forms; these are particularly great consumers of surface litter. The majority of endogées were also eubiotic forms, while the majority of épigées were mesobiotic forms that live at the soil surface in litter layers, or under the bark of trees, i.e., they live in media that are rich in little-decomposed organic matter.

Among seven lumbricid species, in an English woodland, Phillipson *et al.* (1976) distinguished *Dendrobaena octaedra* and *Lumbricus rubellus* that inhabit soils with C : N ratio > 8, from *Aporrectodea caliginosa, A. rosea, L. terrestris, L. castaneus* and *Satchellius mammalis* that inhabit soils with C : N ratio < 8. The data of Phillipson *et al.* (1976) relate to only one forest area, and it is of interest to compare their findings with those of Bouché (1972) who sampled earthworms from > 1400 sites in forests, grasslands and some cultivated lands throughout France. Minimum, mean and maximum C : N ratios of soils from which the seven species listed by Phillipson *et al.* (1976) were collected by Bouché (1972) are listed in Table 3.

The range of tolerance of C : N ratios recorded by Bouché includes that found in the investigations of Phillipson *et al.* (1976) for all species except *L. castaneus,* for which the minimum C : N ratio was marginally higher (8.34) than that recorded by Phillipson *et al.* It is apparent that the common lumbricids have very wide tolerance of variation in C : N ratio, with little

Table 3. Minimum, mean and maximum C:N ratios of soils from which seven lumbricid species were collected, based on measurements from forest, grassland and cultivated sites in France sampled by Bouché (1972). (Data from M. B. Bouché).

Species	C:N ratios of soils		
	Minimum	Mean	Maximum
Aporrectodea caliginosa	7.30	11.69	47.81
Aporrectodea rosea	7.30	11.08	24.38
Lumbricus castaneus	8.34	10.76	20.42
Lumbricus rubellus	8.53	12.44	21.21
Lumbricus terrestris	7.72	11.39	47.81
Dendrobaena octaedra	9.05	14.29	27.27
Satchellius mammalis	7.98	10.80	28.32

variation between species, except in the case of *D. octaedra*, which is a species characteristic of relatively undecomposed surface litter layers.

Near Bangalore, in India, Kale and Krishnamoorthy (1981) related the abundance of common earthworm species and population diversity in pastures, arable lands and gardens to C : N ratio and to the ratio of humic acid to fulvic acid (H : F ratio) of soils. Of seven species known in the district only three were abundant at most sampling sites. Over a range of C : N ratios *c.* 2–18, abundance of *Lampito mauritii* and *Pontoscolex corethrurus* increased steadily, while that of *Perionyx excavatus* decreased steadily with increasing C : N ratio. Over a range of H : F ratios *c.* 0.6–2.5, abundance of each of the three species varied with increasing H : F ratio in the direction opposite to that for the C : N ratios. The H : F ratio is a measure of the degree of humification (i.e., stage of degradation) of soil organic matter; the lower the H : F ratio the greater is the humification. *Lampito mauritii* and *P. corethrurus* were most abundant in well humified soils with relatively high C : N ratios, while *P. excavatus* was most abundant in poorly humified soils with low C : N ratios.

II. pH

It was demonstrated by Hurwitz (1910) that *Eisenia fetida* withdrew when its anterior end was dipped in acid solutions, and that the lower the pH of the solution the more rapidly *E. fetida* withdrew. It is also well established that earthworms are rare in soils with pH < *c.* 4.0–4.5, are generally absent where pH < 3.5, and that there are considerable differences between species in their preferred pH range. Laverack (1961) recorded action potentials in segmental nerves of *Aporrectodea longa, Lumbricus terrestris* and *L.*

rubellus when the body surface was flooded with buffered solutions and showed that the threshold for nerve response was between pH 4.6-4.4, pH 4.3-4.1, and at pH 3.8 respectively for the three species. Further experiments (Laverack 1961) showed that these species would not burrow when placed on the surface of soils with pH below the threshold indicated by the buffer experiments. Electrolytes other than hydrogen ions are important to earthworms (see following paragraphs) and it has been suggested that the inhibitory effects of low pH may be due to low concentrations of other ions, especially calcium ions. However, the neurophysiological and behavioural data of Laverack (1961) indicate that low pH does itself inhibit earthworm activity.

Satchell (1955a) related occurrence in the field of 13 species of lumbricids to the pH of surface litter, and distinguished three categories; (a) acid-tolerant species (*Allolobophora eiseni, Dendrodrilus rubidus, Dendrobaena octaedra*) found where litter pH was in the range 3.7 to 4.7, (b) ubiquitous species (*Lumbricus rubellus, L. terrestris, L. castaneus, Octolasion cyaneum, D. rubidus*) over the pH range c. 3.7-> 7.0, and (c) acid-intolerance species, confined to the pH range of c. 4.7-> 7.0.

The distribution of 67 taxa of Lumbricidae in France was related to soil pH (measured in aqueous suspension) by Bouché (1972). Most of the French taxa were found in the pH range 5.0-7.4; 26 taxa were found in soils with pH < 4.0; four species were found only in soils with pH > 6.6. On the basis of the mean soil pH where they occurred Bouché distinguished 18 acidophilic taxa, whose preferred pH was < 6.0, 40 neutrophilic taxa (preferred pH 6.0-7.0), and nine basiphilic taxa (preferred pH > 7.0). Many papers on the effects of soil pH on lumbricid earthworms are reviewed by Edwards and Lofty (1977); the preferences and tolerances of most species appear to fall in the ubiquitous range (of Satchell 1955a).

There is little information on the pH preferences and tolerances of non-lumbricid earthworms. Lee (1959) found no megascolecids in New Zealand soils with pH < 4.0; the pH of most New Zealand soils is in the range 5.0-7.0, and overall, Lee found no obvious relationship between species distribution and pH. Madge (1969) tested the reactions of the Nigerian eudrilids *Hyperiodrilus africanus* and *Eudrilus eugeniae* to dipping the anterior portion of the body in a series of buffered solutions with intervals of 0.4 pH unit, between pH 2.0 and 10.0. At pH 2.0-4.0 all worms (30 per test) contracted instantaneously and remained contracted until washed in water, and at pH 4.4 nearly all contracted; at pH 4.8 c. 50% withdrew, while at pH 5.2 nearly all did not react; at pH 5.6-9.2 none contracted, but they became increasingly agitated at pH > 8.4; at pH 9.6 and 10.0 the worms contracted spasmodically. Madge concluded that the optimum pH range for the two species was between 5.6 and 9.2. Emiliani *et al.* (1971) in Argentina, recorded the ocnerodrilid species *Eukerria halophila* as the only earthworm species in saline soils of pH 9.4–9.8, and found it also in more acid soils (pH 6.3).

III. Electrolyte Concentrations

A. Osmotic and Ionic Regulatory Mechanisms

The permeability of the cuticle to water (p. 35) and to dissolved ions limits the ability of earthworms to control the osmotic pressure of their internal fluids and, although some control of transfer rates of electrolytes is achieved, earthworms are particularly sensitive to high concentrations of electrolytes in the soil solution. Darwin (1881) was aware that lumbricid earthworms are unable to survive prolonged immersion in sea water, and was puzzled by the presence of earthworms (*Microscolex kerguelarum*) at Kerguelen Island (southern Indian Ocean). The Kerguelen species and some others that are known from remote islands are euryhaline and are found in the intertidal zone as well as in soils (Lee 1968, and p. 65), but most earthworms are confined to habitats with low salinity.

Oglesby (1969, 1978) has discussed the concentrations of electrolytes and non-ionic compounds that contribute to the osmotic pressure of the blood and coelomic fluid of earthworms; the reader is referred to these sources for details of their concentrations and of the physiological mechanisms involved in their regulation. The nephridial system, the body surface, the gut wall and the calciferous glands are involved in maintaining ionic equilibria. Their contributions to the system are discussed briefly below.

Salts are resorbed as urine passes down the lumen of the nephridia of *Lumbricus terrestris*, with Na^+ being actively resorbed against the electrochemical gradient, while Cl^- is passively absorbed. Some organic molecules may also be absorbed; in fully hydrated *L. terrestris* the urine is strongly hypo-osmotic to the coelomic fluid and the blood. In some megascolecid earthworms (especially in the *Pheretima* s.l. group of genera) the nephridial system discharges into the gut; water is resorbed (p. 34) and it is likely that some ions are also exchanged across the gut wall. Dietz and Alvarado (1970) calculated that 36% of the total voided Na^+ and 32% of the total voided Cl^- in hydrated *L. terrestris* was excreted into the gut, and that some metabolically produced ammonium, NH_4^+, was voided by exchange with Na^+ in the gut. The importance of Ca^{2+}, absorbed from the posterior region of the gut and transferred to the calciferous glands where it is excreted as crystalline calcium carbonate, is discussed on pp. 226–227. Bouché (1983) has demonstrated that Fe^{3+}, Mn^{2+}, and Na^+ are also taken up in the gut and recycled through the calciferous glands in *Aporrectodea velox*, and this may be the case in other earthworms. Ions diffuse through the body wall of *L. terrestris* between the coelomic fluid and the external environment, but the body wall is differentially permeable to different ions. For example, in worms exposed to a range of salt concentrations Dietz and Alvarado (1970) calculated that the equilibrium flux of Cl^- was about 15 times that of Na^+. The concentration of Ca^{2+} has an important influence on Na^+ flux, which is much increased in the absence of Ca^{2+} (Dietz and Alvarado 1970). Richards (1978) suggests that the acid mucins, produced by epidermal gland cells, that provide the respiratory film on the body surface act as ion barriers, either in a selective capacity or by allowing abrupt changes in the ionic environment to be buffered.

B. Significance of Electrolyte Concentrations in Field Conditions

Khalaf El-Duweini and Ghabbour (1965a) related abundance and the mean weight of individuals of *Aporrectodea caliginosa* to a range of soil properties, including total soluble salt and Cl^- concentrations in two transects in Beheira Province, Egypt. Total soluble salts ranged from *c.* 0.3% to 0.8% and Cl^- from *c.* 0.06% to 0.45%. Over these ranges there was a general decline in both abundance and average weight with increasing total salt and Cl^- concentrations, more closely related to Cl^- than to total salt content. Average weight ranged from *c.* 0.85 g at 0.06% Cl^- to *c.* 0.13 g at 0.45% Cl^-, while abundance fell from $22/m^2$ to $1/m^2$ over the same range. In an earlier investigation Khalaf El-Duweini and Ghabbour (1964) had determined the survival in experimental conditions of *A. caliginosa, Metaphire californica* and *Alma stuhlmanni* after 24 hours of immersion in solutions of $NaCl$, Na_2CO_3, $NaHCO_3$, KH_2PO_4, $Ca(NO_3)_2$, $CaCl_2$, $(NH_4)_2SO_4$, NH_4Cl and NH_4NO_3, in ascending steps of 0.1% concentration. The salts used in the tests are those that are most common in Egyptian soils. Five earthworms were used for each test; maximum concentrations tolerated without mortality and concentrations that resulted in 100% mortality are listed in Table 4. In all cases *A. caliginosa* tolerated the highest concentrations, equalled only by *M. californica* in the case of NH_4NO_3. Sodium chloride is tolerated in highest concentrations in all species, except that $Ca(NO_3)_2$ is more tolerated by *A. caliginosa*. Sensitivity to $(NH_4)_2SO_4$ is less than to most of the other salts tested; it is widely used as a fertilizer and has been claimed to be particularly lethal to earthworms. *Aporrectodea caliginosa* is the most common earthworm in the generally alkaline soils of Egypt and this may be at least partly due to its tolerance of high electrolyte concentrations.

Table 4. Maximum concentrations tolerated without mortality and minimum concentration causing 100% mortality of *Aporrectodea caliginosa*, *Metaphire californica* and *Alma stuhlmanni* when immersed for 24 h in solutions of electrolytes in ascending steps of 0.1% concentration. Five individuals used for each test. (Data from Khalaf El-Duweini and Ghabbour 1964).

Electrolyte	*Aporrectodea caliginosa*		*Metaphire californica*		*Alma stuhlmanni*	
	Maximum tolerance %	100% mortality	Maximum tolerance %	100% mortality	Maximum tolerance %	100% mortality
NaCl	1.4	1.5	0.8	1.2	0.8	0.9
Na_2CO_3	0.3	0.4	<0.1	0.1	0.1	0.2
$NaHCO_3$	1.0	1.1	<0.1	0.2	0.5	0.6
KH_2PO_4	0.5	0.7	0.1	0.3	<0.1	0.2
$Ca(NO_3)_2$	1.9	2.0	0.6	0.8	0.7	0.8
$CaCl_2$	0.9	1.0	0.8	0.9	0.4	0.7
$(NH_4)_2SO_4$	1.0	1.1	0.5	0.6	0.2	0.4
NH_4Cl	0.7	0.9	0.1	0.3	0.1	0.2
NH_4NO_3	0.4	0.6	0.4	0.6	0.3	0.4

Piearce and Piearce (1979) and Piearce (1982) studied lumbricid populations at two farmland sites in northwest England that were flooded by sea water in 1977. *Aporrectodea* spp. and *Allolobophora chlorotica* were dominant in all the flooded soils, with *Lumbricus* spp. and *Dendrodrilus rubidus* and *Satchellius mammalis*. Total populations, biomass and diversity of species declined following flooding, but *Aporrectodea* spp. and *L. terrestris* were less affected than other species. Experimental immersion in sea water (2.9% salt content) was fatal to eight species tested, but *A. longa* and *L. terrestris* survived longer than others; *L. castaneus* and *S. mammalis* were least resistant. In an experiment with choice of salinity of substrates, the threshold of avoidance was 0.7% salt content, similar to the threshold for nervous excitation of the body wall reported by Laverack (1960). The viability of cocoons deposited before the flooding was apparently unaffected, and was probably a major factor in the almost complete recovery of the earthworm population recorded in 1981 (Piearce 1982).

Bouché (1972) related the distribution of French lumbricids to soil carbonate CO_3^{2-} content, which is present mainly as $CaCO_3$, but also as $MgCO_3$. He distinguished 13 taxa (calcifuges) that were confined to soils with mean CO_3^{2-} content < 1%, 16 taxa (calcicoles) confined to soils with mean CO_3^{2-} content > 10% and a majority of 39 taxa from soils with mean CO_3^{2-} content of 1%–10%. Phillipson *et al.* (1976) similarly distinguished acidophilic species, *D. rubidus* and *L. rubellus* in soils with $CaCO_3$ < 4% from basiphilic or neutriphilic species, *A. caliginosa, A. rosea, L. castaneus, L. terrestris* and *S. mammalis* in soils with $CaCO_3$ > 4%.

IV. Oxidation-Reduction Potentials

The oxidation-reduction potential, or redox potential (E_h), of the soil solution is a measure of its tendency to accept or to donate electrons, i.e., of its reducing or oxidizing capacity. Redox potentials are measured as the potential difference between a platinum electrode inserted in the soil and a saturated calomel electrode, converted by reference to the normal hydrogen electrode and expressed as E_h (mV potential difference). Low E_h values indicate reducing and high E_h values oxidizing conditions. Redox potentials are particularly dependent upon soil moisture, ease of diffusion of oxygen, organic matter content, and probably provide some synthesis of factors that affect the favourableness of soils as an earthworm environment, especially where the climate is moist. They have not been much used by earthworm biologists, nor by soil zoologists in general.

Nomura and Usuki (1951) compared redox potentials of soils at sites in the field inhabited by *Amynthas communissimus* and *Eisenia fetida*. The redox potential of the soil at the *A. communissimus* site was E_h = 523-566 mV, while that at the *E. fetida* site was E_h = 181-463 mV. Soil pH at the two sites was 6.4-7.2 and 6.8-7.4 respectively. Nomura and Usuki (1951) and Usuki (1955) concluded that E_h measurements were superior to pH measurements and probably to most soil chemical data in discriminating the habitats of the two species. *Eisenia fetida* lives in habitats with unusually high organic matter content, while *A. communissimus* is widespread in central Japanese farmlands; Nomura and Usuki related the lower E_h of the *E. fetida* soil to its higher organic matter content, consequent more active microflora and tendency to a less aerobic condition compared with the *A. communissimus* soil.

In a *Calluna*-dominated heathland at Silpho Moor, in northeastern Yorkshire, Satchell (1980c) related extremely low earthworm numbers in depressions, where water accumulates after heavy rain, to low E_h (174.2 mV) at 4 cm depth in the soil. He compared populations in these saturated microhabitats with others nearby, where there was no standing water, E_h was 279.2 mV, and earthworms were about four times more numerous, and suggested that the differences were related to low oxygen tensions in the saturated sites, as reflected in the differences in E_h.

Anaerobic sewage sludges are toxic to earthworms, and Kaplan *et al.* (1980a) related the change in E_h of sludge when it is aged in an aerobic environment to its ability to provide an environment suitable for survival and growth of *E. fetida*. They found that *E. fetida* survived only when E_h ⩾ 250 mV, and related the attainment of suitable E_h levels to mixing initially anaerobic sludge with soil, or to placing it on a substrate (soil) that allowed free drainage.

V. Earthworms in Aquatic Habitats

Most earthworms can survive for some time in oxygenated fresh water, and many species may occasionally be found in aquatic habitats. There are a few species that are known exclusively from aquatic habitats or that appear to be able to move freely between aquatic and terrestrial habitats; they should not be confused with the specialized aquatic families of Oligochaeta (see Brinkhurst and Jamieson 1971), but are representatives of primarily terrestrial oligochaete families that have adopted aquatic habitats. Their habitats may be classified as follows:

(a) in mud and under stones of stream, pond and lake bottoms: *Eiseniella tetraedra* is most commonly found living under stones on the bottom of streams or lakes and occasionally in the water of swamps (Cernosvitov 1945, Roots 1955, Lee 1959); it has been recorded from many such locations in Europe, North America, New Zealand and elsewhere. Cernosvitov (1945) recorded *Allolobophora chlorotica* from a depth of several metres in the waters of Lake Windermere, in England. Two New Zealand megascolecids, *Rhododrilus aquaticus* and *Diporochaeta aquatica*, were recorded by Lee (1959) under stones in a fresh water stream and *R. aquaticus* also from brackish water, and they are not known except from aquatic habitats.

(b) in the water of lakes: Benham (1903) collected *Diporochaeta aquatica* and *Pontodrilus lacustris*, swimming in the waters of Lakes Manapouri and Wakatipu in New Zealand. Nothing is known of the life cycles of these species, and they may have been washed into the lakes from streams and not be permanent inhabitants of lakes.

(c) in swamps: A few species of several earthworm families live an aquatic or semi-aquatic life in swamps. Some have special respiratory adaptations, with heavy vascularization of caudal segments that enable them to live in burrows under water (p. 50). Others have no special respiratory adaptations, e.g., the New Zealand megascolecids *Decachaetus violaceus, Diporochaeta chathamensis, Eodrilus paludosus* and *Perionyx helophilus*, which were recorded by Lee (1959) from swamps. It is likely that these species were living only temporarily in surface water, as two of the species were collected also from soil.

(d) intertidal zone or in brackish water: Most earthworms die if immersed in sea water (p. 60), but a few megascolecid species live in the intertidal zone of sea shores under stones and debris or in brackish water.

Eight species of *Microscolex* are found on the isolated subantarctic islands of the South Atlantic, Southern Indian and South Pacific Oceans (Lee 1968). Of these, *M. aucklandicus* and *M. campbellianus* are known to be euryhaline and are found on ocean beaches, in brackish water and in soils on Auckland, Campbell and Snares Islands, south of New Zealand. It is not known whether the other six *Microscolex* spp. are euryhaline, though it might be presumed from their close morphological and ecological similarities and from the occurrence of individual species on islands that are widely separated that they are euryhaline and have achieved their distribution by drifting with plant debris from island to island, moved by the west-east ocean current that circulates around the Antarctic continent.

Two species of *Rhododrilus, R. cockaynei* and *R. leptomerus*, are also found on subantarctic islands south of New Zealand. *Rhododrilus cockaynei* is recorded from the intertidal zone, from fresh water streams and from soils on Auckland, Campbell and Snares Islands and also from a saline meadow on the southern New Zealand coast. A third species, *R. aquaticus*, is known from brackish water and from fresh water streams on the southwest coast of New Zealand (Lee 1959).

Pontodrilus matsushimensis is widely distributed on the shores of Pacific islands (Lee 1959), usually living in beach sand under stranded plant debris at the high-water line. In the Solomon Islands Lee (1969) found *P. matsushimensis* sporadically on the shore. Fresh water springs are common along the shore close to high-water mark and *P. matsushimensis* may consequently not be living in a strongly saline environment. Lee (1969) took *P. matsushimensis* from sand at the high-water line on the island of Makira and put them into a pool of sea water among coral outcrops. All specimens showed immediate signs of irritation, writhing and exuding coelomic fluid from dorsal pores and nephridiopores, but within ten minutes they relaxed and burrowed into the sand on the bottom of the pool. Takeuchi (1980) tested the capabilities of *P. matsushimensis* to regulate the osmotic pressure of its coelomic fluid to that of its external environment. Specimens that were kept under sea water in sand from a Japanese beach were transferred to sand under water that ranged from concentrated sea water (1.2 times salinity of normal sea water) to deionized water. The osmotic pressure of the coelomic fluid increased in a more or less linear relationship to that of the water, and *P. matsushimensis* was shown to be capable of tolerating much higher salinity than other megascolecid earthworms. The coelomic fluid was hyper-regulated as related to external media of diluted sea water and hyper-conformed to media of normal and concentrated sea water. Body weight was inversely related to salinity, which would explain the observations of Lee (1969) of loss of coelomic fluid. Takeuchi (1980) compared the range of tolerance of salinity levels of *P. matsushimensis* to that of the euryhaline polychaete *Nereis diversicolor*. Sodium ion concentration is greater than chloride concentration in most earthworms (p. 60), but it was shown that in *P. matsushimensis* the opposite is true, as in nereids, reflecting adaptation to the concentration of these ions in sea water.

4
Phenology

I. Life Cycles

The physiology of reproduction in annelids has been reviewed by Olive and Clark (1978). They distinguished three basic reproductive strategies among animals; (a) *monotelic*, in which a species breeds only once during life, (b) *polytelic*, in which breeding occurs at several discrete times during life, and (c) *semi-continuous/continuous*, in which species breed several times during life, and release gametes in a number of small broods over an extended breeding season within one or more years. Earthworms were grouped among *semi-continuous/continuous* breeders.

Earthworms are generally hermaphroditic, with separate testes and ovaries that are simultaneously functional. In many species, as described in classical texts, sperm cells are interchanged between pairs of copulating individuals, and are stored in spermathecae for subsequent fertilization of the ova. An alternative mode of fertilization, in which sperm cells are transferred between individuals in spermatophores, to achieve direct fertilization of ova as the cocoons are produced, was observed by Bouché (1975d) in *Bimastos antiquus michalisi*, from Albania. Bouché listed more than twenty species of Lumbricidae, including European and North American representatives of the genera *Aporrectodea, Bimastos, Dendrobaena, Dendrodrilus, Eisenia, Eisenoides, Kritodrilus, Lumbricus, Murchieona, Octolasion* and *Satchellius*, in which he or Ljungström (1968) had recorded spermatophores. Reproduction through direct transfer of spermatophores may be facultative or obligate. It provides an alternative means of interchange of sperm between individuals, and its implications have not been adequately considered.

Uniparental parthenogenesis, with self-fertilization, is known in some species. It may be facultative or obligate, and is most commonly recorded in widely distributed peregrine species (pp. 164–165). Parthenogenesis has sometimes been inferred on the basis of absence or retrogression of some secondary sexual organs (spermathecae, prostates). The absence of these organs is often due to their destruction by gregarines or other internal parasites, so that their absence does not necessarily imply parthenogenesis (R.W. Sims, pers. comm.). Biparental reproduction may be achieved after the loss of secondary sexual organs by transfer of sperm cells in spermatophores, as described above (Bouché 1975d).

The physical processes of copulation, the formation of cocoons, and the embryology of lumbricids are described in classical text books of zoology; they do not differ greatly in other families of earthworms, and will not be discussed here. There are no larval stages. Hatchlings emerge from the cocoons and pass through a stage of growth, which may extend from a few weeks to more than a year, to attain sexual maturity and enter the reproductive phase of the life cycle. The duration of reproductive life of individuals varies greatly between species and may extend from a few weeks to several years.

A. Life Cycles of Lumbricidae

1. Cocoon Production

Cocoons may be produced at any time of the year. Production rates of field populations are highest in Europe in late spring to early summer, with a lesser peak in autumn, and lowest in winter (Gerard 1967). In regions with hot dry summers, e.g., Australia, South Africa and much of New Zealand, lumbricids are inactive during summer, and cocoon production ceases.

Evans and Guild (1948) recorded numbers of cocoons produced per adult per year over a two year period, by 11 lumbricid species, kept in optimum conditions in laboratory cultures. Analysis of their data by Satchell (1967) showed that the deep-burrowing species *Aporrectodea caliginosa, A. longa* and *Octolasion cyaneum* produced 3–13 cocoons per year, topsoil species (*A. chlorotica*) produced 25–27 cocoons per year, while *Lumbricus rubellus, L. castaneus* and *Dendrodrilus rubidus*, which live close to the soil surface, produce 42–106 cocoons per year. Satchell (1967) related these differences to the differing severity of environmental hazards, especially heat, drought and predation, and consequent chances of embryonic and juvenile survival, encountered by the species in the field; the higher the risk of mortality in early life, the higher the rate of cocoon production.

Phillipson and Bolton (1977) explained seasonal variation in cocoon production by *Aporrectodea rosea* in an English woodland in terms of temperature variation. Annual production was shown to be 3.13 cocoons per individual per year; as mean soil temperature at 5 cm rose from winter to summer, from 6.0°C to 12.6°C, cocoon production increased from 0.06 to 1.6 per individual in the first three months compared with the second three months; as the temperature fell from summer to winter the rate fell again to 0.4 per individual in the second three months. These changes were interpreted as evidence for differential partitioning of resources between body tissue growth and production of reproductive tissue.

Cocoon production also reflects variability in food quality. Evans and Guild (1948) compared rates of cocoon production of *Allolobophora chlorotica* and *Lumbricus castaneus* supplied with various kinds of organic matter as food (Table 5), and found them to vary by factors of 10 times to 20 times depending on the food type.

Nowak (1975) collected earthworms from Polish pastures, under normal grazing, from an area that had formerly been used as a sheep fold and so must have had large inputs of excreta, and from an area currently in use as a sheep fold. In laboratory cultures, earthworms from these three habitats produced respectively 26, 35 and 42 cocoons per individual per year. From numbers of cocoons collected in the field she calculated slightly lower rates of 23, 30 and 38 cocoons per individual per year. She concluded that individual fecundity depends on the input of organic matter to the soil. Total cocoon production per unit area was shown to depend on this factor and numbers of sexually mature individuals; in the habitats investigated by

Table 5. Rates of cocoon production of *Allolobophora chlorotica* and *Lumbricus castaneus* in laboratory cultures, with various kinds of organic matter as food. (Data from Evans and Guild 1948).

Organic matter supplied	Number of cocoons produced per individuals per month	
	A. chlorotica	L. castaneus
Well decayed		
sewage sludge	0.05	0.56
farmyard manure	0.01	0.57
Partially declayed		
pasture peat	0.67	1.88
horse manure	0.84	3.68
sheep manure	0.93	5.07

Nowak, the total number of cocoons (per square metre per year) produced was 115, 1503 and 520, corresponding to mean populations of 5, 50.1 and 13.7 sexually mature individuals per square metre respectively.

The proportion of reproductive individuals in field populations was shown by Nowak (1975) to be 2%–23% for *A. caliginosa*, 21%–42% for *A. rosea*, and 18%–67% for *Octolasion lacteum* in three pasture soils. *Aporrectodea caliginosa* was the dominant species in each case; though this species included the lowest proportion of individuals able to reproduce, it had the highest individual fecundity.

Litter and its decomposition products contain little nitrogen; there is evidence that earthworms derive much of their nutritional requirements from fungi and microorganisms that are ingested with detritus (p. 23), so overall rates of cocoon production, and seasonal variations in rates, may relate more to rates of production of fungal and microbial tissue on decaying detritus than directly to physical parameters of the soil environment.

Cocoon size is closely related to adult size. Lavelle (1981a) plotted fresh weight of cocoons against fresh weight of adults for 11 European lumbricids and 14 tropical species of several families and found a high level of correspondence ($r = 0.98$, $P < 0.01$).

The most prolific of all earthworms is probably *Eisenia fetida*. Given luxury rates of energy-rich and nitrogen-rich detritus as food, and optimum moisture, cocoon production in *E. fetida* is almost entirely dependent only on population density and ambient temperature. The natural habitat of this species is probably under the bark of dead tree trunks (M.B. Bouché, pers. comm.), but it is most commonly found in animal dung or accumulations of decaying plant material and it has been spread by European peoples and become successfully established throughout much of the world. Hartenstein et al. (1979b) measured the fecundity of *E. fetida*, kept in laboratory cultures with horse manure or activated sewage sludge as food. Maximum fecundity was achieved at a constant temperature of 25°C. The worms became clitellate and began to produce cocoons at age four to six weeks; from five to ten weeks the rate of cocoon production increased and from ten to 27 weeks the rate declined; some cocoons were produced at up to 45 weeks, when the experiments were terminated.

Effects of population density were examined, using a standard volume of 300 mL of horse manure as a medium, in a dish with surface area 78 cm^2, and with *E. fetida* populations that included a range of 4, 8, 12 and 16 adults per dish. Maximum cocoon production was attained at age 9–11 weeks, at rates of 5.5, 5.0, 3.4 and 2.4 cocoons per worm per week for 4, 8, 12 and 16 worms per dish respectively; production rates then fell fairly uniformly, to rates of *c.* 1.5, 1.6, 0.4 and 0.2 per worm per week respectively at age 27 weeks. The total number of cocoons produced over the age range 5–27 weeks varied from c. 70 at a density of four per culture dish to c. 26 at a density of 16 per culture dish. The rate per worm decreased over the full range of increasing populations, but the maximum cocoon production per culture was at a population of eight worms per culture. Reinecke and Kriel (1981) kept *E. fetida* at constant temperatures of 10°C, 15°C, 20°C and 25°C and at temperatures that fluctuated diurnally, with means as for the constant temperature treatments, to simulate winter, autumn and two summer regimes in South African conditions. Mature worms (9–15 weeks old) were used, and were introduced to petri dishes, at a rate of two worms to five grams of cow manure. Production rates of cocoons were measured over 20–24 days. They increased with increasing temperatures; maximum cocoon production was attained at constant 25°C and was 3.4 per worm per week, while minimum production, at constant 10°C was 0.13 per worm per week. Graff (1974) reported a cocoon production rate of 3.8 per worm per week for *E. fetida* at 25°C. In contrast, Evans and Guild (1948) recorded a production rate of only 11 cocoons per year for *E. fetida*, but these were maintained at low temperatures in an unheated cellar (J.E. Satchell, pers. comm.).

Cocoon production of *E. fetida* has special significance, since this species is recognized as the most suitable candidate as an agent of disposal of organic wastes (p. 316), and as a potential producer of useful protein (Chapter 16). It is apparent from the data above that rates of cocoon production, and eventually of protein production in the form of *E. fetida* tissue, can be manipulated in cultures by varying population densities, temperatures and food sources. Almost nothing is known of the possibilities of similar manipulation of production in the 3000 or more other species of earthworms, except for *Eudrilus eugeniae* (p. 317), and the unexplored potential of the majority of earthworms as protein producers is an open field for research.

In summary, cocoon production is influenced by a variety of characteristics of populations, particulary population density, biomass and age structure, and by external factors especially soil temperature and moisture and the energy content of available food.

2. Incubation Time

Incubation time varies widely, mainly in response to soil temperature, but hatching is delayed also when soil moisture content is low.

Gerard (1960) found that *Allolobophora chlorotica* cocoons, kept in the laboratory with adequate moisture, hatched in c. 36 days at 20°C, 49 days at 15°C and 112 days at 10°C. Evans and Guild (1948) measured incubation times in field populations of *Lumbricus castaneus* and *Aporrectodea caliginosa* in Scotland, from cocoons produced throughout most of the year; the incubation time varied from seven to 14 months for *L. castaneus*, and from 5.5 to 11 months for *A. caliginosa*. For *A. caliginosa*, Nowak (1975) estimated that the minimum hatching time in optimum culture conditions was 1.5 months; the incubation period of cocoons produced in the field (in Poland) was, however, at least three months, and cocoons produced late in autumn could persist through the winter and have incubation periods of up to eight months.

For *Eisenia fetida*, Tsukamoto and Watanabe (1977) showed that hatching rate was negatively and linearly correlated with temperature, up to a maximum of 25°C; at 10°C mean hatching rate was 85.5 days, while at 25°C it was 19.2 days. Grant (1955b) reported that the upper lethal temperature limit for *E. fetida* in eastern USA, was c. 25°C. It seems likely (M.B. Bouché, pers. comm.) that the earthworms studied by Grant (1955b) included more than one species and his results have not been confirmed by other workers, who have found the upper lethal temperature of *E. fetida* to be > 30°C (Table 2).

Persistence of cocoons through periods of summer drought is illustrated in soils formed on marine sediments, uplifted 50 years ago during the Napier (New Zealand) earthquake of 1931. *Aporrectodea longa* is common during spring, winter, and autumn in shallow soils that support some grasses and overlie a saline water table regulated by the tide, but in summer, when the surface soil, above the saline water table, becomes very dry and the grasses wilt and cease to grow, no active *A. longa* can be found, but cocoons of this species are found in the 0-10 cm layer (Lee 1959). Adult worms would be unable to aestivate in deep soil layers, which are saturated with sea water, and the earthworms are trapped in the shallow layer of dry soil at the surface. *Aporrectodea longa* apparently persists in this environment by the delayed hatching of cocoons, deposited during the spring in the superficial soil layers when they are moist from rain, surviving through the summer drought, when adult worms die out, and hatching in autumn when there is rain. The same strategy is recognized by Bouché (1977b) as typical of litter-inhabiting species (épigées) when subjected to seasonal desiccation. He has termed this habit "cocoonization" and regards it as equivalent to encystment in many other animal groups.

3. Hatching Rate

Wilson (quoted in Stephenson 1930) counted 10-60 ova in individual cocoons of *Eisenia fetida* but found that no more than 10-12, and frequently only one or two, developed into embryos. Counts of one or two ova per cocoon were recorded by Stephenson (1930) in *Lumbricus rubellus, Dendrobaena* spp. *Aporrectodea caliginosa*, and several ova per cocoon have been recorded in *L. terrestris* by Walsh (1936) and Gates (1963).

It is generally accepted that, in most species, only one juvenile successfully hatches from each cocoon. Evans and Guild (1948) kept 14 lumbricid species in culture and found that only *E. fetida* commonly produced > one juvenile per cocoon, but subsequent investigations have shown that *A. caliginosa, Eiseniella tetraedra* and *Octolasion lacteum* cocoons produce one or two juveniles and those of *Dendrodilus rubidus* up to four (Reinecke and Visser 1981). In *Bimastos tumidus* Vail (1974) recorded hatching rates from 100 cocoons; numbers and frequency distribution of juveniles hatched per cocoon were: 0(19%), 1(6%), 2(17%), 3(28%), 4(24%), 5(3%), 6(3%), with a mean of 3.1 per cocoon. Cocoons of South African populations of *A. caliginosa* and *A. rosea*, collected from the field, produced up to four and three juveniles respectively per cocoon, with two hatching from c. 80% of cocoons and a mean of 1.8 juveniles per cocoon in both species.

The hatching rate of cocoons of *E. fetida* has attracted attention because of the use of this species for waste disposal and its potential as a source of protein for animal food. Reynolds (1973) found that, of 251 cocoons, 199 hatched and produced 516 progeny, with a range of 1-6 cocoons and a mean of 2.6 per cocoon. Tsukamato and Watanabe (1977) hatched *E. fetida* cocoons at temperatures of 10°C, 15°C, 20°C and 25°C. About equal numbers of juveniles, with a mean c. 2.1 per cocoon, were produced at 10°C and 15°C, while at 20°C and 25°C the mean was c. 1.9 and 1.7 per cocoon. Hartenstein *et al.* (1979b) showed that the number of hatchlings produced by *E. fetida* was proportional to the weight of the cocoon, and that cocoon weight related closely to the weight of the adult that produced the cocoon. In general a cocoon weight of 6 mg was necessary to produce one hatchling, and for each 4 mg increase in cocoon weight above the 6 mg limit, one extra hatchling was highly probable; hatchlings weighed 2-3 mg, and the maximum number of hatchlings was 11.

The mechanism of cocoon production, resulting from secretion of a ring of albuminous material by gland cells in the clitellum, implies that the size of cocoons must increase with increase in the dimensions of adult worms, and it therefore follows that for a particular species the number of hatchlings per cocoon will increase with increasing adult size. This was confirmed by Hartenstein *et al.* (1979b), who showed that c. 1 mg of cocoon weight was produced per 100 mg live adult weight of *E. fetida*; the minimum cocoon weight for a viable hatchling was shown to be c. 6 mg and this correlated well with adult weight. Phillipson and Bolton (1977) showed a similar relationship between cocoon weight and adult weight for *Aporrectodea rosea*.

4. Growth

Lumbricus terrestris cocoons hatch mainly in spring, and the peak established in the size class distribution of field populations can be followed in successive samples to establish a growth curve for about the first year (Lakhani and Satchell 1970; Satchell 1971). After about a year, variation in individual growth rates obscures the peak and these authors used a field culture method, in which large numbers of newly hatched worms were individually confined in nylon bags of worm free soil and litter, buried in the soil, recovered at monthly intervals and weighed. By comparing the rate of growth of the culture worms with those in free-living field populations, up to the last point at which a peak in the size class distribution could be traced from the spring peak in hatching, and applying a correction factor to compensate for the difference in growth rate, they showed that it was possible to extend the period of growth rate estimates beyond that recognizable in field populations alone.

Nowak (1975) followed the growth of *A. caliginosa* in field cultures, from a mean live weight at emergence of 27 mg to a mean live weight of c. 325 mg, 20 months after hatching. She distinguished a rapid pre-productive phase of growth, to a mean weight of c. 210–260 mg attained c. 13 months after hatching, followed by a phase of steadily decreasing growth after attainment of sexual maturity. Michon (1957) and Avel (1959) distinguished a third, post-productive phase, of very slow growth and finally decreasing weight, in senescent individuals. Phillipson and Bolton (1977) kept *A. rosea* in large pots of soil, buried in the soil, removing and weighing them monthly, and then constructed a generalized growth curve for an "average" individual grouping the data at bimonthly intervals according to arbitrary weight classes. Growth per individual per unit time was at a maximum in "large immatures", 100–180 mg live weight, over the range c. 1.5–2.3 years, exceeding that of "small immatures", < 100 mg and up to c. 1.5 years, and of adults, > c. 180 mg. The generalized growth curve was imperfectly sigmoid up to c. four years and c. 260 mg live weight, similar to that for *A. caliginosa* (Nowak 1975) up to c. 2.7 years and c. 360 mg live weight. Adult *A. rosea*, from c. four years old and c. 260 mg live weight, showed a negative mean growth rate, indicating that there were more senescent than actively growing individuals in this age class; individual *A. rosea* in the field commonly reached 330–350 mg, and one individual, which weighed 439.4 mg, was estimated, by extrapolation of the generalized growth curve, to be c. eight years old.

Similar patterns of growth are known from other lumbricids, e.g., *Lumbricus terrestris*, which Lakhani and Satchell (1970) showed to grow rapidly, though with short seasonal pauses in mid-summer and mid-winter, for about three years. At three years of age, the mean live weight of *L.*

terrestris was *c.* 9.5 g, and it changed little for the next four years, though few earthworms survived for seven years. Sexual maturity in *L. terrestris* is usually reached in < one year (Evans and Guild 1948; Wilcke 1952), but the duration of the pre-productive phase is strongly influenced by environmental factors.

The importance of substrate quality in regulating growth rate was indicated by experiments of Murchie (1960) on the North American species *Bimastos zeteki. Bimastos zeteki* is a woodland species, commonly found in rotting logs, litter layers and soil A horizons immediately under litter layers. Murchie introduced newly hatched *B. zeteki*, with a mean fresh weight of 30 mg, to laboratory cultures with substrates that included organic matter-rich loamy sand, leaf litter from aspen forest in several stages of decay, soil from a sandy B horizon under aspen forest, and rotting wood from a *Betula* log. Increases in weight of the earthworms were measured over a period of 109 days. Those cultured on B horizon sand and on rotting *Betula* wood did not survive to the end of the experiment. Weight increased *c.* 80-fold on the organic matter-rich loamy sand, 15–30-fold on four litter substrates, and analysis of variance showed a significant relationship ($P < 0.01$) between weight increase and substrate type, which may reasonably be attributed to the nutritional quality of the substrate.

Eisenia fetida has a sigmoid growth curve similar to that of *L. terrestris* through phase 1 and phase 2, but the rate of growth is much faster and the time required to reach maturity much shorter than in the other lumbricids described above. Tsukamoto and Watanabe (1977) kept *E. fetida* in compost-filled petri dishes in the laboratory; rate of increase of live weight was shown to be much affected by culture temperature, with worms that had a mean weight of 5 mg at hatching attaining 100 mg mean live weight after *c.* 150 days at 10°C, *c.* 70 days at 15°C, *c.* 46 days at 20°C, and *c.* 27 days at 25°C. Comparison of mean weights 70 days after hatching at the four experimental temperatures shows that 5 mg hatchlings had attained weights of *c.* 27 mg at 10°C, 100 mg at 15°C, 190 mg at 20°C, and 650 mg at 25°C, i.e., the growth rate at 25°C had been *c.* 24 times that at 10°C. Hartenstein *et al.* (1979a, b) have shown that *E. fetida*, kept at 25°C on a diet of horse manure or activated sewage sludge, becomes clitellate and begins producing young *c.* 35 days after hatching, and that maximum cocoon production is attained *c.* 70 days after hatching. The implications of these rates of growth and reproduction for the use of *E. fetida* as a source of useful protein are discussed in Chapter 16.

5. Length of Life

The longevity of Lumbricidae in field conditions is not easily determined. In laboratory conditions, specimens of *E. fetida, L. terrestris* and

Aporrectodea longa have lived for 4.5, 6.0 and 10.25 years respectively, but Satchell (1967) estimated an average longevity of < two years for ten common species, ranging from 1.25 years for *Allolobophora chlorotica* to 2.6 years for *A. longa*, and an average of *c.* 1.25 years with a maximum of eight to nine years for some individuals in field populations of *L. terrestris*. Lakhani and Satchell (1970) found that growth of *L. terrestris* virtually ceases after three years.

Phillipson and Bolton (1977) followed the growth of individuals of two cohorts of *Aporrectodea rosea* that hatched in May and September respectively in an English woodland soil. After approximately two years it was no longer possible to separate the two groups on the basis of individual weights. It was estimated that the largest individuals found in the field were from five to six years old, and it is likely that a few *A. rosea* lived longer than this.

B. Life Cycles of Non-Lumbricid Earthworms

The most comprehensive studies of life cycles of non-lumbricid earthworms are those of Lavelle (1971b, c; 1974; 1978; 1979), who studied a mixed population of Megascolecidae and Eudrilidae in the soils of savannas and gallery forests at Lamto, Ivory Coast. Species for which data were obtained were the megascolecids *Millsonia anomala, M. ghanensis, M. lamtoiana, Dichogaster agilis, D. terrae-nigrae*, and *Agastrodrilus opisthogynus*, and the eudrilid *Chuniodrilus zielae*. The annual climatic cycle at Lamto includes two wet seasons, the "big rains" and the "small rains", separated by two dry seasons, the "big dry" and the "small dry"; earthworm life cycles relate closely to the seasons. The species studied in most detail was *M. anomala* and the sequence of events in its life cycle during 1968–1970 is summarized in Figure 7 and related to the seasonal climatic cycle.

1. Cocoon Production

Lavelle (1978) showed that *M. anomala* may produce cocoons at any time of the year, but production is concentrated mainly in two peaks, which coincide with soils beginning to dry out at the end of each of the wet seasons (Fig. 8), and result in two distinct generations per year. The other five megascolecids have only one period of cocoon production per year, and their approximate timing in relation to the seasons is summarized in Figure 8. The only period during which there appeared to be few or no cocoons produced by any species was at the end of the big dry, extending into the beginning of the big rains. For various reasons it was difficult to define any seasonality of cocoon production by *C. zielae* and other eudrilids, but reproduction was observed in nearly every month (Lavelle 1978).

4. Phenology 77

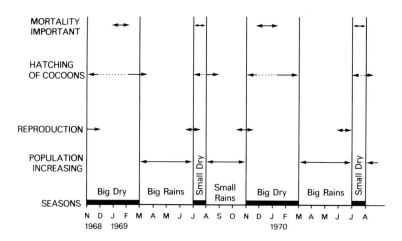

Fig. 7 Life cycle stages of *Millsonia anomala* related to wet and dry seasons in the soils of savanna lands at Lamto, Ivory Coast. (After Lavelle 1971c).

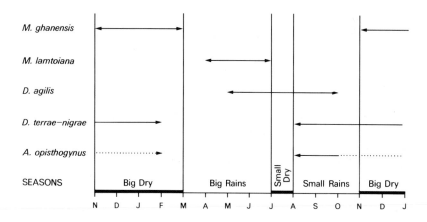

Fig. 8 Periods of cocoon deposition in five species of megascolecids at Lamto, Ivory Coast. (Data from Lavelle 1978).

The numbers of cocoons produced per adult, for each of the species studied, is listed in Table 6. These estimates are based on field collections; observations of hatching rate, also included in the table, are from counts made on cocoons collected from the field and hatched in the laboratory. The rates of cocoon production are comparable to and lie within the low to medium range of cocoon production rates of European lumbricids (p. 68). Population densities are also within the low to medium range of European lumbricids (see Table 7). Increasing population density is accompanied by decreasing fecundity; this is illustrated in Figure 9 (from Lavelle 1981a).

The incubation time of *M. anomala* is *c.* three weeks, but if cocoons have not hatched before the onset of a dry season development is delayed, and hatching takes place at the beginning of the next wet season. In the field, Lavelle (1971c) estimated that the hatching rate for those that develop without interruption and hatch in *c.* three weeks is *c.* 90%, but cocoons that pass through a dry season become partially dehydrated and some fail to hatch, so that the overall hatching rate is *c.* 70%. Incubation times of lumbricids in the field are similarly delayed by the intervention of cold temperatures or drought, and the cocoons hatch when temperatures or soil moistures are more favourable (p. 71).

Lavelle (1978) calculated, from demographic data, the life expectancy of seven species from the Lamto savannas, and these are included in Table 6.

2. Growth

Millsonia anomala populations seem usually to include five generations. Cohorts within the population, representing separate generations, are recognizable, and Lavelle (1971c) followed the development of four such cohorts from monthly sampling of a field population for two years. He was able to draw curves representing growth and survival rates and to construct a diagram of the age structure of the population at the end of the sampling programme (Fig. 10). Estimates derived from means for 11 cohorts (Lavelle 1978) showed that sexual maturity was attained at 8-14 months, with a maximum at *c.* ten months, when mean fresh weight of individuals was *c.* 1.25-2.5 g; the time taken to reach maturity varied with rainfall during the growth period. Growth of adults continued up to *c.* 20-21 months, when mean weight was *c.* 5 g, and maximum weight 5.8 g.

In a population of *Millsonia lamtoiana*, sampled in April, Lavelle (1978) could readily distinguish three cohorts, comprising the very small juveniles of the 1972 generation, a group of large immature worms, with individual weights *c.* 10 g, which represented the one year old generation hatched in 1971, and a group of reproducing adults, with individual weights > 20 g, up to 32 g and 50 cm in length, representing generations two or more years old.

Table 6. Life cycle data for earthworms from savanna habitats at Lamto, Ivory Coast. (Data from Lavelle 1978).

Species	Cocoon production		Life expectancy at hatching (months)	Mean duration of growth (months)			Mean survival rate. (Adults/Juveniles) %	Maximum longevity years
	Mean no. per adult per year	Hatching rate (%)		Juvenile	Adult	Total		
Dichogaster agilis	7.5–15.8	91	3.3	9	6	15	5	2–2.5
Chuniodrilus zielae	c. 5–9[a]	c. 70[a]	3.4	10	8	18	7.5	1.5–2
Millsonia anomala	5.6–14.2	95	6.2	10	10	20	29	2–2.5
Millsonia lamtoiana	2.4–2.8	100	7.5	12	12	24	28	2–3
Agastrodrilus opisthogynus	0.4–1.9	43	11.1	10	14	24	60	3–4
Dichogaster terrae-nigrae	1.3–2.4	95	11.6	22	14	36	35	4
Millsonia ghanensis	0.8–1.4	71	10.6	12	30	42	18	4–5

[a]These data are for several species of Eudrilidae, including *C. zielae*, which are not readily distinguishable from the morphology of their cocoons.

80 Ecology

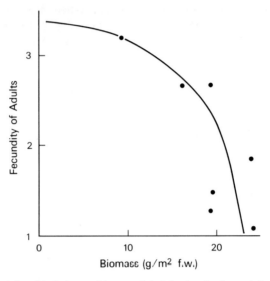

Fig. 9 Inverse relationship between biomass (g/m² fresh wt) of populations and fecundity (number of cocoons per adult per period of reproduction) in *Millsonia anomala*, at Lamto, Ivory Coast. (From Lavelle 1981).

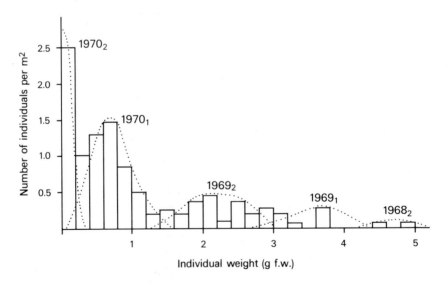

Fig. 10 Demographic constitution of a population of *Millsonia anomala* in plateau soils of savanna lands at Lamto, Ivory Coast, recorded in July, 1970. (After Lavelle 1971c).

One month later, the new generation of immatures was recognizable as a cohort, but the remainder were all adult, and ranged upward from c. 14 g individual weight. Maximum weight was attained at c. two years old.

The duration of the growth period for seven species from Lamto is summarized in Table 6. During periods of growth all the species sampled displayed a cyclic pattern, with the rate of increase in individual weight highest during the two wet seasons, lowest and sometimes becoming negative during the two dry seasons. The cyclic pattern was most marked in *M. lamtoiana*. There is a close parallel between this cyclic pattern and the growth pattern of *Lumbricus terrestris*, which grows for about three years, with short seasonal pauses in summer and winter (p. 74).

3. Length of Life

The maximum longevity of seven species from Lamto, their life expectancy at hatching and the mean survival rate of juveniles that grow to maturity were estimated by Lavelle (1978), and are listed in Table 6. There is a clear relationship between rate of cocoon production, life expectancy at hatching, duration of the growth period, rate of survival to adulthood, and maximum longevity, and these factors also relate to the depth in the soil at which the species live and to individual size. The smallest species live closest to the soil surface, have the lowest life expectancy and survival rate to adulthood, the least duration of growth and the shortest length of life, while the largest species live deepest in the soil and stand at the opposite end of the scale in each of the other characters listed above. The concept of "demographic profiles", to relate these factors to eachother, was developed by Lavelle (1978, 1979), and is discussed on pp. 116-119.

Knowledge of the life cycles of other non-lumbricid earthworms is scanty. *Hyperiodrilus africanus* (Eudrilidae) has three generations per year in Nigerian grassland soils (Madge 1965). The first generation, which hatches at the beginning of the wet season (April–May), matures and produces a second generation later in the wet season (July–August); the second generation matures and produces a third generation in September–October, which survives in small numbers through the dry season to produce the first generation at the beginning of the next wet season.

Amynthas hupeiensis (Megascolecidae), a cosmopolitan species, has one generation per year in Japanese pastures (Watanabe 1975). Hatching begins in April and reaches its peak in May (spring); the worms are active in the surface soil (0–10 cm) through the summer and into the autumn, from April to October, and then hibernate through the winter. *Amynthas sieboldi*, which is restricted to upland forested areas in southern Japan, appears to have a generation time of two years (Sugi and Tanaka 1978a). Hatching occurs in early summer (May–June), and the worms live in the litter layer on forested slopes through the summer and into autumn until November, when they are still immature. They then migrate and concentrate in the soils of small valleys, where they hibernate in the soil or under stones. In rainy weather in late April they emerge from hibernation and return to the slopes, where they mature in late spring and early summer, deposit their cocoons in the litter, and die by August. The cocoons apparently remain in the litter, to hatch in the following April.

Neuhauser *et al.* (1979) studied the life history of *Eudrilus eugeniae* (Eudrilidae) and its suitability as a decomposer of waste products. Maximum production was attained at 24°C–29°C in horse manure or activated sewage sludge. Longevity was *c.* 12 weeks, with maximum cocoon production of *c.* 0.7 cocoon per week at 7-12 weeks, and a total cocoon production of from four to five cocoons throughout its life. At 25°C cocoons hatched in *c.* three weeks. At the population densities in the experiments, sexual maturity was attained in five to seven weeks, so that three to four generations could be produced each year. Production of *E. eugeniae* does not approach that of *Eisenia fetida*, which hatches and grows to maturity in about the same time (p. 316) but produces cocoons at rates up to five per week at optimum population density, and continues to reproduce at high rates up to 27 weeks (p. 70).

Perionyx excavatus, which is common in compost heaps in southern India, was similarly studied by Kale *et al.* (1982). The earthworms were fed on sheep, cow or horse dung. Cocoon deposition began at an earlier stage (90–150 days after hatching) in cultures on sheep dung than in those on cow dung (150–210 days after hatching) or horse dung (162–195 days after hatching). Cocoon production per pair of earthworms in the three months following commencement of deposition was 41.5 on sheep dung, 28.9 on cow dung and 37.1 on horse dung, but the percentage that hatched in the three culture media was 35.0%, 67.5% and 33.0%, so the effective production of juveniles was respectively 15, 20 and 12. Growth rates of earthworms on cow or horse dung were similar, mean weight steadily increasing *c.* two-fold to three-fold to a maximum of *c.* 0.60–0.65 g in the period 20–360 days after hatching. Those fed on sheep dung grew much faster, mean weight increasing *c.* three-fold to a maximum of *c.* 1.7 g at 145 days, and thereafter falling steadily to *c.* 0.8 g at 360 days; their comparatively rapid growth and decline was related to the earlier attainment of maturity and cocoon deposition.

Soil temperature and moisture directly affect rates of cocoon production, incubation times and hatching rates in the Indian megascolecoids *Lampito mauritii, Octochaetona surensis,* and moniligastrids *Drawida calebi* and *D. willsi* (Dash and Senapati 1980, 1982). Threshold conditions for cocoon production were temperature ≥ 20°C and soil moisture ≥ 7%. Incubation time decreased and hatching rate increased with increasing temperature, e.g., hatching time of *O. surensis* cocoons is 12 weeks at 20°C, 15% soil moisture, four weeks at 25°C, 15% soil moisture. Temperatures ≤ 15°C and ≥ 35°C, and soil moisture ≤ 5% or at saturation cause sharp increases in mortality, but these Indian species generally retreat to deep soil layers and enter a resting stage when temperature > 30°C and soil moisture < 10%.

II. Age Structure of Populations

Cocoon production and hatching rates of earthworms exceed the requirements for maintenance of population size, and juvenile mortality rates are high. Lakhani and Satchell (1970) constructed a set of survivorship curves for *Lumbricus terrestris*, assuming potential longevities of 2400, 2500, or 2600 days. Their curves were based on data from field populations and a field-culturing technique, and from a model that best fitted the observed data. Numbers decline rapidly in early life, with about 20% mortality after c. 120 days, when individual mean live weight for field populations is c. 0.3 g, 40% after c. 400 days, at a mean live weight of c. 1.75 g, and 50%–60% after c. 600–800 days, at a mean live weight of c. 3.3–5.0 g.

Over a period of six years, in Holland, van Rhee (1965) found that a population of *Aporrectodea caliginosa* always consisted mainly of immature worms, with the ratio of immature : mature individuals ranging from c. 2 to c. 25. In mixed populations of *Lumbricus terrestris, L. castaneus, A. caliginosa, A. rosea* and *Allolobophora chlorotica* in orchard soils, he found (van Rhee 1967) that, over five successive years, immatures always outnumbered mature worms, except for *L. castaneus*, where immatures were usually in the majority.

In a mixed population of *A. caliginosa* and *A. rosea*, in pastures at Potchefstroom, South Africa, Reinecke and Ljungström (1969) took monthly samples from March to November of one year, and found that the ratio immature : mature varied between c. 0.6 and 24 for *A. caliginosa* and between 2 and 19 for *A. rosea*. Samples of a population of *A. caliginosa* and *A. chlorotica*, taken monthly for a year by Gerard (1967) in an English pasture, showed that for *A. caliginosa* immatures outnumbered mature individuals from April to December, while the reverse was true from January to March, while for *A. chlorotica* immatures were in the majority from June to December, and the reverse was true from January to May.

The only comprehensive studies of the age structure of non-lumbricid earthworm populations are those of Lavelle (1971b, c; 1974; 1978) for several

84 Ecology

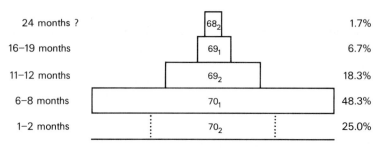

Fig. 11 Pyramid of ages of cohorts in a population of *Millsonia anomala* in plateau soils of savanna lands at Lamto, Ivory Coast, recorded in July, 1970. Recruitment of the generation 1970_2 was incomplete at the time of sampling. (After Lavelle 1971c).

species from the soils of savannas of the Ivory Coast. The species for which most data were accumulated was *Millsonia anomala* (Megascolecidae). *Millsonia anomala* is apparently able to produce cocoons throughout the year, but in most cases there are two well defined maxima, one in July–August, and one in December–January, and these result in addition to the population of two recognizable cohorts of juveniles per year. It was possible to follow the growth of these cohorts and to recognize five cohorts within some field populations, on the basis of their body weights. The structure of a population of *M. anomala*, sampled in August, 1970, is illustrated in Figs 10 and 11. The worms become adult at a mean age of 230 days (*c.* 7.5 months), at a live weight of *c.* 2 g, but continue to grow, more or less continuously, to a maximum mean weight of *c.* 5 g. From the pyramid of age-classes in Fig. 11 it can be seen that, at the time of sampling, nearly half the population was in the six to eight months age-class; recruitment of the cohort 1970_2 was at this time incomplete and it comprised only 25% of the total population, but it is apparent from Figure 11 that when its recruitment was complete it would greatly outnumber the next youngest cohort. The mortality rate between hatching and attainment of sexual maturity of *M. anomala* was estimated as 75%–85% by Lavelle (1974).

Analyses of the age structure of populations of other species from the Ivory Coast may be found in Lavelle (1978), but the data on which they are based are less comprehensive than those for *M. anomala*.

III. Rhythmic Behaviour Patterns

Well defined rhythmic seasonal patterns and some circadian patterns of behaviour are known among earthworms.

A. Seasonal Rhythms

Nearly all earthworms have the ability to go into a resting stage, in response to drought or low temperatures, retreating deep into the soil where they become motionless and do not feed. Primary and secondary sexual characteristics of adults regress during the resting period. These resting periods coincide with seasonal changes and are often described as aestivation or hibernation. Olive and Clark (1978) distinguished two kinds of resting behaviour, *diapause* and *quiescence*, which they defined in behavioural and physiological terms.

Diapause was defined as a response to drought. The earthworms cease feeding, empty the alimentary canal, and construct a spherical mucus-lined chamber in the soil in which they coil into a tight ball, with only one earthworm per chamber. During diapause earthworms lose weight but do not suffer tissue dehydration. Diapause is considered to be obligatory in some species but facultative in others. It has been postulated that obligatory diapause is controlled by neurosecretions; once initiated, usually by drought, but sometimes by a severe injury such as caudal amputation (Saussey 1966), its duration is not affected by artificial manipulation of soil moisture content (Michon 1957, Saussey 1966). Saussey (1966) followed the classical definition of diapause, as used by entomologists, and confined use of the term diapause in earthworms to the obligatory state, distinguishing it from the facultative state which he termed para-diapause. In the sense used by Saussey, facultative diapause, or para-diapause, differs from obligatory diapause only in that it may terminate at any time when soil moisture conditions become suitable for renewed activity.

Quiescence was defined as a response to drought or to low soil temperature. The earthworms cease feeding and go into a torpid state, but do not excavate resting chambers and line them with mucus; quiescent species often suffer severe tissue dehydration. Bouché (1972) distinguished three types of quiescence among lumbricid earthworms, as follows:

(a) anhydrobiosis, a response to dehydration, often associated with congregation of earthworms into dense masses. Lumbricid earthworms can survive water loss up to *c.* 75% of the water content when fully hydrated (Michon 1957);

(b) hibernation, a response to low soil temperatures, usually initiated at temperatures of *c.* 2°C–4°C. Earthworms are usually found, inactive and tightly coiled, in small soil spaces just below the superficial frozen soil layer in seasonally frozen soils (Bouché 1972);

(c) reaction to toxicity, resulting from accumulation of metabolites from microorganisms in the drilosphere. This is rare. Some evidence that it may induce quiescence was presented by Doeksen (1964) and Bouché

(1972) has observed it where worms are living in long-established burrows during long periods of fairly constant environmental conditions. Quiescence is terminated when physico-chemical conditions change so that there is stimulation of bacteria that can decompose the toxic metabolites.

There is evidence that the stated differences between obligatory diapause, facultative diapause and quiescence recognized in European lumbricids are site-specific, and that in the more general case there is a continuous gradation in the intensity of the response within species, dependent on the severity of the stress to which individual populations are subjected, and also varying between populations representing separate genetic stocks within species.

Oglesby (1969) concluded that obligatory diapause is confined to two species of *Lumbricus* (*L. festivus* and *L. rubellus*) and to some species of *Allolobophora* s.l. and *Eophila*. Satchell (1967) recorded obligatory diapause only in sexually mature *Aporrectodea longa* and *A. caliginosa*, contrasting with quiescence in juveniles of these species and in all life stages of *A. rosea*, *A. caliginosa* and *Allolobophora chlorotica*. Avel (1959) listed *Lumbricus* spp., *Eisenia fetida*, *Eiseniella tetraedra*, *Dendrodrilus rubidus*, *Octolasion cyaneum* and *O. lacteum* as species that become quiescent, but have no diapause. Mazantseva (1982) recorded *A. caliginosa* in quiescence from mid-October to mid-June in the Moscow region, when soil temperatures are continuously below 4°C. The Japanese species *Eisenia japonica* has been recorded in "true" diapause during summer in Sapporo (Nakamura 1968).

However, in Sweden, Nordström (1975) found that *A. longa* (adults and juveniles) has no obligatory diapause as defined above, but enters a quiescent stage in summer, as do *A. caliginosa* and *A. rosea*. In New Zealand, Martin (1978) recorded *A. caliginosa* coiled tightly in chambers in the soil through the summer, while *Lumbricus rubellus* remained active. I have similarly found *A. caliginosa* from early summer to early autumn, in many parts of New Zealand, tightly coiled, with the gut empty, with no apparent characters of sexual maturity, in spherical mucus-lined chambers at depths of *c.* 20–40 cm in the soils of pasture lands, while *L. rubellus* could be found in a relatively inactive stage, but not coiled in chambers, at similar depths. The chambers left by *A. caliginosa* in these soils are filled with humus-stained casts from the vertical burrows made by the earthworms when they abandon the resting stage and move back nearer to the surface in autumn. The spherical masses of recognizable casts so formed (many hundreds of them per square metre of surface area) are prominent features of the B horizon of many New Zealand pasture soils. Reinecke and

Ljungström (1969) recorded summer diapause especially in *A. rosea*, and also in *A. caliginosa* in South African pastures.

The contradictory evidence, within species and between species, indicates that the behavioural and physiological adaptations correlated with "obligatory diapause" are not so restricted as has been supposed, but are within the capabilities of a variety of species, in response to severe moisture stress. Some species are probably incapable of "obligatory diapause", but it is uncertain which these are.

It is known that many non-lumbricid earthworms survive adverse seasonal conditions by retreating to deep soil layers and becoming inactive, and that generally non-lumbricids are active for not more than about six months of the year. However, there is little information on their physiological state during periods of inactivity. Watanabe (1975) recorded hibernation of *Amynthas hupeiensis* in Japan from about November to March. Dash and Senapati (1980) similarly recorded "diapause" of *Octochaetona surensis, Lampito mauritii* and *Drawida willsi* in pastures and cultivated lands in India from about January to March, when soil moisture levels close to the surface fall to < 10%. Lavelle (1978, 1983a) in savanna lands at Lamto, Ivory Coast, recorded a decline in earthworm activity as soil moisture levels decreased, first noticeable at pF = 3, and with activity virtually ceasing at pF = 4.2. The number of months per year when > 50% of the earthworm population was inactive ranged from zero in grass savannas to two in shrub savannas and three or four in unburnt shrub savannas. An unusual form of resting behaviour was reported by Sugi and Tanaka (1978a) for *Amynthas sieboldi* in upland forest areas of Kyushu Island, Japan. During the warm months from April to November *A. sieboldi* is active at the litter-soil interface, while in the colder months many of the earthworms die, and those that survive move downhill into valley bottoms and hibernate in the soil or under stones on the ground surface. In the spring the survivors move uphill and are active again at the litter-soil interface. Gates (1941, 1948) recorded "diapause" in some megascolecids and glossoscolecids, while Madge (1969) found that the eudrilid *Hyperiodrilus africanus* in Nigeria assumed a resting state deep in the soil from about November to May.

Bouché (1977a) regarded diapause as an important attribute of anéciques, species that live in burrows in the soil, but feed at the surface. Subsurface conditions do not give a sure guide to conditions at the surface, and Bouché considered, though without evidence, that diapause, controlled by a "biological clock", is an important survival mechanism, imposing aestivation, with reduction of metabolic rate by *c.* 50%, during periods when normal activity would endanger survival. There is little evidence from the real situations described above to support the existence of a rhythmic pattern controlled by a "biological clock".

Byzova (1977) investigated the sources of energy used by earthworms (*Aporrectodea caliginosa*) during aestivation for periods of one to three months. Glycogen reserves, stored in chloragogen tissue, are rapidly used up, and are reduced almost to nil in *c.* two months; total erythrocruorin falls *c.* 10% after one month and *c.* 40% after three months. Body weight falls *c.* 60% during three months' aestivation, so that total erythrocruorin per unit body weight increases. It seems likely that erythrocruorin is used to provide energy.

B. Circadian Rhythms

Ralph (1957) recorded diurnal rhythmic patterns of activity in *Lumbricus terrestris*, with a maximum at 0600 h, when the earthworms were active for 32 min/h, falling quickly to a minimum at 1000 h of 25 min/h, rising again quickly to *c.* 29.5 min/h at 1300 h, and fluctuating between 29 min/h and 31.5 min/h from 1300 h until it reached the maximum again at 0600 h. Measurement of oxygen consumption through the diurnal cycle demonstrated a maximum at 0800 h, two hours after the maximum of activity, indicating that *L. terrestris* may incur an oxygen debt during periods of maximum activity.

Edwards and Lofty (1977) demonstrated that *L. terrestris* showed maximum activity at night, from *c.* 1800 h to *c.* 0600 h, although there were seasonal variations. The rhythmic activity continued for more than seven days when *L. terrestris* was transferred to conditions of constant temperature and darkness.

Circadian rhythms of surface-casting activity by *Millsonia anomala* were recorded by Lavelle (1974) at Lamto, Ivory Coast, and were correlated with ambient temperatures. Maximum rates were observed in the early morning hours, with a sharp decline about the middle of the day. When soils were dry, in April, there was usually only one peak per day, but in October, when soil moisture levels were higher, there was a regular but much smaller peak in late evening. In an Indian grassland Dash and Patra (1977) noted a regular downward movement of *Lampito mauritii* to 15–20 cm in summer when ambient temperatures are at a maximum, from about noon, with upward migration to surface layers with falling temperatures in the evening. No such movements were observed in winter.

5
Populations and Associations

I. Abundance and Biomass

Some estimates of abundance and biomass of earthworms in selected habitats throughout the world are listed in Table 7. Few authors have estimated biomass as dry weight, and though this would be preferable for consistency (p. 347) the figures in Table 7 are fresh weights.

The data in Table 7 illustrate the wide range of habitats and geographical localities where earthworms are found; they are entirely lacking only in regions of extreme cold or drought, and they usually dominate the biomass of soil animals (see, e.g., Kitazawa [1971] for a summary of soil invertebrate population data in forest soils).

About 15 species of the palaearctic family Lumbricidae have been spread by European man throughout temperate regions of the world. In the northern regions of Europe, Asia and North America Lumbricidae are the earthworms of forests, grasslands, sown pastures, croplands and gardens. Populations range from $< 10/m^2$ to several hundreds per square metre, not usually exceeding $c.$ $400/m^2$, but occasionally reaching higher numbers, $500-1000/m^2$ in forest, or in pastures where very heavy rates of fertilizers are applied (Table 7). Lumbricids that have been introduced to temperate regions of the southern hemisphere are confined mainly to sown pastures, cultivated lands and gardens, and in pastures and orchards often attain abundances of $> 1000/m^2$, occasionally up to $> 2000/m^2$, with biomass (fresh weight) sometimes $> 300 g/m^2$ (Table 7).

In the forests and natural grasslands of tropical and southern temperate lands the endemic earthworms are not lumbricids; abundance varies widely but does not attain the very high levels of lumbricids in pastures.

Table 7. Abundance and biomass of earthworms in selected habitats from various parts of the world.

Habitat	Location	Collection method	Earthworm taxa	Abundance no./m²	Biomass g/m² Fresh wt	References
Sown pastures	New Zealand	Hand sorting	Lumbricidae	208–775	60–241	Sears and Evans 1953
				740–1235	146–303	Waters 1955
				690–2020	305 (mean)	McColl and Lautour 1978
Sown pastures	South Australia	Hand sorting	Lumbricidae	460–625	62–78	Barley 1959a
Sown pastures	South Africa	Hand sorting	Lumbricidae	72–1112	—	Reinecke and Ljungström 1969
Fertilized pasture	Argentina	Hand sorting	Lumbricidae, Megascolecidae and Glossoscolecidae	27	—	Ljungström and Emiliani 1971
Pastures with heavy rates of fertilizers	Ireland	Hand sorting	Lumbricidae	400–500	100–200	Cotton and Curry 1980a
Old pasture	Sweden	Hand sorting	Lumbricidae	109	59	Nordström and Rundgren 1974
Old pasture	England	Hand sorting	Lumbricidae	390–470	52–110	Svendsen 1955
Old pasture	Wales	Hand sorting	Lumbricidae	646	149	Reynoldson 1966
Old pasture	France	Washing/sieving	Lumbricidae	288	125	Bouché 1977a

5. Populations and Associations

Habitat	Location	Method	Family	Population	Biomass	Reference
Fallow	South Australia	Hand sorting	Lumbricidae	210–460	16–76	Barley 1959c
Fallow	Wales	Hand sorting	Lumbricidae	226	79	Reynoldson 1966
Cropland	South Australia	Hand sorting	Lumbricidae	20–25	2–2.5	Barley 1959c
Cropland	Rumania	Hand sorting	Lumbricidae	5–100	0.5–20	Gruia 1969
Natural grassland	Rumania	Hand sorting	Lumbricidae	200 (mean)	10–60	Gruia 1969
Natural grassland	Wales	Hand sorting	Lumbricidae	22	8	Reynoldson 1966
Natural grassland	Tennessee, USA	Hand sorting	Lumbricidae	13–41	3.2–7.5	Reynolds 1970
Natural grassland	South Africa	Hand sorting	Glossoscolecidae	74	96	Ljungström and Reinecke 1969
Natural grassland	India	Hand sorting	Megascolecidae and Ocnerodrilidae	64–800	6–60	Dash and Patra 1977
Tropical savannas	Ivory Coast	Hand sorting and washing/sieving	Megascolecidae and Eudrilidae	230	49	Lavelle 1974
Natural grassland	New Zealand	Hand sorting	Megascolecidae	250–750	—	Lee 1958
Orchard	Netherlands	Hand sorting	Lumbricidae	300–500	75–122	van Rhee and Nathans 1961
Orchards	Australia	Hand sorting	Lumbricidae	150	—	Tisdall 1978
Mulched and irrigated orchards	Australia	Hand sorting	Lumbricidae	2000	—	Tisdall 1978

Table 7—continued

Habitat	Location	Collection method	Earthworm taxa	Abundance no./m²	Biomass g/m² Fresh wt	References
Garden	Egypt	Hand sorting	Megascolecidae	420	153	Khalaf El-Duweini and Ghabbour 1965a
Gardens	Argentina	Hand sorting	Lumbricidae, Megascolecidae and Glossoscolecidae	73	—	Ljungström and Emiliani 1971
Taiga	Finland		Lumbricidae	17.4	2.8	Zajonc 1971c
	Siberia	Hand sorting		23.0	8.4	Zajonc 1971c
	USSR			3–7	—	Perel' and Karpechevsky 1971
Northern European and Asian coniferous forests	Finland		Lumbricidae	14–68	—	Huhta et al. 1967
	Sweden	Hand sorting		103–167	30–35	Persson and Lohm 1977
	USSR			12	—	Ustinov 1962
	Japan			27–72	—	Kitazawa 1971
Spruce forest with lime topdressing	USSR	Hand sorting	Lumbricidae	1000	—	Brauns 1955
European deciduous forests	England		Lumbricidae	118–138	—	Phillipson et al. 1976
	USSR	Hand sorting		136	68.3	Zajonc 1971c
	Czechoslovakia			106	98.1	Zajonc 1971b
North American deciduous forests	Canada		Lumbricidae	240–780	38–109	Maldague 1970
	Tennessee, USA	Hand sorting		2–96	1.3–14	Reynolds 1970
	Indiana, USA			14–124	26.3–280.3	Reynolds 1972

Dry schlerophyll forest				7–38	1.3–25.5	
Wet scherophyll forest	Australia	Hand sorting	Megascolecidae	34–76	12.3–47.9	Wood 1974
Sub-alpine woodland				15–106	5.7–35.7	
Gallery forests	Ivory Coast	Hand sorting and washing/sieving	Megascolecidae and Eudrilidae	70–130	3.4–6.8	Lavelle 1978
Tropical Forest	Nigeria		Eudrilidae	34	10.2	Madge 1969
Tropical forest	Nigeria	Hand sorting	Eudrilidae	61.7	2.5	Cook et al. 1980
Lowland dipterocarp forest			Moniligastridae and Megascolecidae	37–92	0.7–1.3	
Lower montane forest			Moniligastridae and Megascolecidae	55	3.1	
Upper montane forest	Sarawak	Hand sorting	Moniligastridae and Megascolecidae	47–108	1.8–2.7	Collins 1980
Upper montane low forest			Megascolecidae	2–24	0.2–2.1	

II. Species Associations

The number of species that make up a community is a simple measure of diversity, basic to the consideration of spatio-temporal relationships, ecological strategies, niche partitioning and sharing of resources among the species that are associated in communities. Numbers of earthworm species recorded from a wide variety of geographical localities and vegetation types are listed in Table 7. The range is from one to 11 species, most commonly c. two to five, with a remarkable consistency that is more or less independent of taxonomic groups and major geographical regions.

Lumbricid associations in Europe and Iceland include fewer species in colder northern than in warmer southern latitudes, are relatively species-poor in coniferous forests, peat lands, heath lands and arable lands and species-rich in non-coniferous mixed woodlands, permanent pastures and gardens. The species-poor communities usually include *Dendrobaena octaedra* and/or *Dendrodrilus rubidus* which are very small litter-dwelling species; with increasing diversity there are progressively increasing proportions of larger, soil-dwelling species of *Allolobophora* s.l. and *Lumbricus* spp.

Lavelle (1983b) contrasts the small number of species in earthworm populations with the species-richness of insect and mite populations in soils; he comments on the lack of increasing diversity in a meridional sequence of populations from Sweden to the Ivory Coast, which appears to be at variance with the general rule of increasing diversity proposed by MacArthur and Wilson (1967a) and confirmed for many animal groups. Lavelle (1983b) points out that although species-richness shows little variation, there is a wider size spectrum among earthworms closer to the equator. The *Dendrobaena* and *Dendrodrilus* spp. of cold Swedish soils weigh < 1 g (fresh weight); in temperate European forests and grasslands fresh weights of adults of some species reach > 10 g; in the tropical savannas at Lamto (Ivory Coast) weights of adults range from c. 100 mg to 30 g; Németh (1981), quoted by Lavelle (1983b), found species with weights of adults ranging from 10 mg to 20 g in the forests of Amazonia. Lavelle considers the increasing diversity of size, and therefore of functional diversity in the ecosystem, as corresponding to the more general increasing species-diversity of other animal groups.

It is doubtful that this argument can be sustained for all earthworms. One of the largest of New Zealand earthworms, *Maoridrilus montanus* (up to 300–350 g fresh weight), is found only in the mountains of the southern North Island, where it lives in areas that are snow-covered in the winter and is associated with other species of earthworms whose adult weights are c. 1 g; the smallest New Zealand species, *Diporochaeta minima* (c. 100 mg) and

D. punctata (*c.* 200 mg) are commonly found with *Octochaetus multiporus* (*c.* 20-25 g) and several other species, in species associations of three to five species (data from Lee 1959). There is some evidence that the maximum, if not the minimum, size of earthworm species may sometimes relate to ecological factors that vary meridionally, but in general large size is better explained in terms of geographical isolation and evolutionary sequences within related earthworm groups, predatory pressures, especially as they affect the need for rapid escape, and exploitation of food sources with low energy content, which requires a large gut-surface area (Table 8).

From a consideration of models that attempt to relate the complexity (i.e., number of species) of communities to their dynamic stability, May (1979) concluded that stable or predictable environments permit species-richness and complexity, while unstable or unpredictable environments require dynamically robust communities and therefore relative simplicity and species-poverty. This conclusion assumes that communities are assembled randomly, but this is not so; they are "the winnowed products of the long workings of evolutionary processes" (May 1979). May (1979) discussed the concept of "connective stability" developed by Siljak (1974, 1975a, b) and further refined by Goh (1978), which proposes that if a system can be redefined as an assemblage of loosely coupled subsystems it may be seen to be stable, even when there may appear to be strong interactions within the system, or stability may result from time-dependent changes in the interactions between the components of the system. The concept of "connective stability" was developed in the context of relationships that encompass complete ecosystems, but it seems likely that it applies also to the interactions between species associated together in mixed species populations of a group such as earthworms.

Phillipson *et al.* (1976) attempted to quantify the significance of species associations that they found among ten lumbricid species, collected over a two year period of intensive sampling in an English beech wood (Brogden's Belt). Using Sorensen's quotient of similarity and Cole's coefficient of similarity (see Southwood [1966] for methods), two species associations were shown to be particularly frequent. These were: (a) *Aporrectodea rosea, Lumbricus terrestris, L. castaneus, Satchellius mammalis*; (b) *A. rosea, A. caliginosa, L. terrestris, S. mammalis,* (Fig. 12). The dominant species in all collections was *A. caliginosa*. The validity of these two associations was confirmed by Fager's method of recurrent grouping. Using all three of these methods together, association (a) above was distinguished as the more significant. This was further confirmed by the construction of a three-dimensional model of the nature of species groupings (Fig. 13), using Cole's C_{AB} index and the significance of associations determined from the x^2 values of the 2 × 2 contingency tables used to calculate C_{AB}. Although *A.*

Table 8. Number of species in earthworm populations from a selected range of habitats in various geographical regions.

Geographical location	Vegetation type	Number of species[a]	Taxonomic groups	Reference
Europe				
Southern Sweden	Heath	2	Lumbricidae	
	Spruce plantations	3–4	Lumbricidae	
	Pine plantations	4	Lumbricidae	
	Alder-pine wood	3	Lumbricidae	
	Beech woods	6–8	Lumbricidae	Nordström and Rundgren (1973, 1974)
	Alder-birch wood	8	Lumbricidae	
	Alder wood	10	Lumbricidae	
	Elm-ash wood	9–10	Lumbricidae	
	Elm wood	11	Lumbricidae	
	Meadow and grassland	5–7	Lumbricidae	
Denmark	Mixed deciduous woodlands	4–7 (7)	Lumbricidae	Bornebusch (1930)
France	Pastures	9–14	Lumbricidae	
	Mediterranean forest	4–5	Lumbricidae	Bouché (1976, 1978)
	Deciduous forest	9	Lumbricidae	
	Pine forest	8	Lumbricidae	
	Pastures	5	Lumbricidae	
Poland	Meadows	6–9	Lumbricidae	Nowak (1975)
	Oak-hornbeam low forest	8	Lumbricidae	Rozen (1982)
	Oak-hornbeam high forest			
Czechoslovakia	Oan-hornbeam forest	6	Lumbricidae	Zajonc (1971b)
Scotland	Pastures	4–10 (7–8)	Lumbricidae	Guild (1951)
England	Beech wood	10	Lumbricidae	Phillipson *et al.* (1976)
	Moorland	2–7	Lumbricidae	Standen (1979)
Wales	Pasture	7	Lumbricidae	
	Ley	5	Lumbricidae	Reynoldson (1966)
	Arable	5	Lumbricidae	
Ireland	Hay field	15	Lumbricidae	Baker (1983)

5. Populations and Associations

Location	Habitat	Number	Families	Reference
Spain	Cultivated fields	1–7 (2–3)	Lumbricidae and Megascolecidae	Diaz Cosin et al. (1981)
	Uncultivated field	0–5 (2)		
Iceland	Bogs	1–2 (2)	Lumbricidae	
	Shrub heaths	0–2 (0)	Lumbricidae	
	Grassy heaths	0–2 (2)	Lumbricidae	
	Grass meadows	3	Lumbricidae	
	Herb meadows	3–4 (3)	Lumbricidae	Bengston et al. (1975)
	Birch woods	2–4 (2)	Lumbricidae	
	Farm lands	0–4 (3)	Lumbricidae	
	Gardens and plantations	2–5	Lumbricidae	
North America				
Canada	Various	1–6 (1–2)	Lumbricidae	Reynolds (1975)
Indiana	Woodlands	2–4 (2)	Lumbricidae and Megascolecidae	Reynolds (1972)
Tennessee	Woodlands	1–3 (2)	Lumbricidae and Megascolecidae	Reynolds (1969)
	Grassland	2–3 (2)	Lumbricidae	
South America				
Venezuela	Tropical rainforest	8	Glossoscolecidae	Németh and Herrera (1982)
Africa				
Egypt	Various	1–2 (1)	Lumbricidae and Megascolecidae	Khalaf El Duweini and Ghabbour (1965a)
Ivory Coast	Savannas	4–9	Megascolecidae and Eudrilidae	Lavelle (1978)
	Gallery forest	4–11	Eudrilidae and Megascolecidae	Lavelle (1983b)
Nigeria	Grassland	> 3		Madge (1969)

Table 8—continued

Geographical location	Vegetation type	Number of species[a]	Taxonomic groups	Reference
South Africa	Pasture	2	Lumbricidae	Reinecke and Ljungström (1969)
India	Grazed pasture	5	Moniligastricae	Senapati and Dash (1981)
	Ungrazed pasture	5	Megascolecoidea	
Japan	Peat bogs	1–2	Lumbricidae	Nakamura (1967)
	Grasslands	5–6	Lumbricidae and	Tsuru (1975)
	Grasslands	4–11	Megascolecidae	
Australia	Savanna woodland	4		
	Low eucalypt forest	2–5 (3)		
	Tall eucalypt forest	4–6 (4)	Megascolecidae	Wood (1974)
	Subalpine woodland	2–5		
	Alpine herbfield	3–5 (4)		
	Gardens and clearings	2–4	Megascolecidae and Lumbricidae	Wood (1974)
	Grassed orchard	4	Lumbricidae	Tisdall (1978)
New Zealand	Native forests	1–7 (2)	Megascolecidae (and Lumbricidae)	
	Native shrublands	1–3 (2)	Megascolecidae (and Lumbricidae)	
	Pine plantations	1–3 (3)	Megascolecidae (and Lumbricidae)	Lee (unpublished data)
	Native grasslands	1–5 (2)	Megascolecidae (and Lumbricidae)	
	Pastures	1–4 (2)	Lumbricidae (and Megascolecidae)	

[a] Numbers in brackets are those most frequently found.

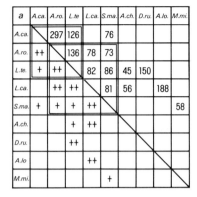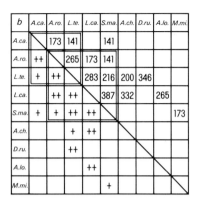

Fig. 12 Species associations illustrated in the form of trellis diagrams; (a) according to Sorensen's quotient, (b) according to Cole's index. Two crosses indicate significance at $P < 0.10$; one cross indicates significance at $P < 0.05$. A.ca. = *Aporrectodea caliginosa*, A.lo = *A. longa*, A.ro. = *A. rosea*, A.ch. = *Allolobophora chlorotica*, D.ru. = *Dendrodrilus rubidus*, L.ca. = *Lumbricus castaneus*, L.te. = *L. terrestris*, M.mi. = *Murchieona minuscula*, S.ma. = *Satchellius mammalis*. (From Phillipson et al. 1976).

caliginosa was the numerically dominant species in the study sites, it showed a lower level of correlation with *L. terrestris*, *A. rosea* and *S. mammalis* than did *L. castaneus*, which make up group (a) above. The occurrence of species together in a recognizable association indicates only that they do occur together regularly, and does not show that they necessarily have similar ecological requirements. To detect associations whose requirements are similar, Phillipson et al. (1976) used the ordination method of Mountford (1962), and this indicated close similarities between *D. rubidus* and *L. terrestris*, *S. mammalis* and *L. castaneus*, *A. rosea* and *A. caliginosa*.

The two levels of association recognized by Phillipson et al. (1976) might reasonably be regarded as subsystems within the overall system, in the sense of May (1979).

Phillipson et al. (1976) used the dominance data summarized by Nordström and Rundgren (1973) to compare the earthworm faunas of 36 beech woods, 37 deciduous woods dominated by species other than beech, 38 coniferous woods and 43 permanent pastures. For each earthworm species in each of the four major habitat types they calculated number of times the species was dominant/total number of times the species was found as an index of the relative importance of the species. Of the ten species recorded from the Brogden's Belt site of Phillipson et al. (1976) the

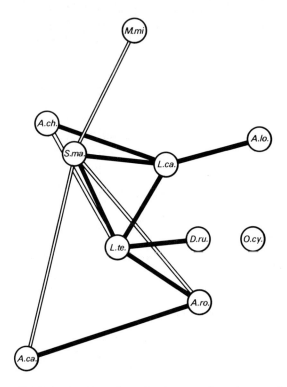

Fig. 13 Diagram illustrating the three dimensional nature of species groupings according to Cole's C_{AB} index, based on data from the earthworm community at Brogden's Belt, England. A.ca = Aporrectodea caliginosa, A.lo. = A. longa, A.ro. = A. rosea, A.ch. = Allolobophora chlorotica, D.ru. = Dendrodrilus rubidus, L.ca. = Lumbricus castaneus, L.te. = L. terrestris, M.mi. = Murchieona minuscula, O.cy. = Octolasion cyaneum, S.ma. = Satchellius mammalis. Solid lines indicate associations significant at $P < 0.01$ and double lines at $P < 0.05$. (From Phillipson et al. 1976).

calculated index showed that *A. caliginosa, A. rosea, D. rubidus* and *L. castaneus* can be considered as predominantly beech wood species, *L. terrestris* as a deciduous woodland/pasture species, and *A. longa, Allolobophora chlorotica* and *O. cyaneum* as most characteristic of permanent pastures. *Satchellius mammalis* and *Murchieona minuscula* did not dominate in any of the habitats. The Brogden's Belt site, with *A. caliginosa, A. rosea* and *L. castaneus* together comprising 68%–76% of total earthworm numbers, may thus be considered typical of many beech woods. On the other hands, this site must be regarded as atypical in that *L.*

rubellus, D. octaedra and *O. lacteum*, all of which are common in other beech woods, were not found at Brogden's Belt, while *Murchieona miniscula* was common at Brogden's Belt but has not been reported from many other beech woods. There are other major differences between the associations recognized by Phillipson *et al.* (1976) and those recognized by Nordström and Rundgren (1973).

The concept of correlating species associations with vegetation types has some validity, but Phillipson *et al.* (1976) conclude that differences in soil characteristics, including pH, quantity and quality (C : N ratio) of organic matter, calcium carbonate content, and moisture regimes, have direct effects on earthworm associations that are more significant than similarity in vegetation. Soils vary over short distances, horizontally and vertically, in their physical and chemical characteristics, and in the quantity and quality of their organic matter content. Such variations are sometimes extreme; e.g. in the north of New Zealand, deep podzols with strongly acidic (pH < 4), saturated and near anaerobic litter up to > 1 m deep and acidic A_2 horizons > 1 m deep may be found in isolated areas up to *c.* 20 m diameter under individual kauri (*Agathis australis*) trees, in forests consisting predominantly of mixed broadleaved species on soils that were formed from the same parent materials as those under the kauris, with mull-type litter overlying neutral to slightly acid deep friable loams. There are virtually no earthworms under the kauri trees, while under the broadleaved trees there are associations of three to six megascolecid species, with high population densities (Lee 1959). More generally, soils vary over small distances within less extreme limits, related to geological, topographic, vegetational and hydrological variations. Soil is a product of its environment, and though earthworms affect the nature of soils (see Part II of this book), there is an over-riding reciprocal effect of soils on earthworms that determines the parameters within which the abundance and diversity of earthworm populations may vary.

Phillipson *et al.* (1976) used indices of species diversity and equitability of abundance to indicate the ecological "fullness" or maximum attainable ecological diversity, comparing the earthworms of their British and two Swedish beech woods. They concluded that the ten earthworm species recorded from the Brogden's Belt beech wood constitute only 70%–80% of the total species that might theoretically be accommodated in this habitat, while the two Swedish beech woods, with six and eight species respectively, might be considered ecologically "full" with respect to earthworms.

Using the same methods of data analysis, Rożen (1982) concluded that the six or eight species of earthworms she had found in Polish oak–hornbeam forests represented ecological "fullness".

Species associations reflect taxonomic diversity and a two-dimensional view of spatial relationships between co-habiting species. They are an essential first step towards the understanding of the ecological strategies adopted by species to minimize interspecific competition and ensure their reproductive success, and thus their survival in multi-species associations.

III. Ecological Strategies

Most earthworms are terrestrial animals that live in the soil, including superficial litter layers, and in some cases in above-ground habitats, especially in very wet forests and peat lands, where they may be found in decaying logs, moss clumps and in a variety of arboreal habitats. A small number of species have adopted aquatic habitats, including the water of streams, ponds, lakes, swamps and the intertidal zone of sea shores. I have limited the discussion of ecological strategies to the terrestrial species.

A. Vertical Stratification

Earthworm communities, i.e., species associations in relatively small areas that are recognizable from their vegetation and soils as ecosystems or catenary sequences, are usually stratified vertically. In addition, individual species display morphological, physiological, reproductive and behavioural differences that are related to their positions in the vertical stratification. The species in a vertically stratified association can be classified on the basis of these attributes, but the classification does not usually relate closely to taxonomic categories, and similar associations are recognizable in superficially similar ecosystems, e.g., temperate rainforests, grasslands, throughout the world.

Satchell (1980e) summarized early work of Evans and Guild (1947–1955), who recognized among British Lumbricidae surface-dwelling species that have no burrows, produce no recognizable casts, and do not aestivate, and distinguished them from species that live deeper in the soil in well-defined burrow systems, cast at the surface or in spaces in the soil, and generally aestivate in summer. Graff (1953a) further defined these two groups, recognizing that the surface-dwelling species are usually red in colour, produce many cocoons, mature rapidly, and have several generations per year, while the soil-dwelling species are usually unpigmented, have relatively low rates of cocoon production, mature slowly, usually have only one generation per year, and occupy all agricultural habitats except compost.

Lee (1959) recognized three main ecological groups among New Zealand megascolecoids. He distinguished them primarily in relation to the soil horizons in which they are most commonly found, into leaf mould (litter or O horizon) species, topsoil (A horizon) species, and subsoil (B and/or B/C horizon) species, and further defined the groups on the basis of shared morphological, behavioural and physiological features, their food preferences, susceptibility to predation, geographical distribution, and reaction to changes in land use patterns. Characteristics of the three main groups are summarized in Table 9.

Classification on the basis of soil profile morphology enabled Lee (1959) to correlate the ecological groups with the distribution and classification of soils and landscape types. Using the zonal/intrazonal/azonal classification of New Zealand soils developed by Taylor (1948, 1952), he related the earthworm groups to Taylor's soil groups (Fig. 14) and to the five soil forming factors—parent material, climate, topography, vegetation and age—that Taylor had adopted from the Russian school of "genetic" soil classification and developed to a high level of refinement in his soil classification. Darwin (1881) had recognized the significance of lumbricid earthworms in the genesis of soils, and his findings were confirmed and extended from the evidence provided by New Zealand megascolecoid earthworms.

Lee (1959, 1969, 1975a, b) further recognized that a minority of "earthworms" lived outside the litter/topsoil/subsoil strata of soil profiles, and classified these into eight categories, four terrestrial and four aquatic, on the basis of their habitats. The four terresrial categories were as follows:

(a) under logs and stones: a variant of the topsoil or litter habitat, with the cover of a log or stone providing protection from most predators, from extremes of temperature and humidity, and sometimes enabling earthworms that are intolerant of salinity to colonize soils on rocky shorelines, where they would otherwise be excluded by wind-blown sea spray, or to live at the surface of some peaty soils from which they would otherwise be excluded by anaerobiosis or extreme acidity;

(b) under bark of trees: accumulation of organic debris in cracks and under loosely attached pieces of bark of trees provides an environment with protection from light, moderation of extremes of temperature and moisture, and protection from predators, which enables small litter and topsoil earthworms to live above the ground in standing trees, or close to the ground surface under the bark of fallen logs. This habitat acquires a special significance in the extremely wet environment of tropical rainforests. In the superhumid climate of mountainous regions in the Solomon Islands and in Vanuatu (Lee 1969; 1975a, b; 1981) where deep layers of saturated acidic peat (pH usually < 4) lie on the ground surface, small earthworms were commonly found under bark of living and dead trees.

Table 9. Ecological grouping of New Zealand megascolecoid earthworms, classified in relation to soil horizons in which they are most commonly found. (After Lee 1959).

Distinguishing characters	Litter species	Topsoil species	Subsoil species
Burrows	None	Permanent — in mineral soil, opening to surface, used primarily as retreat.	Extensive systems, often extending deep below surface, constantly extended. Large diameter, up to 20 mm.
Casts	No recognizable casts	Cast at surface or in soil spaces adjacent to burrows.	Cast in burrows or other soil spaces, rarely at surface.
Size of mature individuals	Small; range 15–180 mm length; most 20–50 mm. Ratio length:diameter 8–17.	Medium-large; range 25–300 mm length; most 75–200 mm. Ratio length:diameter 15–40.	Medium–very large or long and slender; range 32.5–1400 mm length; most 100–400 mm. Ratio length: diameter 20–130.
Pigmentation of body	Heavy; usually ventral and dorsal; red, brown, purple, or green, often with spots of contrasting colour (yellow, white) around setae. Effectively camouflaged.	Medium; usually only dorsal, unpigmented ventrally; red or brown. Little camouflage.	Unpigmented or lightly pigmented only on dorsal surface. No camouflage.
Muscular development	Very thick muscles in body wall.	Medium thickness of body wall muscles.	Body wall muscles weakly developed.
Reaction to touch	Initially no movement; continued stimulation results in lashing of body and leaping.	Immediate very rapid longitudinal contraction to withdraw into burrow.	Limited rapid contraction of extremities, but generally sluggish movement.

Structure of gut	Simple narrow tube	Simple tube, but more convoluted than in litter species.	Very wide thin-walled tube; occasionally narrow tube with complex spiral morphology.
Source of food	Fragments of decomposing litter at surface; little or no soil ingested.	Decomposing litter collected from surface and drawn into burrows; some soil ingested.	Much soil ingested, but selection of organic-rich material; some probably dead root-feeders.
Predatory pressure	Very high, especially from ground-feeding birds, small mammals, centipedes.	Predation from ground-feeding birds, mammals, centipedes, less severe than for litter species.	Minimal; some large nocturnal ground-feeding birds.
Geographic distribution of individual species	Usually very restricted, some exceptions.	Usually restricted; may be widespread in regions with fairly uniform vegetation and soils.	Sometimes restricted; often very widely distributed.
Reaction to change in land-use patterns	Extinction	Usually extinction, but some species survive change from forest to low-intensity use for grazing.	Often survive extreme changes of land use.

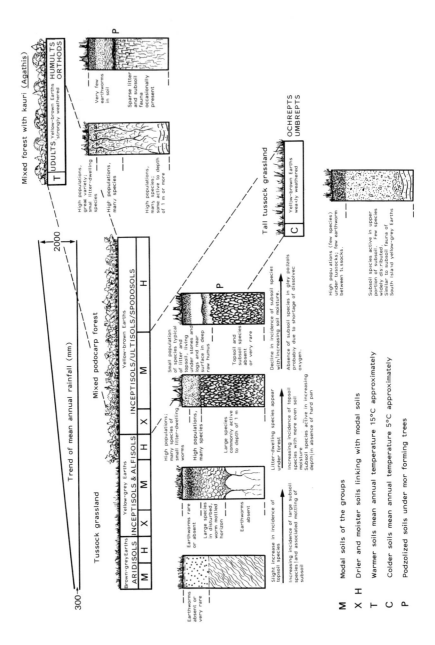

(c) in rotting logs: earthworms are not known to be primary invaders of fallen logs, but they are commonly found in logs that are in an advanced stage of decomposition. In New Zealand Lee (1959) found four log-inhabiting species, and only one (*Megascolides suteri*) was considered to be primarily a log inhabitant; all four were found also in forest litter or topsoil. In the wetter tropical rainforests of the Solomon Islands and Vanuatu, the larger species of earthworms were most commonly found in rotten logs (Lee 1969, 1975a, b). Many other invertebrates, e.g., geophilid and cryptopsid centipedes, millipedes, scorpions, that are more usually found in litter and topsoil horizons were also common in rotten logs and rare in soil horizons. In this environment it seems that rotten logs provide a habitat that is acceptable as a substitute for the upper soil horizons.

(d) at the bases of epiphytes and in leaf axils of understorey forest trees: accumulations of litter at the bases of epiphytes (e.g., *Astelia* spp. *Asplenium* spp.) in rainforests provide a habitat similar to that of forest litter, and small earthworms similar to those in the litter layer are commonly found there. Some understorey plants, especially Pandanaceae, accumulate litter that falls from the forest canopy, and the spaces between their leaf bases and the plant stem are filled with a slurry of decomposing organic debris. A wide variety of small invertebrates, including small earthworms, adult and larval stages of many groups of insects, small land crabs, copepods and other crustaceans, were found (Lee 1969, 1975a, b) in leaf bases of Pandanaceae in the superhumid tropical rainforests of the Solomon Islands and Vanuatu; for earthworms in these tropical island rainforests this was the favoured habitat.

Bouché (1971, 1972, 1977b), independently of Lee (1959) recognized three major morpho-ecological groups among European Lumbricidae. He distinguished the groups primarily on the basis of morphological characters that he considered to have functional significance, not as Lee had, on the basis of the soil horizon in which they were most commonly found, but the resultant groups of Lumbricidae were remarkably similar to Lee's groups of megascolecoids. Bouché's groups, as defined in his paper of 1977b, were:

(a) *épigées:* species that live above the mineral soil surface, typically in the litter layers of forest soils. True litter species were distinguished as a subgroup of *straminicoles*, corresponding closely to Lee's *litter species*. Other related subgroups were *corticoles* that live under bark of trees and

Fig.14 Relationships between the endemic megascolecid earthworms of New Zealand and main soil groups, with reference to climate and vegetation. (After Taylor 1952 and Lee 1959, with assistance from Dr M. L. Leamy).

feed on decomposing phloem, *pholéophiles*, in cracks and crevices above ground, *détritiphages*, in temporary accumulations of organic matter (e.g., dung or rubbish heaps), and *coprophages*, that live in mammalian faeces.

(b) *anéciques:* species that live in burrows in mineral soil layers, but come to the surface to feed on dead leaves, which they drag into their burrows. They are important in burying surface litter. Most species would be included in the *topsoil species* of Lee (1959), but some make burrows that extend deep into the subsoil. The anéciques are a major component of the European Lumbricidae, but are not common among New Zealand megascolecoids; Lavelle (1979) similarly found that anéciques are a minor components of the megascolecid and eudrilid earthworms of the Ivory Coast (p. 112).

(c) *endogées:* species that inhabit mineral soil horizons, feeding on soil more or less enriched with organic matter. Bouché (1977b) distinguished *hypoendogées*, the typical deep-dwelling endogées that correspond closely with Lee's *subsoil species*, from *épiendogées*, which live closer to the surface and often resemble some *épigées*, especially when they are specialized feeders on dead roots (*saprorhizophages*). *Épiendogées* would probably be included among *topsoil species* in Lee's classification.

Bouché's (1977b) definitions of groups were based on characters that included most of those used by Lee (1959), but Bouché used some additional characters, particularly:

(a) "digging muscles": in some species the body wall and septa of anterior segments are strongly muscular. Bouché related this feature to burrowing behaviour, applying the term "digging muscles" to such muscular development. In his three main ecological groups of European Lumbricidae "digging muscles" were absent or only slightly developed in épigées, very strongly developed in anéciques, and feebly developed in endogées. The situation is rather more complex in megascolecoid earthworms (Lee 1959), and I believe it is also more complex in lumbricids than is described in Bouché's scheme. Most megascolecoid litter-inhabiting species have, in proportion to their size, very thick body wall muscles and rather muscular septa throughout their length. This is correlated with their ability to move rapidly, in a habitat where they are exposed to high predatory pressure. Topsoil species vary, usually having well developed body-wall musculature throughout their length, but with extra body wall thickening and associated strongly muscular septa in the 10–20 most anterior and most posterior segments. This is correlated with the ability of these species, when feeding at the surface with their tails in the burrow, to contract the anterior end of the body and retreat very rapidly into their burrows if disturbed or attacked (the withdrawal reflex, p. 53), and also to hold on very strongly to the walls of the burrow with the tail if caught anteriorly by a bird or other predator. In subsoil species the body wall musculature is generally very weak. In large species (e.g., *Octochaetus multiporus, Hoplochaetina* spp.) some anterior and posterior septa are strongly muscular, the anterior series resembling overlapping, backward pointing cones when relaxed. Contraction of these conical septa must provide a powerful forward thrust, which probably assists in burrow formation and so might be important in burrowing; these muscles may, however, be more important simply to enable movement in very large earthworms.

(b) moistening of cuticle: Bouché (1977b) noted relatively copious secretion of mucus by epidermal gland cells in épigées, and that the nephridiopores were not in straight longitudinal lines in his épigées and anéciques, while they generally were in endogées. He considered that these two factors would help to maintain a moister cuticle in épigées and anéciques than in endogées, which are less liable to desiccation. All lumbricids are holonephric, but many megascolecoids are meronephric, and some excrete urine into the gut where much of the water is resorbed. Meronephric and holonephric species, some with regularly aligned and some with irregularly arranged nephridiopores, may be found in any of the recognized ecological categories of Lee (1959), sometimes sharing the same forest soil habitats, with no apparent advantage to those with one or other nephridial form.

(c) reproductive rates of épigées are higher than in anéciques and endogées, with épigées growing to maturity rapidly compared with the other groups. Bouché related these differences to the more extreme variability in physical environment at or near the surface and greater susceptibility to predation of surface species compared with subsurface species.

(d) resting stages: among lumbricids Bouché (1977b) distinguished the anéciques, that survive summer drought and high temperatures by entering a state of diapause, endogées that enter a quiescent stage, not a true diapause, and the épigées, with their short life cycles, that survive through seasonal drought as cocoons even if the adults die. Distinctions based on differences in over-summering behaviour can not easily be sustained (pp. 85-88).

(e) intestinal transit time: Bouché considered that the time for food to pass through the intestine was relatively slow in épigées, variable in anéciques, and rapid in endogées. For comparable megascolecoids I believe the opposite to be true, especially in that the endogées, which live in and feed on soil that contains little organic matter and have a long intestine with a large surface area, are specialized to maximize the time of intestinal transit of their food.

Lee (1959) and Bouché (1977b) recognized that although most species could fairly readily be assigned to one of the three groups each had defined,

110 Ecology

many species had some characters intermediate between those regarded as typical for a group. Rundgren (1977), Persson and Lohm (1977) and Pokarzhevskij and Titisheva (1982) have shown that the depth distribution of some European lumbricid species varies seasonally, and sometimes diurnally.

Bouché (1977b) proposed that the characters of individual species might be imagined as lying within a triangular space whose apices represent the extreme examples of épigées, anéciques and endogées (Fig. 15). Similarly, Lee's (1959) scheme could be represented in the same way, with corresponding apices representing extreme examples of litter, topsoil and subsoil species, as represented in Figure 15. Bouché (1977b) represented the variation in individual characters on separate triangles, pointing out that the position of a species in the over-all triangular framework represents a summation of the positions of individual characters on the same basic framework. Figure 16 represents a comparison between the litter/topsoil/subsoil categories of Lee (1959) and the épigées/anéciques/endogées categories of Bouché (1977b), set out on a series of triangular frames, for eight diagnostic characters. Agreement in the assignment of values to corresponding ecological categories is close; where there are slight disagreements the corresponding triangles are black.

Bouché (1977b) further proposed that the combinations of characters that define the apices of his triangles represent evolutionary as well as ecological

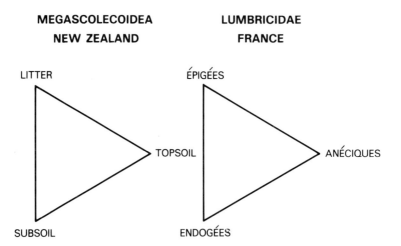

Fig. 15 Corresponding major ecological groupings of New Zealand Megascolecoidea, according to Lee (1959), and of Lumbricidae of France, according to Bouché (1977b), represented on a triangular base.

5. Populations and Associations 111

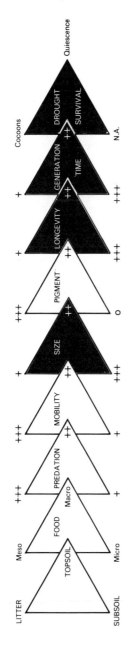
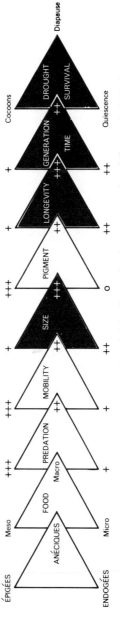

Fig. 16 Representation on a series of triangular frameworks of eight diagnostic characters used to define ecological characters in the systems proposed for New Zealand Megascolecoidea (Lee 1959) and for French Lumbricidae (Bouché 1977b). Where there are slight disagreements between the two systems the corresponding triangles are black.

strategies. Convergence of ecological strategies, as recognized by Lee and Bouché within genetically isolated earthworm stocks, is paralleled in most, if not all, major groups of animals and plants. It might be regarded as a significant isolating mechanism for sympatric species within restricted geographical regions, but it should not be confused with the more basic unity of the larger groups, at the level of families, subfamilies and probably most genera, whose coherence derives from continuous evolution within geographically isolated gene pools.

Nordström and Rundgren (1974), Phillipson *et al.* (1976), Persson and Lohm (1977), and others have recognized the épigée, anécique, endogée classification of Bouché in the lumbricids of Sweden, England and elsewhere in Europe. Németh and Herrera (1982) distinguished épigées, anéciques and endogeés among the glossoscolecid earthworms of tropical rainforests in the upper Rio Negro basin of Venezuela. Wood (1974), who studied megascolecoid earthworms in the mountains of southeastern New South Wales, distinguished "topsoil" species that live in surface litter and the A horizon, but do not make burrows that penetrate the B horizon, "subsoil" species that live in burrows in the B horizon, penetrating into the A horizon, but not into the litter, and "other" species that live in semi-permanent vertical burrows opening at the surface and penetrating into the B horizon. Wood's groups correspond to Bouché's épiendogées, hypoendogées and anéciques.

Studies of the earthworms of the tropical savanna lands of the Ivory Coast led Lavelle (1979) to question the applicability to tropical earthworms of Bouché's (1977a) ecological classification into épigées, anéciques and endogées. The soils of the Lamto savannas, where Lavelle worked, are very impoverished, as are those of many tropical regions. Thirteen earthworm species were found at Lamto and of these, seven were abundant everywhere. The seven dominant species included two that feed on surface litter mixed with some soil, and might be classed as intermediate between épigées and endogées, and five that are strictly geophagous (soil-ingesting) and would be classed as endogées. Lavelle noted the lack of anéciques among the dominant species, and contrasted this situation with that in the Lumbricidae of temperate regions, where anéciques often make up 50%-75% of the total earthworm biomass (Bouché 1975c). There are some tropical species that might be classified as anéciques, but Lavelle (1979) found them only as a minor component of the total population, and concluded that the anécique group may be important only among the Lumbricidae. The groups of species recognized by Lavelle might more profitably be compared with those of Lee (1959) (see Table 9, p. 104). The depth-distribution of species shown by Lavelle (Fig. 17) indicates that *Dichogaster agilis* might be recognized as a litter species, *Millsonia lamtoiana, M. anomala* and *Chuniodrilus zielae* as

5. Populations and Associations   113

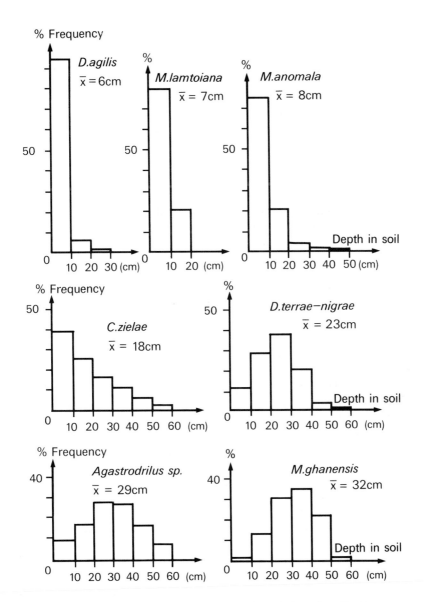

Fig. 17 Frequency distribution as a function of mean depth in the soil profile of seven species of earthworms collected over the period August 1971–August 1972 in the soils of savanna lands at Lamto, Ivory Coast. (After Lavelle 1979).

topsoil species, *D. terrae-nigrae* as an intermediate topsoil/subsoil species and *Agastrodrilus opisthogynus* and *M. ghanensis* as subsoil species in Lee's classification.

Perhaps the problem, recognized by Lavelle, of fitting ecological groups of earthworms to patterns derived from the European Lumbricidae results from the more general inadequacy, also illustrated in the humus former/humus feeder classification of Perel' (see below), of regarding all earthworms in the light of the accumulated knowledge of the Lumbricidae of Eurasia. These species include only about 10% of all earthworms. Most of the non-taxonomic literature of earthworms derives from work on Lumbricidae, and furthermore it deals almost exclusively with only a few of the most common lumbricid species. These few species are those that successfully recolonized the newly developing soils of the uninhabited deserts exposed by the retreat of the polar ice sheets that covered northern Europe, northern Asia, and much of North America only 12 000–15 000 years ago. Their genetic and ecological diversity is not comparable with that of the non-lumbricid earthworms, nor even with that of the remainder of the Lumbricidae. Most earthworms have continuously evolved in soils and vegetation types that have themselves evolved over long periods of time, in the major continental land masses and islands of the world that were not denuded of life by the polar ice sheets of Pleistocene and Recent geological periods.

B. Humus Formers and Humus Feeders

Perel' (1977) classified lumbricid earthworms into *morpho-ecological* categories. She distinguished *humus formers*, earthworms that eat plant debris that is only slightly decomposed, from *humus feeders*, that eat plant debris that is already much decomposed, and participate in the further transformation of organic matter into humic materials. The terms humus formers and humus feeders correspond respectively with the detritivores and geophages, whose food preferences are discussed on pp. 18–22.

The *humus formers* have a comparatively simple intestine, with a small typhlosole that has no secondary folds, and the intestine is moniliform (see Semenova [1966] for details); this form of intestine was associated by Perel' with the ability to propel discrete food masses along the intestine. Other characters that Perel' associated with humus formers include purple or dark brown pigmentation, a flattened tail, closed epilobic or tanylobic prostomium which can be readily manoeuvred to grasp plant debris from the ground surface and remove it to the burrow, high mobility, which Perel' related to a rather complex pennate arrangement of the longitudinal muscles, rapid response to physical stimuli (e.g. Stolte [1962] recorded that

Lumbricus terrestris, a humus former, responds to UV light at 64%-69% of the lethal limit), and gas exchange rates higher than in humus feeders (Byzova 1965b).

The *humus feeders* have a more complex intestine, essentially cylindrical but with a large involuted typhlosole, often with secondary folds that increase the surface area available for absorption. Other characters that Perel' associated with humus feeders include little or no pigmentation, no flattening of the tail, open epilobic or proepilobic prostomium with little independent manoeuvrability, low mobility, which Perel' related to arrangement of the longitudinal muscles in a more simple "bundle-like" disposition, relatively delayed response to physical stimuli (e.g., Stolte [1962] recorded that *Aporrectodea caliginosa*, a humus feeder, responds to ultraviolet light at 78%-80% of the lethal limit), and gas exchange rates lower than in humus formers (Byzova 1965b).

The simple typhlosole structure, with its associated characters, was found by Perel' to be especially typical of species that inhabit forest litter and decaying wood, including *Dendrobaena attemsi, D. hortensis, Dendrodrilus rubidus, Allolobophora parva* and *Eisenia lucens*, while the more complex lobed typhlosole structure was characteristic of soil-inhabiting species, including *Aporrectodea caliginosa, Allolobophora chlorotica, A. bartolii, A. sturanyi, Cernosvitova biserialis* and *Kritodrilus calarensis*.

Perel's *humus formers* include litter dwellers, e.g., *Dendrobaena octaedra*, that are confined to surface litter layers, soil-litter dwellers, e.g., *Eisenia nordenskioldi, Lumbricus rubellus, Perelia diplotetratheca*, that are active in the A horizon and feed on dead roots and surface debris, making burrows that penetrate to 20-30 cm, but may penetrate deeper when they are in resting stages, and deep-burrowing, surface-feeding species, that live in deep burrows up to 1 m or more, but feed at the surface, e.g., *Aporrectodea longa, Lumbricus terrestris*. Her *humus feeders* include topsoil-inhabiting species that live permanently in the humified A horizons, middle-layer species that live in the humified A or non-humified B horizons and deep-layer species that live in the B and C horizons of soils.

Perel's morpho-ecological groups cut across the ecological groups recognized by Lee (1959) and Bouché (1971, 1972, 1977b). The morphology of the intestine and of the longitudinal muscles are probably important in determining and limiting the assimilation efficiency and mechanical capabilities of earthworms, and Perel' has shown that there are recognizable patterns in these characters that relate to life forms and behavioural patterns. There is, however, no evidence that these characters have any special significance that distinguishes them as more important than the much wider range of characters used to define ecological strategies in the groups recognized by Lee and Bouché.

C. Demographic Profiles and Demographic Indices

The striking difference in life forms, behaviour, food quality and quantity ingested, phenology and productivity of earthworms that inhabit the soils of the savanna lands of the Ivory Coast led Lavelle (1979) to propose that selection may have favoured the development of distinct "demographic profiles", characteristic of groups of species best fitted to distinct subdivisions of the soil environment.

Analysis of population structure, based on monthly sampling for four years of the earthworms of five different savanna associations, convinced Lavelle (1979) that the demographic profiles of the seven dominant species could be described in terms of three parameters (Table 10): (a) total duration (i.e., maximum) of growth period; (b) life expectancy at hatching; and (c) number of cocoons produced per adult earthworm.

From the demographic parameters in Table 10 Lavelle concluded that duration of the growth period is shorter for species that live close to the surface than for deeper dwelling species, that life expectancy at hatching is directly related to the product of mean maximum weight and mean depth of activity ($W.\bar{p}$), and that the number of cocoons produced per year is at a maximum in the small species *Chuniodrilus zielae* and *Dichogaster agilis*, at a minimum in the large geophagous species *Millsonia ghanensis*, *Agastodrilus opisthogynus* (possibly a carnivore [Lavelle 1983c]) and *D. terrae-nigrae*, and intermediate in *M. anomala* and *M. lamtoiana*. The smallest species usually live closest to the surface, so $W.\bar{p}$ also relates reasonably closely to annual cocoon production (Fig. 18). Lavelle (1979) found that each of his three demographic parameters related reasonably well to $\log(W.\bar{p})$. He proposed a formula, to apply to individual species

$$D = 10^3.F / (C . Ev)$$

where F is the number of cocoons per year; C duration of growth in months, Ev, life expectancy (in months) at hatching, and D a demographic index that provides a measure of the capability of a species for population growth. A minimum D (2–5) occurs for deep-dwelling geophagous species (*D. terrae nigrae, A. opisthogynus, M. ghanensis*), a maximum (200) for small surface-dwelling species (*D. agilis, C. zielae*), and intermediate for *M. lamtoiana* and *M. anomala* that live at shallow depths (Table 10).

Lavelle (1979) extended his concept of demographic profiles by examining the relationships between the demographic index (D) and energy utilization of individual species. He related D to **P/B̄** and **P/I**, where **P** is annual production, **B̄**, mean biomass and I, total assimilation of food of the population of the species. He further considered the relationship **P_R/P**,

Table 10. Some characters of the dominant earthworm species of the Lamto savannas (Ivory Coast), their ecological classification according to Bouché (1977a), and mean values of demographic parameters. (After Lavelle 1979).

Species	Ecological group of Bouché (1977a)	Source of food	Maximum weight (g)	Mean depth of activity of adults (cm)	Ratio length: diameter	Demographic parameters		
						Duration of growth (months)	Life expectancy at hatching (months)	Cocoon production (no./y)
Dichogaster agilis	Intermediate	Litter	0.5	6	15	15	3.4	10.7
Millsonia lamtoiana	épigées/endogées	Litter	32.0	7	15	24	7.5	3.1
Millsonia anomala	épiendogées	Soil	6.0	8	17	20	6.2	6.2
Chuniodrilus zielae		Soil	0.2	18	35	18	3.3	13.0
Dichogaster terrae-nigrae		Soil	29.0	23	40	36	11.6	1.9
Agastrodrilus opisthogynus	hypoendogées	Soil (possibly carnivorous or root feeder)	3.0	29	56	24	11.1	1.3
Millsonia ghanensis		Soil	16.0	32	32	42	10.6	1.3

5. Populations and Associations 117

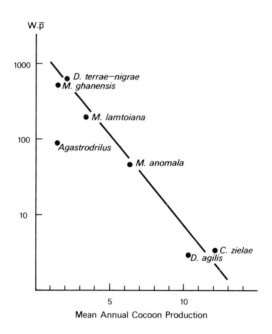

Fig. 18 Relationship between cocoon production (mean number of cocoons produced per adult per year) and $W.\bar{p}$ (the product of mean maximum weight and mean depth of activity in the soil) for seven species of earthworms in the soils of savanna lands at Lamto, Ivory Coast. (After Lavelle 1979).

where P_R is the component of P devoted to reproduction. These relationships are summarized in Table 11. The value of D was found to correlate closely with $P/\bar{B}$ and with P_R/P, i.e., species whose production per unit biomass is high invest more energy in reproduction than those with low production per unit biomass. Extreme examples are *C. zielae*, where D is 200, $P/\bar{B}$ is 3.2 and 9% of production goes into reproduction, and *M. ghanensis* with a D value of 2.9, $P/\bar{B}$ of 1.3 and only 0.5% of production going into reproduction.

The value of P/I tends to vary inversely with $P/\bar{B}$, but close comparisons are difficult because of differences in the type and nutritive quality of food ingested. The large geophagous species *M. anomala*, *D. terrae-nigrae* and *M. ghanensis* feed on soil with a low organic matter content and therefore are obliged to expend much energy to ingest 10–30 times their body weight of soil per day, while the litter-feeding species *D. agilis* and *M. lamtoiana* have access to food with a high organic matter content and need to expend little

Table 11. Demographic indices (D), relationship between maximum weight and mean depth of activity ($W.\bar{p}$), and ratios between the energy equivalents of annual total production (P), mean biomass (B), ingesta (I) and reproduction (P_R) for seven species of earthworms in a population at Lamto, Ivory Coast. (From Lavelle 1979).

Species	D	W.$\bar{p}$	P/$\bar{B}$	P/I%	P_R/P%
Chuniodrilus zielae	200	3.6	3.2	1.2	9.0
Dichogaster agilis	200	3.0	2.1	0.4	7.2
Millsonia anomala	50	48.0	2.1	0.3	4.1
Millsonia lamtoiana	17	224.0	1.7	1.8	2.1
Dichogaster terrae-nigrae	4.5	644.0	1.0	0.9	2.1
Agastrodilus opisthogynus	4.9	116.0	1.0	2.7	2.5
Millsonia ghanensis	2.9	576.0	1.3	0.9	0.5

energy in feeding. The deep soil-dwelling species *Agastrodilus opisthogynus* is anomalous; Lavelle (1979) presented some evidence that it may be a root feeder, or (Lavelle 1983c) a predator on other earthworms, and may thus have access to high energy food resources. Species with high potential for population growth (high P_R/P) have high $P/\bar{B}$ and low P/I, and take advantage of food resources that have high nutritive value (litter) but are not permanently available, while species with low P_R/P have low $P/\bar{B}$ and high P/I and are thus able to take advantage of food resources that have low nutritive value (soil) but reliable availability. Lavelle (1979) related these two groups of earthworms to the more general existence of such ecological strategies recognized by Pianka (1970) and Calow (1977).

Lavelle's recognition of distinct demographic strategies in a tropical earthworm community is an important contribution to understanding of community structure and resource partitioning in earthworm populations. It relates the vertical stratification recognized by Lee (1959) and Bouché (1971, 1972, 1977b) to the ecological energetics of the individual species that make up species associations, and deserves wider investigation. Its relationships to some more general ecological concepts are discussed below.

D. *r*- Selection, *K*- Selection and *A*-Selection and the Habitat Templet

1. r-Selection and K-Selection

The concept of *r*-selection and *K*-selection was introduced to ecological and biogeographical theory by MacArthur and Wilson (1967a, b), although some of its basic ideas had been discussed earlier (Dobzansky 1950). The terms *r* and *K* originated from the Verhulst–Pearl equation

$$\partial N/\partial t = r(1 - N/K)N$$

which relates population, *N*, increase over time to the intrinsic rate of natural increase, r, and to the carrying capacity of the environment, K. Macarthur and Wilson distinguished two basic selection processes, expressed in reproductive strategies, that could be used to relate the establishment and maintenance of viable populations of a species or other taxon to biotic and abiotic factors of the environment. These were *r*-selection, selection for maximum population growth in uncrowded populations, and *K*-selection, selection for competitive ability in crowded populations. They applied the distinction particularly to the establishment and subsequent reproductive success of the elements of animal populations on islands, distinguishing strategies that favoured successful colonization in disturbed or unpredictable habitats (*r*-selection) from strategies that favoured success in more stable and predictable habitats (*K*-selection). On this basis they examined the distribution and constitution of some island faunas and propounded a theory of island biogeography.

The original MacArthur and Wilson *r*-selection and *K*-selection concept, based on a few criteria that relate to reproductive success, led to many new investigations of ecological and evolutionary strategies. Ecological, phenological, behavioural, physiological and other criteria have been correlated with the *r*-selection to *K*-selection spectrum, and the concept has also been applied to plants and microorganisms and more broadly to communities and to ecosystems.

2. A-Selection

Greenslade (1972) proposed a selection process additional to the *r*- and *K*-selection processes that operates in consistently and predictably harsh environments, where those few species that can maintain viable populations are faced with little interspecific competition. He first called this "beyond *K*" selection and used it to distinguish the ecological strategies of some species of beetles (*Priochirus* spp.), that are found in high montane areas of tropical islands, from the strategies of species of the same genus that are found at lower altitudes. Greenslade subsequently adopted the term adversity-selection, proposed by Whittaker (1975), as identical with his "beyond *K*", and abbreviated it to *A*-selection.

The A-selection concept has been further developed by Greenslade, especially as it applies in arid environments (Greenslade 1982), and more generally in arid, cold, or resource-limited environments (Greenslade 1983). Selection is first for adaptation to the harsh environmental conditions, then for conservation of adaptation. It is correlated (Greenslade 1982) with low population densities and species-richness and little interspecific competition in communities, low reproductive rates and rates of population increase, long generation times, commonly with parthenogenesis and with dormancy, and often with restricted migratory ability.

Grime (1974, 1977, 1979) has proposed three basic selection processes and associated strategies that relate to ecological and evolutionary success in vascular plants. He distinguished strategies related particularly to disturbance, to competitive success in relatively undisturbed habitats, and to environmental stress, which correspond fairly closely with Greenslade's r-selection, K-selection and A-selection strategies.

Some of the parameters used in defining ecological strategies of earthworms, discussed previously, are shared with those used in defining the r–K or r–K–A selection spectra, and there have been a few attempts to integrate the two approaches. Lavelle (1979) proposed that, of the species commonly found in the Lamto savannas, those with high potential for population growth (p. 116) should be regarded as r-selected while those that have low potential for population growth (p. 116) should be regarded as K-selected. Bouché (1977b) linked r-selection with small size, which implies inability to make extensive burrows and necessitates feeding at the surface on small particles of plant detritus, and linked K-selection with large size, ability to make burrows in the soil and to remove large pieces of plant detritus from the surface into burrows, or to seek their food beneath the soil surface. On this basis, Bouché's épigées would be regarded as r-selected, his endogées as K-selected, and his anéciques as K-selected or intermediate between r-selected and K-selected.

Satchell (1980e) began from the attributes used by Bouché (1977b) to distinguish épigées, anéciques and endogées, and with some additional material distinguished 22 criteria that could each be interpreted in terms of correlates of r-selection and K-selection as defined by Pianka (1970). Then he graded 17 common peregrine species of Lumbricidae, on the basis of his 22 criteria, along a scale ranging from extreme r-selection to extreme K-selection. His conclusions were that surface-feeders (predominantly épigées), including those that find their food in litter layers, animal dung, and in semi-aquatic habitats, are r-selected, those that feed below the soil surface (predominantly endogées) are K-selected, while some species that move between the soil and surface to find their food are intermediate between r-selected and K-selected forms.

It should be appreciated that the peregrine species classified along the r–K spectrum by Satchell (1980e) and those included in Bouché's (1977b) treatment are predominantly representatives of the small group of Lumbricidae that followed the retreating polar ice sheets back into northern Europe in the last 12 000–15 000 years and have spread with European man throughout the temperate regions of the world. In the context of all earthworms they are opportunistic species that are most successful colonizers of disturbed and uncrowded habitats and so should all be considered as r-selected, though some may be at the more extreme limit of r-selected strategies than others. They are probably most comparable to the euryhaline *Microscolex* spp. that originated in South America and have repopulated subantarctic islands in the last few thousand years (Lee 1968), or to the predominantly pheretimoid earthworms that have spread, with some help from humanity, through the isolated oceanic islands of the Pacific basin (Lee 1981). A more comprehensive expression of the possibilities of the r–K or r–K–A selection spectrum in earthworms might be expected among the diverse and ancient groups that inhabit continental tropical regions.

Using demographic and ecological data accumulated by Lavelle (1971b, c; 1974; 1978; 1979) for seven dominant megascolecid and eudrilid earthworm species of tropical savannas in the Ivory Coast, and relating these data to criteria of r-selection and K-selection (from Southwood 1977) and to criteria of r-selection, K-selection and A-selection (from Greenslade 1982, 1983), I have compiled Table 12. The available data permit only eight of the 12 contrasting criteria in Southwood's scheme and seven of the 12 in Greenslade's scheme to be used. The criteria listed are as in Southwood's scheme, with the pertinent quantitative data used to assess them in parentheses. Each species is scored for each of the criterion, as follows:
(a) on the r–K scheme of Southwood (1977), r-selected = +1, K-selected = −1, intermediate = 0;
(b) on the r–K–A scheme of Greenslade (1982, 1983) the score for each species is assessed from $\Sigma(r\text{-} + K\text{-} + A\text{-})$ selected characters.

The limits set in the scoring procedure are necessarily arbitrary and therefore arguable. Additional data (from Lavelle 1979) on the source of food, mean depth of activity in the soil of adults and interpretation of their ecological group (of Bouché 1977b) are also listed for each species. There is a continuous gradation between extreme r-selection and K-selection or extreme r-selection, K-selection, A-selection for each of the criterion.

Scoring on the r–K criteria the small litter-feeding species *Dichogaster agilis* emerges as extreme r-selected, one of the large soil-feeding species, *D. terrae-nigrae*, as extreme K-selected, while the remainder are in the middle range (from +2 to −2) and must be regarded as intermediate between r-selected and K-selected. Lavelle (1983a) used demographic data alone to rank

5. Populations and Associations   **123**

some of these species, concluding that *D. agilis* and *Chuniodrilus zielae* are extreme *r*-selected, *D. terrae-nigrae* and *Millsonia ghanensis* extreme *K*-selected, while *M. lamtoiana* and *M. anomala* are intermediate (*r–K*)-selected. Scoring on the *r–K–A* criteria yields a more precise separation, with the litter feeding *D. agilis* emerging as *r*-selected, three species that live in the upper soil layers as (*r–K*)-selected or *K*-selected, two of the deep soil dwellers as *A*-selected and one as intermediate, (*K–A*)-selected.

Some relationship is apparent in Table 12 between the *r–K–A* classification and the ecological categories of Bouché (1977b), but closer affinities can be seen with the litter/topsoil/subsoil categories of Lee (1959). This is reflected in the relationship between mean depth of activity of adults and the position of individual species in the *r–K–A* spectrum (Table 12). Reference to Lavelle (1978) shows that *D. agilis* is a typical litter species, in the sense of Lee (1959), *M. lamtoiana, M. anomala* and *C. zielae* are topsoil species, *D. terrae-nigrae* is an intermediate topsoil/subsoil species, while *Agastrodilus opisthogynus* and *M. ghanensis* are subsoil species. The *r–K* spectrum, excluding *A*, gives a less satisfactory match with the ecological categories of Lee (1959) or Bouché (1977b).

3. The Habitat Templet

Strategies that ensure biological success in the *r–K* (or *r–K–A*) selection spectrum must necessarily be such that they maximize the numbers of species' descendants. Southwood (1977) distinguished three basic parameters, namely, favourableness for reproduction, **r**, survival expectancy, **E**, and variation in favourableness for reproduction, *V*, which modifies **r** and can be represented as $V \pm \mathbf{r}$. He recognized that each of these parameters may vary in time and space, and for each of them drew up a two by two matrix of variability on time and space axes. The summation of these three matrices, also drawn up as a two by two matrix on time and space axes, then yields the reproductive success matrix of the *r–K* (or *r–K–A*) selection spectrum. Further, Southwood (1977) recognized that the parameters that define a habitat also vary in time and space, and thus present a varying framework that affects the possibilities for expression of the biological components of the reproductive success matrix. Variability in habitats may be described by a multi-dimensional model, whose axes represent ranges of levels of temporal and spatial constancy of the various abiotic and biotic parameters that determine *habitat favourableness* and *habit predictability*. Within the framework defined by the model, which Southwood designated the *habitat templet*, the adaptive capabilities of particular organisms could be seen as defining a space, centred around an optimum, within which exist

Table 12. An interpretation in terms of some r-selection and K-selection criteria (Southwood 1977) and some r-selection, K-selection and A-selection criteria (Greenslade, 1982, 1983) of life cycle and energetics data for seven common species of earthworms from the Lamto savannas (Ivory Coast). Data derived from Lavelle (1971a, b; 1974, 1978; 1979). Data for individual species are scored and additional data on source of food, mean depth of activity in the soil, and ecological groupings (of Bouché) are listed.

	Dichogaster agilis	Millsonia lamtoiana	Millsonia anomala	Chuniodrilus zielae	Dichogaster terrae-nigrae	Agastrodrilus opisthogynus	Millsonia ghanensis	Limits of categories
Generation time (months of juvenile growth)	Short (9)	Intermediate (12)	Short (10)	Short (10)	Long (22)	Short (10)	Intermediate (12)	<11 Short 11–20 Intermediate >20 Long
Size (fresh weight, g)	Small (0.5)	Large (32)	Intermediate (6)	Small (0.2)	Large (29)	Intermediate (3)	Large (16)	<1 Small 1–15 Intermediate > Large
Density-independent mortality (survival rate Adults/Juveniles %)	High (5)	Intermediate (28)	Intermediate (29)	High (7.5)	Low (35)	Low (60)	Intermediate (18)	<10 High 10–30 Intermediate >30 Low
Fecundity (No. of cocoons per year)	High (10.7)	Low (3.1)	Intermediate (6.2)	High (13.0)	Low (1.9)	Low (1.3)	Low (1.3)	<2 Low 2–10 Intermediate >10 High
Time or Food/Space efficient (Production/Ingestion = P/I %)	Time (1.2)	F/S (0.4)	F/S (0.3)	Time (1.8)	Intermediate (0.9)	Time (2.7)	Intermediate (0.9)	<0.5 F/S 0.5–1.0 Intermediate >1.0 Time
"Overshoot" of populations? (Reproduction/Production = P_R/P %)	Yes (9.0)	Yes (7.2)	Intermediate (4.1)	No (2.1)	No (2.1)	No (2.5)	No (0.5)	<3 No 3–7 Intermediate >7 Yes

"Boom and bust"? (Production/Biomass = P/B)	Yes (3.2)	Yes (2.1)	Yes (2.1)	Intermediate (1.7)	No (1.0)	No (1.0)	Intermediate (1.3)	0–1 No 1–2 Intermediate >2 Yes
[a] H$_r$ small or large (months/year when cocoons are found)	Small (5)	Small (3)	Large (12)	Large (11)	Intermediate (6)	Intermediate (6)	Small (5)	<6 Small 6 Intermediate >10 Large
SCORE AND SELECTION TYPE								
i Criteria of Southwood (1977)[b]	+8 r	0 r–K	0 r–K	+3 –r	–6 K	–2 –K	–2 –K	—
ii Criteria of Greenslade (1982)[c]	r	r–K	K	r–k	A	A	K–A	—
Source of food	Litter	Litter	Soil	Soil	Soil	Soil (possibly carnivorous or root feeders)	Soil	—
Mean depth of activity of adults (cm)	6	7	8	18	23	29	32	—
Ecological group of Bouché (1977b)	Intermediate épigées/endogées	Intermediate épigées/endogées	Épiendogées	Épiendogées	Épiendogées	Hypoendogées	Hypoendogées	—

[a] H$_r$ is a measure of the length of time for which a habitat provides conditions suitable for breeding, relative to an organism's generation time (Southwood 1977).

[b] Each of the eight characters listed above is scored on the basis: $+1 = r$-selected, $-1 = K$-selected, $0 =$ intermediate, according to the criteria of Southwood (1977).

[c] The listed characters (except H$_r$) are assessed as $\Sigma\,(r + K + A)$-selected, according to the criteria of Greenslade (1982).

conditions that satisfy the requirements of the reproductive success matrix and so permit a population to persist. The concept provides a basis for fusion of the genetically determined parameters of the reproductive success matrix, represented in the r–K or r–K–A selection spectrum, with physical and biological parameters of the environment that define the habitat templet.

The environmental parameters of the habitat templet can be represented diagrammatically on two axes, *increasing favourableness* and *increasing predictability* (Southwood 1977, Greenslade 1982, 1983). The r–K-selection spectrum was represented by Southwood as parallel to the predictability axis, with adversity selection (Whittaker 1975) parallel but in the opposite sense to the favourableness axis (Fig. 19a). Southwood (1977) considered Greenslade's (1972) "beyond K-selection" (A-selection) to be at the lower limit of the favourableness axis, where those species that do adapt will be relatively free of interspecific competition, but he did not examine this type of selection in detail. Greenslade (1982, 1983) has developed Southwood's concept of the habitat templet to include A-selection, arguing that the K-selection strategy requires habitat favourability and so can not include strategies that are necessitated by extremely unfavourable habitats. Southwood and Greenslade introduced a third vector, which Greenslade (1982) named "biotic unpredictability", to represent trophic and spatial complexity, increasing simultaneously with increasing habitat favourableness and predictability (Fig. 19b).

In Figure 20 the seven species classified in Table 12 on the basis of r–K or r–K–A selection criteria are placed on the habitat templet, as represented by Southwood (1977) in Figure 20a, and as represented by Greenslade (1982, 1983) in Figure 20b, as an illustration of the usefulness of these concepts.

Effects of habitat variability on a population comprising ten lumbricid species, in the soils of an English beech wood, were studied by Phillipson et al. (1976). Monthly estimates of abundance of each species through one year and three-monthly estimates for the next year, in 44 adjacent 10 m × 10 m plots, were related to a variety of soil and vegetation characteristics. Multiple regression analyses showed that 42.4% of the variance in the total population was explained (P < 0.01) by variability in soil depth (D = < or > 25 cm), soil bulk density (weight per unit volume), maximum soil moisture content in the June–August (summer) period, and litter standing crop, in the listed order of significance. For the individual species the percentage of variance explained by measured habitat variables was as shown in Table 13 (from Phillipson et al. 1976). It should be noted that much of the variance remains unexplained, even with the best combinations of measured variables, reflecting the present inadequacy of understanding of earthworm ecology.

5. Populations and Associations 127

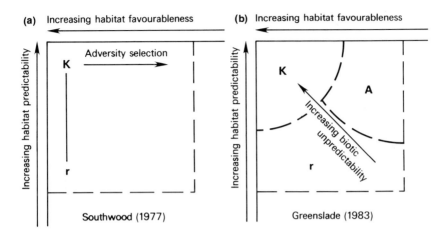

Fig. 19 The habitat templet, as represented (a) by Southwood (1977) and (b) by Greenslade (1982).

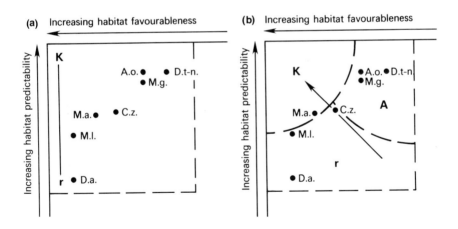

Fig. 20 Approximate positions on the habitat templet, as represented (a) by Southwood (1977) and (b) by Greenslade (1982), of seven species of earthworms from the savannas of Lamto, Ivory Coast, based on an interpretation of data from Lavelle (see Table 11). A.o. = *Agastrodrilus opisthogynus*, C.z. = *Chuniodrilus zielae*, D.a. = *Dichogaster agilis*, D.t-n. = *D. terrae-nigrae*, M.a. = *Millsonia anomala*, M.g. = *M. ghanensis*, M.l. = *M. lamtoiana*.

Table 13. Significance of habitat variables in explaining variability in abundance of ten lumbricid species in an English woodland. (Data from Phillipson et al. 1976).

Species	Habitat variables (in order of significance)	Variance explained (%)	Probability
Épigées			
Satchellius mammalis (litter)	i soil depth ii decrease in soil moisture (June–July) iii litter standing crop	28.7	0.01
S. mammalis (soil)	None	—	—
Lumbricus castaneus (litter)	i maximum soil moisture (June–August) ii decrease in soil moisture (June–July) iii litter standing crop	32.1	0.01
L. castaneus (soil)	i soil depth ii soil bulk density	26.9	0.01
Dendrodilus rubidus	i decrease in soil moisture (June–July)	10.5	0.05
Anéciques			
Lumbricus terrestris	i soil depth	21.8	0.01
Aporrectodea longa	i decrease in soil moisture (July–August)	12.1	0.05
Endogées			
Aporrectodea caliginosa	i litter standing crop	11.3	0.05
A. rosea	i soil depth ii soil bulk density	32.5	0.01
Murchieona minuscula	i litter standing crop	26.3	0.001
Octolasion cyaneum	i decrease in soil moisture (July–August)	10.6	0.05
Intermediate			
Allolobophora chlorotica	None	—	—

The set of habitat variables used in the regressions did not include any effect of the ground flora, but by using 2 × 2 contingency tables Phillipson et al. (1976) were able to show that the distribution of *Aporrectodea longa* and *A. caliginosa* was significantly correlated with particular species in the ground flora ($P < 0.01$ or 0.001), and both of these species had low correlations ($P < 0.05$) with the other site variables used in the regressions. The variables used can be interpreted as indices of habitat favourableness or habitat predictability and can therefore be related to the habitat templet (Fig. 19a, b). One of the variables, soil bulk density, is probably considerably affected by the earthworms themselves, while the availability of litter on the ground surface is greatly affected by the presence or absence of large litter-feeding earthworm species (especially *Lumbricus terrestris*, which is capable of burying all litter by late spring in many woodland soils). However, the results of the analysis illustrate the significance of the habitat templet and demonstrate the value of this approach to the analysis of population structure. Addition of further variables, including biotic pressures from predators or competitors as well as further physical parameters, e.g., soil temperature and soil organic matter content, might be expected to increase the percentage of variance explained.

Greenslade and Greenslade (1983) have considered the soil fauna as a whole in relation to the habitat templet. They recognize a sequence that parallels increasing depth in soil profiles, from predominantly *r*-selected taxa in the energy-rich but environmentally unpredictable surface litter horizon, through predominantly *K*-selected taxa in the energy-rich and environmentally more predictable deep litter to surface soil horizons, to predominantly *A*-selected taxa in the energy-poor and environmentally highly predictable deep soil horizons.

E. Niche Partitioning and Resource Utilization

Separation of earthworms into the litter/topsoil/subsoil categories of Lee (1959), the épigée/anécique/endogée categories of Bouché (1971), the humus former/humus feeder categories of Perel' (1977), and their spatial separation in relation to horizontal variations within a site of soil depth, soil moisture, soil bulk density, litter standing crop and ground flora (Phillipson *et al.* 1976) illustrate some mechanisms of niche partitioning in earthworm communities. Differences between species are also apparent in timing through the year of peak breeding periods (see Lavelle 1978), and maximum population density (Phillipson *et al.* 1976), and in the selection of components of litter, roots or soil organic matter as preferred food (Phillipson *et al.* 1976, Ferrière 1980). Direct interspecific competition for resources is probably not a major constraint on most earthworms, although it must be important in some rather special cases, e.g., where a large litter-feeding species like *Lumbricus terrestris* may bury all the autumn litterfall by early spring in some deciduous forests, thus depriving any other litter-inhabiting species of a suitable habitat for much of the year.

Lavelle et al. (1980) proposed that the ecological niche of earthworms can be defined by six principal variables. These are: three spatio-temporal variables, namely, vertical distribution in soil, horizontal distribution related to plant associations, and seasonal cycles of abundance; and three trophic dimensions, namely, species size, demographic profile, and energy content of ingested food. Taking eight earthworm species of the Lamto (Ivory Coast) savannas, and using the three spatio-temporal variables above, Lavelle et al. (1980) calculated spatio-temporal niche overlaps (Ojk indices) using the method of Pianka (1974). The Ojk index is a similarity coefficient whose value increases from 0 to 1 with increasing overlap. Ojk indices were calculated for each pair of species for each variable, to yield a matrix of indices. The three matrices were then combined by multiplication into a single matrix for the three variables (Table 14). The Ojk indices vary from 0.01 to 0.80 ($\overline{ojk}$ = 0.35), and indicate two groups whose spatio-temporal niches overlap. These are the *Dichogaster agilis–Millsonia lamtoiana–Millsonia anomala* group that inhabit litter and upper soil layers, mainly in shrub savannas, and the *Agastrodrilus opisthogynus–Millsonia ghanensis–Chuniodrilus zielae–Stuhlmannia porifera* group that inhabit the deep soil layers, mainly in open savannas. Species in the two groups were then plotted on three axes representing the trophic dimensions; Lavelle et al. (1980) were thus able to show that differences in size, demographic profile and energy content of ingested food permitted the species to use different components of the available energy resources that ensure a considerable degree of separation of niches.

The food of earthworms is discussed elsewhere (p. 17); it has been shown that co-habiting species may select different fractions of the food apparently available to them, and that in some species juveniles take different foods from adults, so that the resources of a habitat are divided in subtle ways between and within species. Differences in food preferences are paralleled by differences in vertical distribution between species and between life-stages within species, which also contribute to niche separation.

Periods of activity and inactivity or diapause are characteristic of many species, so demands on resources may be separated (p. 84).

Abrupt changes in behaviour are common in individual species and communities, triggered by seasonal, sometimes abrupt, changes in litter fall rates and the nutrient content of the litter fall, e.g., differences in nitrogen and other nutrients of litter from winter to summer in Australian eucalypt forests (Lee and Correll 1978), and in soil moisture content and soil temperature, especially near the surface. Lavelle (1983b) discussed the functional plasticity of earthworm species and communities, distinguishing

Table 14. Spatio-temporal niche overlap (Ojk indices) of eight species in earthworm communities of the Lamto (Ivory Coast) savannas, calculated by multiplication of matrices of Ojk indices for each of three variables, vertical distribution, horizontal distribution, and seasonal cycles of abundance, according to the method of Pianka (1974). (From Lavelle et al. 1980).

	D.a.	M.l.	M.a.	D.t.	C.z.	S.p.	A.o.
M.l.	0.71						
M.a.	0.62	0.65					
D.t.	0.01	0.21	0.34				
C.z.	0.33	0.40	0.46	0.40			
S.p.	0.19	0.11	0.47	0.25	0.41		
A.o.	0.04	0.05	0.24	0.23	0.62	0.79	
M.g.	0.01	0.02	0.18	0.15	0.36	0.64	0.80

D.a. = *Dichogaster agilis*, M.l. = *Millsonia lamtoiana*, M.a. = *Millsonia anomala*,
D.t. = *Dichogaster terrae-nigrae*, C.z. = *Chuniodrilus zielae*, S.p. = *Stuhlmannia porifera*,
A.o. = *Agastrodrilus opisthogynus*, M.g. = *Millsonia ghanensis*.

seven different functional stages of earthworm populations in the Lamto savannas related to changes in the structure of size classes, levels of activity and vertical distribution in the soil. These changes are reflected in seasonal and spatial variations in parameters of the ecological function of populations, including the quantity of soil ingested, production of surface casts and quantities of organic matter decomposed. He related the ability of earthworms to modulate their behaviour, and consequently their requirements for resources, to the low species richness of earthworm communities and the apparently anomalous lack of relationship between species richness and latitude (p. 94). Because of their high level of ecological plasticity earthworm communities contrast with soil microarthropod communities, which include a great variety of species, each highly specialized in its requirements of the habitat. Competition and predatory pressure are probably less important regulators of earthworm populations than is the variability of their physico-chemical environment (Chapter 1).

IV. Models of Earthworm Populations

A. Carbon Flow through an Earthworm Population

Carbon flow in a deciduous forest ecosystem at Oak Ridge was modelled by Reichle (1971a). For the decomposer subsystem of the model, Reichle made a quantitative assessment of the carbon flux, especially the portion due to earthworms. Five compartments were distinguished, namely, the O_1 horizon (surface litter), the O_2 horizon (subsurface, partially decomposed litter), earthworm biomass, total soil organic matter to 35 cm depth, and extractable earthworm biomass estimated from a soil temperature and moisture function (Lakhani and Satchell 1970). Initial values, based on field data, were assigned to the compartments. Four transfer functions covered earthworm respiration, ingestion of litter, ingestion of soil organic matter, and egestion plus mortality, and rates were assigned to these from field and laboratory data. Transfers from O_1 to O_2, O_2 to soil, and respiratory losses from all compartments were assumed to be linear functions of the source compartment. Rates of litter fall were obtained from field data; inputs of dead roots were considered as constant from year to year. The model consisted of a set of seven simultaneous, non-linear differential equations, and several time-dependent forcing functions, solved using a fourth order Runge–Kutta method with a step size of 0.1 month. Running this model with data for repeated years showed little variation from year to year in the contribution of earthworms to carbon flow, and enabled useful predictions to be made of ingestion, egestion and respiration rates, rates of turnover of organic matter related to standing crop of litter, and rates of turnover of soil surface layers by earthworms. Construction of this model required accumulation of a large data base, which is not commonly available for earthworm populations, but the predictive value of the model indicates the value of this approach to study of the ecological energetics and population biology of earthworms.

B. Allez-Les-Vers: A Single Species Model

Lavelle and Meyer (1977) attempted to model the population dynamics, production and consumption of one species of geophagous earthworm, *Millsonia anomala*, which dominates (in biomass) the earthworm population of savanna lands at Lamto (Ivory Coast). The model comprised four submodels and took into account factors of the physical environment, growth rates and population dynamics, particularly hatching and mortality rates, of *M. anomala*. The model was used to simulate population changes during 1972, using initial data established from field studies up to the beginning of that year, and the result of the simulation was compared with detailed field studies of *M. anomala* populations made by Lavelle in 1972. Population parameters, including mean annual density, biomass, hatching rate and production, were well simulated by the model, but seasonal variations were poorly simulated.

Lavelle and Meyer (1983) have revised and refined their first model, to produce the model Allez-Les-Vers. The revised model is based on instantaneous descriptions of entities, attributes and sets. Each worm or cocoon is described as an entity, in terms of demographic characters and its individual environment, especially moisture and temperature. Each entity can be simulated separately and the system, comprising a number of entities, can be varied according to an event list, with events arranged in a time sequence, and processes. Eight subprogrammes describe the changes resulting from the events and processes. The programme provides a detailed simulation of the population, with information on density, biomass, structure, tissue production and quantity of soil ingested.

The revised model was calibrated against the data for *M. anomala* populations in 1972, used to calibrate the earlier model. Results of the simulation agreed closely with observed values for density, biomass, number of mature adults and hatching rates; they differed in predicting amount of soil ingested and mean depth in the soil of occurrence, but these differences were shown to result from slight imperfections in the model. The model was then used, in the same configuration as in the calibration test, to simulate the *M. anomala* population of a clear shrub savanna in a particularly dry year (1969) for which field data had been obtained by Lavelle. Estimates provided by the simulation were within c. 15% of observed data for all but one parameter; imperfections in the model evident in the initial calibration were also evident in the second test. Further development of this model is proceeding, and it is anticipated that it will be extended to other species and may eventually be generalized to apply to all earthworm species.

C. Worm.For: A Single-Species Model

Mitchell (1983) described a simulation model that predicts how much organic waste can be processed and how much earthworm biomass can be produced by *Eisenia fetida* under specified conditions. The model (Worm.For) included four subroutines that calculate growth, mortality, reproduction and ingestion. Each of these could be varied independently. Environmental factors, especially temperature, substrate quality and availability, toxicity, and the size of individual earthworms were incorporated. An iteration interval of one week was adopted. Data from a wide variety of investigations of *E. fetida* were used to establish the basic functions that were incorporated into the model.

Simulations in which specific parameters were varied showed that the output of the model was particularly affected by changes in mortality rates and in temperature. The model was tested against data and predicted performance of a large scale field trial of sewage sludge disposal using *E. fetida* in the city of Lufkin, Texas, USA. At optimum temperature (25°C), over a ten week period, it was shown that *E. fetida* biomass was capable of increasing to a level that would consume about twice the available energy in the sludge, so that the earthworms would not be able to grow and reproduce at optimum rates.

The model could be improved by incorporating data on assimilation and respiration rates, quality of specific food substances, and more detailed environmental data. It was concluded that the model could be applied to natural populations including more than one species if relevant energetic and population data were obtained for each species.

D. A Comprehensive Ecological and Agronomic Model

Bouché and Kretzschmar (1977) constructed a model, which they named REAL (Rôle Écologique et Agronomique des Lombriciens), which attempts to integrate environmental and demographic data, element fluxes (carbon, nitrogen and others), burrowing and casting activity, effects on soil structure, and interactions with microorganisms, to apply generally to earthworm populations. The concept is discussed further in Bouché (1980, 1982). This is an ambitious project, intended to allow quantitative information derived from detailed studies to be used predictively at new sites. At present it is useful only at a conceptual level, to provide a framework within which data might be accumulated to quantify the great variety of factual data and interactions that would be necessary to make it of practical use.

Modelling of the role of earthworms in ecosystems is at an early stage of development but the above examples, provided by Reichle, Lavelle and Meyer, and Mitchell, show considerable promise that further studies using modelling techniques may be expected to add much to knowledge of earthworm ecology.

6
Ecological Energetics

1. Ecological Energetics of Earthworms

Ecological energetics concern the energy transfers involved in the fixation of atmospheric carbon and its elaboration into organic compounds by primary producers, the subsequent flow and partitioning of fixed carbon between and within trophic levels of consumers, and its eventual return to the atmosphere in carbon dioxide. Energy stored or transferred between trophic levels is usually expressed in terms of its calorific value; it is measured directly as heat produced when test materials are oxidized, or in the case of energy used in respiration by measuring oxygen uptake (19.64 J/mL oxygen) or carbon dioxide output. Calculations based on respiratory carbon dioxide output require determination of a respiratory quotient (R.Q.), i.e., the ratio of carbon dioxide evolved to oxygen consumed. For comparative studies, measurements of respiratory exchange must also be corrected to some standard temperature, using the ratio Q_{10}, i.e., the ratio of increase of oxygen consumption for a temperature increase of 10°C.

Expressed in simple form the basic equation for energy flow through consumer organisms is:

$$I = P + R + E$$
$$(\text{Consumption}) = (\text{Production}) + (\text{Respiration}) + (\text{Egestion})$$

where I is energy contained in ingested food, P is energy assimilated and retained in tissue production and storage materials for growth and reproduction, R is energy assimilated but lost in respiration to maintain the organism, and E is unused energy voided in excreta. The sum of P plus R is the total energy assimilated, A.

Consumer organisms comprise herbivores, that feed on living plant tissues, saprovores, that feed on dead plant tissue and/or the microbial and fungal tissue associated with dead plant tissue, and carnivores, that feed on the tissues of herbivores, saprovores, or other carnivores. Earthworms are primarily saprovores although they also derive some energy from digestion of living and dead animal tissue (p. 16). Production efficiency ($P/A\%$) of saprovores is generally lower than for herbivores and carnivores. For a large number and wide variety of insect species Wiegert and Petersen (1983) have calculated that mean $P/A\%$ for herbivores is 43.6 ± S.E. 2.7, for carnivores (including spiders) 38.0 ± S.E. 5.2, and for saprovores 26.7 ± S.E. 4.8.

A. Energy Budgets and Efficiency of Energy Utilization

The energy available in ingesta, the efficiency of its assimilation and the partitioning of assimilated energy between production and respiration have been determined for relatively few earthworm species. Table 15 summarizes data of Bolton and Phillipson (1976) for the European lumbricid *Aporrectodea rosea* and for a mixed megascolecid-eudrilid species association at Lamto, Ivory Coast (Lavelle 1974). *Aporrectodea rosea* is a geophagous species and the dominant species at Lamto are also geophages. The *A. rosea* data relate to measurements at 10°C and those for the Lamto earthworms to measurements at 26°C. These are realistic soil temperatures when earthworms are active at the two localities; they permit comparison between feeding rates and the allocation of assimilated energy in temperate and tropical environments. Differences in energy budgets between small immature, large immature and adult *A. rosea* illustrate the changing metabolic demands of individuals through their life cycle.

I. Consumption (I)

The rate of soil ingestion (mg/g dry weight per day) of the tropical species is about six times that of *A. rosea* (Table 15). Many tropical soils have low organic matter contents. The soil ingested by the Lamto earthworms has a mean organic matter content of *c.* 1% and a mean energy content of *c.* 235 J/g, while corresponding values for the soil ingested by *A. rosea* are *c.* 18% and *c.* 4500 J/g respectively. The total energy available in ingested soil to the tropical earthworms (1570 J $g^{-1}d^{-1}$) is only about one third to one quarter that available to *A. rosea* (4260–5730 J $g^{-1}d^{-1}$). There is little difference

Table 15. Total daily food intake and daily energy budgets (per gram fresh weight) of three life stages of *Aporrectodea rosea* from an English beech wood, at 10°C (after data of Bolton and Phillipson 1976), and of a population of Megascolecidae and Eudrilidae from a sparsely wooded savanna ecosystem at Lamto, Ivory Coast, at 26°C (after data of Lavelle 1974).

Life stage	*Aporrectodea rosea* at 10°C			Megascolecidae and Eudrilidae at 26°C
	Small immature	Large immature	Adult	All life stages
Fresh weight (mg/individual)	60	180	225	1025
Energy equivalent (J/individual)	777	2447	3018	—
Consumption (I)				
$J\ g^{-1}\ d^{-1}$	5730	4910	4260	1570
Organic matter (mg(d.wt)$g^{-1}\ d^{-1}$)	232	197	194	64
Total soil (mg(d.wt)$g^{-1}\ d^{-1}$)	1265	1165	1125	6700
Assimilation ($A = P+R$)				
$J\ g^{-1}\ d^{-1}$	72	53	38	140
in mg organic matter ingested per day	6.6	4.5	2.6	5.7
Production (P)				
$J\ g^{-1}\ d^{-1}$	35	19	8	10
in mg organic matter ingested per day	3.3	1.7	0.4	0.4
Respiration (R)				
$J\ g^{-1}\ d^{-1}$	37	34	30	130
im mg organic matter ingested per day	3.3	2.8	2.2	5.3
Egestion (E)				
$J\ g^{-1}\ d^{-1}$	5660	4860	4220	1430
in mg organic matter $g^{-1}\ d^{-1}$	227	193	191	58
mg total (d.wt) $g^{-1}\ d^{-1}$	1258	1160	1122	6694
R/P	1.0	1.6	3.8	13.0
P/A %	49	36	21	7
A/I %				
In joules	1.3	1.1	0.9	8.9
of organic matter ingested	2.8	2.3	1.3	
of total ingesta	0.4	0.4	0.2	0.1

between the quantities of soil ingested by large immature and adult *A. rosea*, but small immatures ingest more than either of the older life stages.

2. Assimilation (A)

Assimilation rates of young immature *A. rosea* are almost twice those of adults, but the assimilation efficiency ($A/I\%$) of the three life stages is very low, varying from 1.3% in small immature to 0.9% of the energy in ingested food for adults. For the tropical earthworms, assimilation rates are two to four times those for *A. rosea*, despite the much lower energy content of their food, and this difference is reflected in the much higher assimilation efficiency (8.9%) of the tropical earthworms (Table 15). Their superior efficiency may reflect inherent differences in physiological function, or may be related simply to increased rates of chemical breakdown at the higher temperatures. However, it is essential to their survival in soils with low organic matter content, as is apparent from the discussion below of the partitioning of assimilated energy between production and respiration.

3. Production (P) and Respiration (R)

Production rates ($J\ g^{-1}d^{-1}$) in *A. rosea* of small immatures, large immatures and adults are respectively 39, 19 and 8 with $P/A\%$ respectively 49%, 36% and 21% (Table 15), reflecting the higher demands for tissue production of growing immature individuals compared with the more modest demands of adults, which must be mainly for tissue replacement and energy storage for later use in reproduction and maintenance metabolism during periods of inactivity. Energy used in production by the tropical species (10 $J\ g^{-1}d^{-1}$) is about the same as that for adult *A. rosea*, but because of their high energy requirement for respiration $P/A\%$ is only 7% for the tropical species.

Energy use for respiration does not differ greatly between life stages of *A. rosea*, but is greater in immature than in mature individuals. It seems likely that most of the energy used in respiration by *A. rosea* is expended in burrowing; for the three life stages respiration per gram fresh weight per day is in the ratio 37 : 34 : 30, while total soil ingested (mg [dry weight] $g^{-1}d^{-1}$) from the data in Table 15 is in the closely similar ratio 37 : 34 : 32. *Aporrectodea rosea* is a geophagous species and must ingest much inorganic material, which contains no energy available for digestion, to obtain enough organic material for its metabolic requirements. Energy consumed in respiration by the tropical earthworms (130 $J\ g^{-1}d^{-1}$) is much greater than that used in respiration by *A. rosea*. The difference relates fairly closely to the temperature difference between the two sets of experiments. Assuming a Q_{10} of c. 2*, respiratory oxygen consumption would be expected to increase by c. three to four times for a temperature increase from 10°C to 26°C. It is most likely that the energy expended in respiration by the Lamto earthworms is, like that in *A. rosea*, used mainly in burrowing, as they are predominantly geophagous species and must ingest about six times as much soil per unit biomass per day as does *A. rosea*. The three life stages of *A. rosea* use between 1.0 and 3.8 times the amount of assimilated energy in respiration as in production, while the Lamto earthworms use 13 times as much in respiration as in production (R/P in Table 15). For one of the geophagous Lamto species, *Millsonia anomala*, Lavelle and Meyer (1977) and Lavelle (1981a) calculated that 94.7% of the assimilated energy is used in respiration, leaving 5% for tissue production and 0.3% for reproduction; the proportion used in reproduction is only 0.026% of the total energy available in the soil ingested. If the assimilation efficiency of the tropical species were not much higher than that of the temperate species (*A. rosea*) the population could not persist in the energy-poor tropical soils.

Estimates of respiratory energy use by *A. rosea* were based on measurements of carbon dioxide output, using infra-red gas analysis, and of oxygen uptake, using a Warburg respirometer (Phillipson and Bolton 1976). Respiratory rates of *Lumbricus castaneus, Dendrodrilus rubidus, A. rosea* and *Octolasion cyaneum* were similarly determined monthly for a year by Phillipson and Bolton (1976). Respiratory quotients (R.Q.), i.e., the ratio of oxygen consumed to carbon dioxide produced, at 10°C varied seasonally within and between species. The expected range of R.Q. was c. 0.7-1.0. For adult *L. castaneus* R.Q. was 0.52 in spring, 0.53 in summer, 0.58 in autumn and 0.70 in winter; for *D. rubidus* corresponding values were 0.55, 0.66, 0.68 and 0.99, for *A. rosea* 0.81, 0.72, 0.74 and 0.89, and for *O. cyaneum* 0.47, 0.40, 0.58 and 0.29. Only *A. rosea* consistently had R.Q. in the expected range. Phillipson and Bolton considered the possibilities that differences between the species resulted from size-specific metabolic rates, but found no evidence for this. They proposed that in *L. castaneus, D. rubidus* and *O. cyaneum* some respiratory carbon dioxide may be taken up by calcium ions from the calciferous glands, eventually being excreted from the gut as solid $CaCO_3$ (p. 224).

---

*Q_{10} = 2 is a reasonable average value for earthworms. Phillipson and Bolton (1976) calculated Q_{10} for *A. rosea*, over the range 6°C-15°C, as 1.96 for small immatures, 2.42 for large immatures, 1.42 for adults, with a mean for all age classes of 1.93.

4. Egestion (E)

The general inefficiency of assimilation by earthworms is reflected in the large proportion of ingested energy that is lost in egesta.

Later chapters of this book deal with the important role of earthworms in pedogenesis and soil fertility. Their significance results mainly from effects on soil structure and increased porosity, due to their burrowing and casting, from physical disintegration and mixing with inorganic soil materials of plant detritus, and from their contribution to nutrient cycling, especially of nitrogen. It seems unlikely, with their very low assimilation efficiency, that they could make important direct contributions to energy transfer in ecosystems.

B. Significance of Earthworms in Energy Flow in Ecosystems

Most of the net primary production of terrestrial ecosystems, even of those managed by man for grazing animals, enters the soil system in the form of plant litter, dead roots and plant exudates above and below ground, to become the food and energy source of decomposers. For example, Coleman et al. (1976) estimated that, in a lightly grazed pasture, only c. 1% of the total energy flow of secondary producers was due to above-ground herbivores while 87.6% was due to the soil-inhabiting microflora and fauna. The relative contributions of the microflora and the fauna differ considerably between temperate ecosystems and many tropical ecosystems, and so too does the allocation of energy assimilated by earthworms between production and respiration.

1. Temperate Ecosystems

On the basis of data accumulated by collaborators in International Biological Programme Grassland Biome projects Coleman and Sasson (1978) concluded that in temperate grasslands 90% or more of the total energy flow through decomposer systems was due to the microflora; of the small proportion due to invertebrates, the fraction resulting from earthworms (Lumbricidae) varied between c. 3% in dry steppe lands and c. 6% in moist meadow lands, i.e., as a proportion of total energy flow through decomposer systems earthworms would rarely account for more, and often for much less than 5%–6%.

In a Swedish grassland Persson and Lohm (1977) showed that the earthworms (five species of Lumbricidae), with a mean biomass of 5.9 g/m^2 consumed c. 15.7 L m^{-2}y^{-1} oxygen, equivalent to c. 300 kJ m^{-2} y^{-1}, which was c. 40% of the total energy consumption of animal saprovores and c. 3% of the total available from primary production. This compares with energy consumption by a population of *Aporrectodea caliginosa* of 0.4%-2.9% of the total available in Polish pastures (Nowak 1975). Populations of *Lumbricus terrestris* in two English woodlands were estimated to consume 8% and 10% respectively of the annual litter fall (Satchell 1967, Lakhani and Satchell 1970); the total organic matter input to the soil in these woodlands would greatly exceed that in litter fall alone, and the proportion assimilated by earthworms would probably be not more than 5% of the total input.

2. Tropical Ecosystems

Coleman and Sasson (1978) concluded that in tropical savannas the decomposition process was radically different from that in temperate grasslands. In the savannas of Lamto (Ivory Coast) a biomass of 0.49 g/m^2 of fungus-growing termites (Macrotermitinae) was said to be responsible for the consumption of not less than 1200 g/m^2 dry weight of litter. Reference to the original data of Josens (1974) shows that a mean biomass of 0.49-0.79 g/m^2 of fungus-growing termites assimilate c. 130-160 g m^{-2} y^{-1} dry weight, or about 10% of the total gross annual primary production. The data of Lavelle (1974) show that earthworms at the same site, with a mean biomass of 13.4-54.4 g/m^2 assimilate c. 150-225 g m^{-2} y^{-1} dry weight organic matter, or c. 10%-15% of the gross annual primary production. The organic matter consumed by the Lamto earthworms includes that consumed by litter-feeding species and a substantial proportion of sub-surface organic matter consumed by geophagous species. In addition to the fungus-growing termites there are "humus-feeding" subterranean termites and other species that feed at the surface. The energy assimilated by termites and earthworms may total 30%-40% of the total available from the annual input of plant litter, and with the additional consumption due to other soil animals the total contribution to decomposition and energy flow of the soil microflora may be no more than c. 50% of the total.

Further evidence of the significance of earthworms in energy flow in tropical ecosystems comes from data of Thambi and Dash (1973), Dash *et al.* (1974) and Dash and Patra (1977), who showed that the oligochaete population (including Enchytraeidae) was responsible for assimilation of c. 13.2% of the total energy available from the annual primary production of a pasture at Berhampur, Orissa (India). Estimates of energy utilization were calculated from the sum of (a) oxygen consumption of 60 L m^{-2} y^{-1}, equivalent to c. 1200 kJ m^{-2} y^{-1}; (b) biomass production of 35 g m^{-2} y^{-1} dry weight (mean annual biomass = 8 g/m^2 dry weight), equivalent to c. 680 kJ m^{-2} y^{-1}; and (c) mucus production of 142 g m^{-2} y^{-1} dry weight, equivalent to c. 2400 kJ m^{-2} y^{-1}. Of the total of c. 4820 kJ m^{-2}y^{-1} utilized by the oligochaetes c. 56% went into the production of mucus. About 14% of the total assimilated energy was used in tissue production, and of this it was estimated that 95% was subsequently metabolized. Assuming a more or less steady state of earthworm biomass, the energy pool that the biomass reprsents must be reasonably constant. Living systems are thermodynamically unstable and persist only because of a constant flow of energy through them. Some energy must be stored during favourable periods to permit the system to operate during periods of high energy demand. All of the energy stored in excess of that represented in the mean biomass must eventually be metabolized, in tissue renewal, cocoon production, and in respiration at times when the animals are not feeding.

Dash and Patra (1977) calculated an index of utilization of plant biomass by oligochaetes at their experimental site, i.e., the ratio of mean plant biomass (dry weight per square metre) to mean consumer biomass (dry weight per square metre), proposed by Bremeyer (1974) as an indication of the significance of consumers in energy flow. For the Berhampur grassland site the plant and oligochaete biomasses were respectively 2472 g/m^2 and 8 g/m^2, giving an index of utilization of 309. For temperate grassland Breymeyer had calculated an index for all consumers of 162.7 and for tropical grassland in Panama 818.9. Dash and Patra (1977) considered the index for oligochaetes at Berhampur as confirmation of the importance of oligochaetes in the decomposer system of tropical grassland ecosystems.

C. Pathways of Energy Flow

Bouché (1982) questioned the concept of basing energy flow studies solely on budgets of oxygen uptake or carbon dioxide output and differences in energy content between food and faeces, especially when the calculations are based on measurements over short periods in laboratory conditions. Other pathways of energy flow through earthworms listed by Bouché included the loss of metabolically elaborated materials, especially in the forms of (a) ammonia and urea in urine (p. 213); (b) excreted $CaCO_3$ which may, in some species, account for much respiratory carbon dioxide (p. 224) that would not be included in measurements made with respirometers, (c) protein in mucus secreted onto the body surface or into the intestine (p. 213), (d) cocoons (p. 116), and (e) excess annual biomass production, commonly two to five times the mean annual biomass (p. 217). Bouché (1982) proposed that the total energy assimilated by a population of earthworms might be represented as follows:

$$A = \delta B + E$$

where δB is energy used for growth and maintenance of the mean biomass, and

$$E = El + Ex + R$$

where E is the energy assimilated but lost, comprising El, the energy represented in elaboration of cocoons and of excess biomass (mortality), Ex, the energy represented in elaboration of excreted nitrogen compounds, mucus, $CaCO_3$, etc., and R, the energy dissipated in respiration. The information necessary to quantify all these aspects of metabolism for earthworm populations is generally lacking. Though it is important to have such data, it must be recognized that the only source of energy available to earthworms, for all purposes, is from decomposition of carbon compounds (from dead or living plant, microbial or animal tissue) included in the organic matter ingested, and perhaps from the decomposition of microbial tissue produced in the gut, but this must also be derived from decomposition of carbon compounds in the ingested organic matter. Energy flow must be proportional to the amount of carbon assimilated.

Ferrière, reported in Bouché (1982), has attempted to estimate the metabolic flux from rates of loss of ^{15}N or ^{14}C from earthworms uniformly labelled with the radioisotopes and kept in soil in the laboratory. Preliminary results have shown a nitrogen flux (amount of nitrogen per dry weight of earthworm) of 17 mg g^{-1} d^{-1} for *Aporrectodea longa* at 12°C. Extrapolating from this result to a field population of *A. longa* in a pasture soil, at a mean temperature of 4.2°C, yielded an estimated nitrogen flux through the earthworms of 670 mg g^{-1} y^{-1}. This is about ten times the nitrogen flux estimated by Satchell (1963) for a woodland population of *Lumbricus terrestris*, which itself exceeded the annual input of nitrogen from litter. On the basis of the estimated nitrogen flux, and assuming that the C : N ratio of organic materials actually assimilated would not exceed 5, Bouché (1982) concluded that the carbon flux through *A. longa* would be c. 3 g g^{-1} y^{-1}. Then, taking data from a permanent pasture in northern France, where Ricou (1978) had estimated carbon input from the litter as c. 500 g m^{-2} y^{-1} and the mean earthworm biomass (dry weight) as c. 200 g/m², Bouché calculated the carbon flux through the earthworms as 600–1400 g m^{-2} y^{-1}, i.e., the earthworms metabolize more than the estimated carbon input from the litter. This anomalous result may in part be due to an underestimation of carbon input to the soil, as it takes no account of inputs from lower plants, especially algae, nor of inputs of dead roots and exudates, above and below ground, from higher plants. These may as much as double the total carbon input estimated from litter fall alone. But the contribution of earthworms to carbon flux calculated by Bouché would still require that they assimilate 60%–100% of the annual input of carbon compounds to the soil; this can not be so, as if it were, it would leave little or no food available for the wide variety of other saprophagous animals and microflora in the soil.

Nitrogen flux in the plant–soil–decomposer system involves a fast cycle and a slow cycle (see Coleman et al. 1983). In the fast cycle a small proportion of the total soil nitrogen is cycled through food chains that involve microorganisms and soil animals, and is returned to the soil in waste products, principally as NH_4^+-N and NO_3^--N or in amino acids from dead tissue or mucus, which are readily taken up by plants or microorganisms and recycled. A much larger pool of soil nitrogen is involved in the slow cycle, and is present mainly in low concentrations in organic compounds that are not readily decomposed, or in resting stages of microorganisms. Earthworms excrete and secrete large quantities of NH_4^+-N and amino-N (p. 213) and are an important component in the fast cycling process (see Lee and Ladd, 1984). Nitrogen in the fast cycle may circulate through the plant–soil–decomposer system eight to ten times per year (Coleman et al. 1983); because of this it is not practicable to take gross levels of nitrogen use by earthworms, compared with annual carbon flux in ecosystems, as a basis for assessing the role of earthworms in ecosystems.

Dietz (1979), reported by Bouché (1982), has attempted a direct estimation in the field of carbon flux through earthworms. He mixed ^{14}C-labelled litter with the existing litter of a pasture soil in southern France and found that, after 17 weeks, the earthworms had accumulated 3% of the available ^{14}C in their tissues. On the basis that P/A% is 10%, i.e., 10% of assimilated energy goes into tissue production, Bouché concluded that the earthworms would have assimilated a further 27% of the ^{14}C, but dissipated this in respired $^{14}CO_2$, so that it could be inferred that in 17 weeks the earthworms must have assimilated c. 30% of the carbon available in the litter.

Bouché (1982) pointed to the need for more quantitative work like that of Dietz, with its implication that earthworms may play a more significant role in decomposition processes than is generally supposed. While in no way disagreeing with Bouché's plea for more work of this kind, I believe that the calculations above should be treated with some caution. The ratio P/A% determined by Bolton and Phillipson (1976) for *Aporrectodea rosea* ranges from 21% for adults to 49% for small immatures, and there is little other information available for lumbricids. Assuming a mean of 30% and applying this to Dietz's data, it would indicate that the earthworms at Dietz's site had assimilated c. 10% of the available carbon in the litter in 17 weeks (119 days). Most earthworms, at least in temperate regions, are active for no more than c. 180 days per year, so using Dietz's data and assuming P/A% is c. 30%, the quantity of carbon assimilated in a year would be no more than c. 15% of that available in the litter present during the 119 days of the experiment. Any additional litter that became available during the year, outside the experimental period, would have the effect of reducing the proportion of the total annual carbon flux due to earthworms. The total input of organic carbon to the soil is much greater than is indicated simply in litter fall data (p. 144), so the calculated proportion of the annual carbon flux due to earthworms must be further reduced, and it would probably be considerably less than 10%, which is in agreement with the general finding that earthworms have a low assimilation efficiency.

D. Resource Limitation of Populations

Data on carbon flux through earthworm populations indicate that they do not assimilate more than 10%-15% (usually much less) of the energy available in ingesta. Their populations must rarely, if ever, be restricted by the quantity of food available, but it seems likely that they are limited by food quality, especially by the availability of nitrogen. Earthworms are profligate wasters of assimilated nitrogen (Lee 1983); they lose it in urine as ammonia and urea, in mucoproteins, secreted from epidermal gland cells onto the body surface and from intestinal glands into the gut contents, and in production of excess biomass, which has a protein content of c. 60%-70% of dry weight, corresponding to a nitrogen content of c. 7%-8%. Contributions of earthworms to nitrogen cycling in soils are discussed on pp. 211-221. It seems likely that loss of nitrogen via urine and mucus is about half that lost through mortality. On the basis that mortality is generally two to five times the mean annual biomass (p. 217), earthworm populations must assimilate and lose annually nitrogen equivalent to c. 20%-60% of the mean annual biomass (dry weight) of the population.

Reichle (1971b) related decomposition rates of litter from eight tree species, growing on comparable soils and in similar climatic conditions, to the C : N ratio of the litter. He calculated a regression

$$Y = 113.2 - 1.75x \qquad (P < 0.05)$$

where Y is percentage of annual litter fall decomposed per year, and x is the C : N ratio of the litter, i.e., litter decomposition rate was negatively correlated with C : N ratio, and the value of the C : N ratio accounted for > 50% of the variance in decomposition rate. This relationship has not been calculated specifically for earthworms, but their relatively high demand for nitrogen among decomposer invertebrates implies that their contribution to decomposition processes should have a strong negative correlation with the C : N ratio of their food (p. 56). Satchell (1967) regarded nitrogen content of food materials as the critical factor limiting earthworm populations in heathlands and grasslands in England, and Lee (1983) concluded that the same restriction must also apply to tropical earthworm populations.

7
Predators, Parasites and Pathogens

I. Predation on Earthworms

A. Predators

Earthworms are favoured food of many species of birds, and are also preyed upon by some mammals, reptiles, amphibians, fish, and a variety of invertebrates, including some earthworms, doryline ants, mole crickets, carabid and staphylinid beetles, dipteran larvae, centipedes, carnivorous slugs and snails, turbellarians, large land planarians, and probably many others. There is, however, little quantitative data on the amounts of earthworms eaten, on the effects of predation on populations, or on the significance of earthworms as a component of the diet of their predators. Earthworm tissue has a protein content of c. 60%-70% (dry weight) and its amino-acid content is well matched to the requirements of vertebrates. Predation on earthworms by terrestrial vertebrates has recently been reviewed by Macdonald (1983).

B. Effects on Populations and Biomass

Bengtson *et al.* (1976) counted the numbers of earthworms (*Aporrectodea caliginosa* and *Lumbricus rubellus*) taken by golden plover (*Pluvialis apricaria*) in early spring from a 1.2 km^2 hayfield in Iceland. The birds took c. 4.5 m^{-2} d^{-1}. For 22 days, the birds were excluded by nets from plots; at the end of this period the earthworm population was 238/m^2 in the protected plots and 107/m^2 in adjacent unprotected sites. Numbers of birds feeding on the 1.2 km^2 area were c. 2-5 through most hours of the day, and they apparently reduced the population by > 50% during the 22 days. Predation at this intensity is restricted to periods when the grass cover is short, giving little cover from birds that hunt by sight, and when earthworms are active at and close to the ground surface.

Similar high levels of predation on lumbricids in grassland by two species of gulls and by starlings and "mapgies" were observed by Moeed (1976) in New Zealand. Cuendet (1977) estimated that black-headed gulls took *c.* 30–100 kg/ha of earthworms from autumn-ploughed fields in Switzerland, but that this represented not more than 1%–5% of the total earthworm biomass produced during the year. Reynolds *et al.* (1977) reported that woodcock in eastern North America may eat approximately their own body weight in earthworms per day, but compared this with the very much larger quantities collected by commercial fish bait enterprises in regions also inhabited by woodcock, with no apparent stress on earthworm populations; he concluded that the birds' influence on earthworm populations must be insignificant.

Studies of badgers, foxes, shrews and moles as predators indicate that earthworms are an important food source for these mammals, but give no indication that predation is at a level that could have significant effects on the maintenance of earthworm populations or biomass. Mortality is high and results from many causes including predation, but for most species, and especially for those that live or are intermittently active at the surface and are thus most subject to predatory pressure, productivity is high and expectation of survival to maturity is low (Chapter 5).

C. Significance of Earthworms in the Diet of Predators

More than 90% by weight of the diet of black-headed gulls in western Switzerland consists of earthworms during the autumn when the land is ploughed (Cuendet 1983). Earthworms are hunted by moles and shrews when plentiful and stored for later use, after being paralyzed by mutilation of the anterior segments (see, Rudge [1968] and Skoczen [1970], for example). Other animals take earthworms when they are available, but do not store them for later use. They may be particularly important as a quickly available energy source for female mammals when feeding young. For example, Neal (1977) found that earthworms were the main food of badgers when feeding cubs, while Jefferies (1974) found the stomach contents of a lactating female fox to consist of a mass of large earthworms (probably *Lumbricus terrestris*). Jefferies noted that earthworms had not previously been recorded as an important food of foxes, and considered that, though they had probably always been eaten, they may have become more important in the diet, at least in England, following the reduction of rabbit populations by myxomatosis.

Macdonald (1980) studied the hunting behaviour of foxes that eat earthworms (*Lumbricus terrestris*) in grassland near Oxford, England. Over 10% of the calorific intake of the fox population at the site studied was shown to come from earthworms, and in some months and in some hunting territories it was > 60%. Intensity of foraging for earthworms was positively correlated with earthworm population density. Earthworms were captured at night, when they partly or wholly emerged from their burrows onto the ground surface. The method of capture and withdrawal from the soil of earthworms involves grasping with the mouth and carefully pulling them from their burrows, similar to the technique used by badgers and many birds. Earthworms are detected and precisely located by sound, the fox moving its ears to align itself on the prey, contrasting with badgers, which locate earthworms by scent. Fox cubs are taught how to hunt and capture earthworms by their mothers. The rate of capture of earthworms is about the same for foxes as for badgers in the same hunting territory, but there is some evidence that foxes take cues from the success rate of badgers in selecting favourable hunting areas.

The importance of earthworms in the diet of pigs in the highlands of New Guinea was evidenced by Rose and Wood (1980). In the Tari Basin of central New Guinea, sweet potato (*Ipomoea batatis*) is the staple diet of the local Huli people. Pigs are a major source of protein in their diet and it is the custom to tether them in gardens after the harvest of sweet potatoes and allow them to forage for uncollected tubers. The pigs also eat many earthworms and the local people attribute a recent increase in pig populations and growth rates to the appearance in their lands since about 1960 of the pantropical glossoscolecid *Pontoscolex corethrurus*, which they know as *kaungoe*. Previously the megascolecid *Amynthas corticus*, which is known locally as *pedere*, was present in low numbers, but it now comprises < 0.5% of the total earthworm populations of c. 50–125/m^2 in sweet potato fields where *P. corethrurus* is established (Rose and Wood 1980). Sweet potato tubers contain little protein (mean 2.5%) and observed growth rates of pigs could not be maintained solely by the levels of intake of the tubers (Rose and Wood 1980). The local people collect *P. corethrurus* and transfer them to new pig-foraging areas; swamp lands are favoured grazing areas for uncontrolled pig foraging partly because they have relatively high earthworm populations.

The long-beaked echidna (*Zaglossus bruijnii*) which is now confined to the central cordillera of New Guinea and Irian Jaya, but was formerly (until the Pleistocene) also resident in Australia, apparently feeds almost exclusively on earthworms (Griffiths 1978). Its long tongue has a deep groove, extending from the tip to about one third the length of the tongue, lined for about three quarters of its length with backward-pointing keratinous spines. Earthworms are captured by protruding the tongue, with the groove open, and hooking the head or tail end with the spines; the tongue is then retracted and the earthworm is pulled into the beak in a series of jerks. Griffiths (1978) considered that the extinction of *Zaglossus* in the Australian mainland probably resulted from its extreme dietary specialization; increasing aridity during the late Pleistocene must have caused a decline in earthworm populations and a shortage of food for *Zaglossus*.

In the high forest region of Ghana the diet of driver ants *Dorylus (Anomma) gerstaeckeri* and *D.(A.) nigricans* was shown to consist almost entirely of earthworms, and in Kenyan High Forests earthworms were also an important part of the diet of these ants (Gotwald 1974). Their quantitative significance in the diet of other intervebrate predators is unknown.

At least two species of *Agastrodrilus*, a genus of megascolecoid earthworms known from West Africa, are apparently predators on small earthworms. Lavelle (1983c) has observed an individual of *A. dominicae* ingesting a small eudrilid earthworm, and found a recognizable eudrilid in the fore-gut of an undescribed *Agastrodrilus* sp. Lavelle inferred, from shared morphological, physiological and behavioural peculiarities that the remaining two species of the genus, *A. opisthogynus* and *A. multivesiculatus*, are also predators on earthworms.

II. Parasites and Pathogens

A wide variety of parasitic and pathogenic organisms are known from the tissues, coelomic cavity and blood of earthworms. They include bacteria, fungi, protozoa, rotifers, platyhelminths, nematodes, mites and larvae of Diptera. Some are known to be harmful to their hosts, many are parasites of earthworm predators and in some cases earthworms are secondary hosts in which the parasites complete part of their life cycle, while little is known of the significance of others.

A. Direct Effects on Earthworms

Smirnoff and Heimpel (1961) tested the effect of commercial preparations of the insect pathogen *Bacillus thuringiensis* when added to soil in which *Lumbricus terrestris* was cultured. Large doses, equivalent to bacterial cells at a rate of $10^{15}/m^2$ soil surface area, produced 100% mortality, associated with acute septicemia. At dosage rates normally used in agricultural sprays (c. 10^{11} cells/m² soil surface) Benz and Altwegg (1975) found no evidence of pathogenicity, so although *B. thuringiensis* infection is potentially lethal it is unlikely that it has much practical significance.

The best known parasites of earthworms are gregarine protozoa, especially *Monocystis*, though a number of other genera are also known. They have been recorded from the alimentary canal, coelom, blood, testes, spermathecae, seminal vesicles, and cocoons (Edwards and Lofty 1977). Purrini (1983) recorded a mean infection rate of 25% in lumbricid earthworms from two localities in Germany, but found no gregarines, nor any other parasites, in lumbricids from four localities in Spain and one in Austria.

Of the many nematodes and platyhelminths recorded from earthworms few seem to cause serious harm to their hosts (Edwards and Lofty 1977).

The larvae of the cluster fly, *Pollenia rudis*, parasitize lumbricid earthworms in Europe and North America (Walton 1928). Eggs are laid on the surface or in the soil and when the larvae hatch they bore through the body wall of earthworms and feed on their tissues, causing death.

An anoetid mite, *Histiostoma murchiei*, is parasitic in the cocoons of *Allolobophora chlorotica*, and to a lesser extent of those of *Eiseniella tetraedra* in Michigan (Oliver 1962). It is uncertain how infection is achieved, but it is most likely that the hypopal stage of the mite enters the cocoon from the body surface of the earthworm as the cocoon is being formed. The mites feed on and destroy the developing earthworms; infection rates of cocoons of *A. chlorotica* were shown by Oliver to be consistently c. 40%.

Defence mechanisms against parasites have not been extensively studied. It is known that earthworm amoebocytes will engulf and encapsulate many parasites, including nematodes; gregarines, e.g., *Monocystis*, are enclosed in fibrous capsules which sometimes are subsequently calcified (Dales 1978) and the small round white bodies so formed are often seen in large numbers when earthworms are dissected.

B. Earthworms as Carriers and Hosts of Parasites and Pathogens

1. Nematodes and Other Helminths

Associations between nematodes and earthworms were reviewed by Poinar (1978). Four types of associations were distinguished, as follows:
(a) phoretic, in which the nematode is carried in a non-feeding stage by the earthworm, internally and usually in the excretory system, and does no

harm to the earthworm. The phoretic nematodes known from earthworms are free-living microbivorous species that return to the external environment for development or reproduction, or that feed on the earthworm's tissues after it dies;
(b) paratenic, in which juvenile nematodes are carried by the earthworm but do not develop further until the earthworm is ingested by another (developmental) host;
(c) intermediate, in which the nematode lives in the earthworm's tissues and passes through some stages of its life cycle before being ingested by and developing within the tissues of a final (second) host. There is no evidence that these nematodes are harmful to earthworms;
(d) true parasites, in which earthworms serve as sole hosts for the nematodes. These nematodes include some that are lethal to earthworms, especially those of the family Mermithidae and of the superfamily Drilonematoidea.

Poinar (1978) listed some 150 associations of nematodes and earthworms, involving ten families of nematodes and eight families of earthworms.

Detailed lists of a range of parasitic helminths, earthworms known to harbour them, and hosts are in Rysavy (1969), with an extended summary in Edwards and Lofty (1977).

Nematode and other helminth parasites that have earthworms as intermediate hosts include the cysticercoid stage of cestodes (family Dilepididae) that are parasites of birds of the order Charadriiformes and the family Turdidae, and a wide variety of nematodes of several families that include parasites of the lungs, kidneys and stomach of pigs, the trachea, stomach, intestine and urinary bladder of small mammalian predators of earthworms, and the stomach and intestine of fowls, waterbirds and passerine birds. Ghabbour (1966) recorded an infection of the lungs of a man by a nematode (*Metastrongylus* sp., family Metastrongylidae), apparently derived from earthworms, and infections of the kidneys and mesenteries of man by *Dioctophyme* sp. (family *Dioctophymatidae*) that has earthworms as its intermediate host.

Among parasites that have earthworms as phoretic or paratenic hosts but are not themselves known to be parasites of earthworms, the most common are nematodes that parasitize the trachea of wild and domestic birds, especially fowls.

It is not known how commonly eggs, embryos, and other stages of parasites that live in the soil might be ingested and egested unharmed by earthworms. Rysavy (1969) noted that eggs of the nematodes *Ascaris suum*, a gut parasite of pigs, and *Ascaridia galli*, a parasite of fowls, were still viable after passing through the gut of *Lumbricus terrestris*. They and other parasites may be spread through the soil in this way.

2. Other Parasites and Pathogens

Dhennin et al. (1963) kept *L. terrestris* in culture with a virulent strain of the foot-and-mouth disease virus that had been mixed with chopped plant material and incorporated with soil. They found that the earthworms themselves carried the virus, which was still virulent to cattle three to seven days after ingestion, and suggested that *L. terrestris* may act as a reservoir and a dispersing agent for foot-and-mouth disease in field conditions.

Histomonas sp., a protozoan parasite that causes blackhead in fowls is also reported to have earthworms as its intermediate host.

Waste water effluent and sewage sludge applied to soil may include a variety of human pathogens. Brown and Mitchell (1981) investigated the capability of *Eisenia fetida* to reduce populations of *Salmonella*, a common intestinal pathogen. They introduced *S. enteridis* ser. *typhimurium* to culture dishes containing a sterile earthworm growing medium and compared survival rates for periods up to 28 days after inoculation of the bacterium in cultures with and without *Eisenia fetida*. In cultures with earthworms the *Salmonella* population decreased at rates up to 28% per day compared with a maximum decrease of 14% per day in the absence of earthworms. In trials of four to 28 days' duration *Salmonella* populations were reduced by 97.8%–99.9% in cultures with earthworms compared with those in cultures without earthworms. In a further experiment, bacteria from faeces of *E. fetida* were grown in liquid culture, and *S. enteridis* was added. Forty-eight hours after inoculation with *S. enteridis* there was a reduction of 99.9% in population of the bacterial pathogen. It was concluded that the reduction of *Salmonella* populations in the feeding trials may have been due to competition with or antibiosis of bacteria that inhabit the gut of *E. fetida*.

There are indications that earthworms may be significant agents of spread of a wide range of parasites and pathogens, and the subject needs further investigation. The findings of Brown and Mitchell (1981) provide evidence that, at least for one pathogen, earthworms may be beneficial in the treatment of waste materials, but if earthworms are to find wider use for waste disposal and as a source of protein for domestic animal or even human food (Chapter 16), it is important that their role as potential carriers of parasitic and pathogenic organisms should be more clearly defined.

8
Dispersal

I. Dispersal

Dispersal of animals may be active or passive, i.e., the animals move themselves or are transported by other agents, physical or biological.

A. Active Dispersal

It seems likely that all earthworms emerge from the soil and move about, usually at night; some earthworms emerge frequently, some only rarely. They usually return to their own burrows, but some individuals re-enter the soil elsewhere and much of their dispersal is probably from gradual spread in this way. Bouché (1976) used pitfall traps for a period of 13 months to collect earthworms that were moving on the ground surface in a pasture at Citeaux, in France; of eight species collected two were épigées and made up 43.5% of the total number, two were anéciques (36.6% of total number), and four were endogées (19.9% of total number). Épigées live very close to the ground surface and anéciques emerge to feed at the ground surface, so their capture would be favoured by the use of pitfall traps. Capture of a substantial proportion of endogées indicates that probably all earthworms emerge and then move on the surface. Of the endogées collected in Bouché's traps c. 48% were *Aporrectodea caliginosa*, 21% *A. icterica*, 29% *A. rosea* and 2% *Allolobophora chlorotica*. Bouché (1976) pointed out that *A. caliginosa* was over-represented when compared with the other endogées caught in the traps, as its abundance in the soil was less than that of *A. icterica* or of *A. rosea*, and he related its apparently greater tendency to emerge and move about to its unparalleled success as an immigrant where it has been introduce by humans to many temperate regions of the world.

Mass migrations of earthworms have been recorded. Gates (1972) reported huge numbers of *Perionyx* sp. emerging and then crawling and tumbling down hillsides, always early in the morning on a few days at the beginning of the cold season, in the Chin Hills of Burma, and similar mass movements, but with the earthworms moving uphill, early in the rainy season in India. In the Chin Hill example, Gates recorded that all the earthworms disappeared by nightfall, apparently into the soil, and believed that they were seeking a more favourable environment for the cold season by moving downhill. At Medziphema, India, Reddy (1980) recorded a mass migration of vast numbers of *Amynthas alexandri*, moving downhill, in uncultivated grasslands, beginning early on several mornings at the beginning of the cold season. Late in the morning, as temperatures rose, many worms ceased moving and died or were killed and eaten by ants. It is likely that some found shelter in litter layers and re-entered the soil. Madge (1969) noted occasional mass migrations of earthworms in Nigeria, apparently to escape raiding parties of army ants (p. 150). Mass emergence from burrows onto the surface during and following heavy rain was noted by Darwin (1881) and is widely known. Many earthworms die at the surface after emergence, but Schwert (1980) suggested that some may survive and disperse, especially where surface litter layers provide shelter from light and from predators.

B. Passive Dispersal

Earthworms are known to be moved about by downslope movement of surface water resulting from heavy rain, in streams, by ocean currents, and to be carried by birds and especially by humans.

Atlavinyte and Payarskaite (1962) collected earthworms from soil eroded by surface water after heavy rains from experimental plots on a slope of c. 6°–9°. Numbers of earthworms washed away were consistently large, but varied with different vegetation, with a maximum under fallow and arable soils and a minimum under natural grassland. Atlavinyte *et al.* (1974) also noted that after one intense rain shower more earthworms were carried away in subsequent showers than in the first shower; the increase is probably due to earthworms abandoning flooded burrows after the first shower.

Continuous collection for five months from the water of a stream in southern Ontario, Canada, yielded 308 cocoons (92% viable) of at least six species of lumbricids, all of which had been collected from land surrounding the stream. It was concluded that drifting in streams has probably been an important dispersal mechanism for earthworms that have repopulated the recently glaciated regions of North America (Schwert and Dance 1979, Schwert 1980). In France Bouché (1972) found species that were confined to one drainage basin and concluded that they had been dispersed in stream water.

A few megascolecoid species that are widely distributed on the shores and in inland areas of isolated islands and larger land masses in the subantarctic oceans and in the Pacific are known to be euryhaline and are presumed to have dispersed by drifting with ocean currents (p. 65). Most earthworms, apparently including all common lumbricids, are stenohaline and can not have been transported in this way.

It has frequently been suggested, though I know of no recorded instance, that earthworms or earthworm cocoons might be carried in mud on the feet of wading birds or grazing mammals, and some dispersal may be achieved in this way. Birds carry living earthworms to their nests to feed their young, and are frequently seen to drop them while in flight. Meijer (1972) recorded two lumbricid species from the nesting ground of gulls and not elsewhere in the Lauwerszeepolder, Holland, only 16 months after the land had been drained of sea water. The gulls were seen to forage behind the plough in established fields nearby and were probably responsible for introducing earthworms to the nesting ground. Lee (1959) attributed the dispersal of some New Zealand megascolecoids to offshore islands to birds that forage on the mainland and nest on the islands, while Schwert (1980) concluded that carriage by birds was probably an important dispersal mechanism in Canada.

Much long-range dispersal of earthworms is known to have accompanied human migrations and trade during the last few hundred years, and it may reasonably be inferred that humanity and earthworms have been moving together for many thousands of years. The earthworms that have been spread by humans are known as peregrine species, and are discussed below.

II. Peregrine Species

Nearly 100 species of earthworms, *c.* 3% of all known species, including representatives of all families, are widely distributed in regions isolated from their apparent origins, and are presumed to have been transported by humans across otherwise impassable barriers. These are the peregrine species; they do not include species, e.g., *Pontodrilus* spp. and some *Microscolex* spp., that are widely distributed on isolated islands and land masses but that are known or presumed to be euryhaline and to have been dispersed by drifting with debris in ocean currents.

The association with humanity is not confined to providing transport. The most widespread peregrines are confined, or nearly so, to man-modified habitats; Gates (1970) named these species *anthropochores* and attributed their success to their rather unusual ability not only to withstand but to take advantage of human disturbance of their environment.

The most widespread peregrine species comprise:
(a) about 20-30 species of Lumbricidae, particularly of the genera *Lumbricus, Aporrectodea, Allolobophora, Eisenia, Dendrobaena, Dendrodrilus, Bimastos* and *Octolasion*. These species have been spread from northern and western Europe by man, within the last few hundred years, so that they are now the dominant earthworms of agricultural and pastoral lands and gardens throughout temperate and some upland tropical regions of the world. They can not have been widespread in the European regions for more than c. 10 000-15 000 years so their occurrence there must also be regarded as evidence of their great aptitude for occupying new habitats, and perhaps to their introduction by humanity. Their origin must lie to the south of the Pleistocene ice sheets, perhaps in southern Europe and Asia.
(b) representatives of several megascolecid families, in particular are
  (i) about 15-20 species of the genera that constitute the *"Pheretima"* s.l. group (of Sims and Easton 1972), particularly *Amynthas* spp., that apparently originated from eastern and southeastern Asia, and are now established through most of the tropical regions of the world, with some species also well established in temperate regions;
  (ii) *Microscolex dubius* and *M. phosphoreus*, possibly of South American origin, but now established throughout the southern temperate zone and also in North America and in Europe, mainly in agricultural and pastoral areas;
  (iii) *Dichogaster bolaui, D. saliens* and *Eudrilus eugeniae* probably originally from western Africa, but (especially *D. bolaui* and *E. eugeniae*) now widely distributed throughout tropical and temperate regions;
  (iv) *Ocnerodrilus occidentalis* and *Pontoscolex corethrurus*, probably originally from central America, but now pantropical and sometimes found in temperate regions;
(c) the moniligastrid *Drawida bahamensis*, probably originally from eastern Asia and now pantropical.

A. Mode of Initial Introduction and Dispersal

The shipping of plants between remote parts of the world probably accounts for the majority of earthworm introductions. Smith (1893) attempted to trace the origins and initial spread of Lumbricidae in New Zealand and concluded that they had been introduced mainly from England with soil in

158   Ecology

boxes of plants that were propagated in nurseries and gardens and then spread more widely as the plants were moved with soil to other gardens and plantations or in soil that was brought as ballast in ships and dumped on shore. The fifteen species of Lumbricidae that now dominate the soils of farmlands and gardens in New Zealand (and a few others) are also found in similar situations in the northern and southern temperate zones wherever Europeans have migrated and traded during the last 300–400 years. It is not disputed that they were introduced by humans to New Zealand, southern Australia, Africa, South America, and many remote islands, but it has been proposed that their presence in North America reflects the holarctic distribution of Lumbricidae, with many species in common between western and northern Europe and North America.

Omodeo (1963) considered it likely that most of the oligochaetes now found in Iceland and Greenland survived the last, and perhaps all, Quaternary glaciations *in situ*, in ice-free refuges. Støp-Bowitz (1969) reviewed geological, botanical and some zoological evidence in favour of the survival of vegetation in ice-free zones. Various geologists have found evidence that parts of western and northern Norway, Bear Is., Spitzbergen, Scotland, Ireland, Iceland, Greenland, Labrador, Newfoundland and Nova Scotia were ice-free during the last or earlier glaciations. In Norway there are concentrations of montane plant species in two centres, one in the north and one in the south. Many rare species are found in both centres, some only in one, and the flora includes species whose nearest localities outside Norway are in Siberia, the Urals or the European Alps. The closest relatives of Norwegian lemmings are in the Ob River region of Siberia. Støp-Bowitz concluded that a north–south ice-free strip probably persisted through the glaciations along the western side of Norway and that vegetation and animals, including earthworms, survived *in situ* there and in other northern ice-free zones. This could account for the close similarities in many plant and animal groups between Europe and North America.

Schwert (1979) discovered a cocoon case of the peregrine lumbricid *Dendrodrilus rubidus* at a depth of *c*. 3.3 m in marl of a lacustrine sedimentary series in Ontario, Canada. From ^{14}C dating of organic matter in the marl and by association with pollen stratigraphy at the site, the deposit with the cocoon case was dated as *c*. 10 800 years Before Present (B.P.). There was no evidence of sediment reworking or contamination, the marl was *c*. 90% $CaCO_3$, and *D. rubidus* is a litter-inhabiting species that deposits its cocoons near the soil surface. Schwert concluded that the cocoon case was not from a contemporary population, and was most likely from a population that lived in soils around the lake *c*. 10 000 years ago, long before any possible introduction of lumbricids by Europeans.

Some lumbricids, species of *Bimastos* and *Eisenoides*, are confined to the southeast of North America. Other species of *Bimastos* are known from Europe, while *Eisenoides* is confined to North America. The distribution of these lumbricids provides evidence that the family Lumbricidae is truly holarctic (Gates 1970, Reynolds 1974) but their absence from northern United States and Canada, where the only lumbricids are widespread peregrine species, indicates that earthworms may not have survived the Quaternary glaciations in the north. When the first European settlers arrived in the formerly glaciated portions of North America, as well as in large portions of the Great Basin and Great Plains which were not glaciated, they reported that there were no earthworms, and it seems likely that the species now present were introduced from Europe as settlement and agricultural development proceeded (Gates 1970). Additional evidence in support of a recent European origin of these species comes from Gates' (1976) examination of earthworms intercepted by the U.S. Bureau of Quarantine during the previous 25 years. These comprised a total of 24 lumbricid species and 26 species of six other families (predominantly Megascolecidae, which were represented by 17 species of *"Pheretima"* s.l.). The lumbricids included 21 of the 23 lumbricids known from the United States; all of these 21 species are also known in the United Kingdom, where all but four are widely distributed and common. Most of the earthworms were found in soil with imported plants. Gates (1976) concluded that, with plant importations extending over the last 400 years, the evidence is strong that the common peregrine lumbricids have been repeatedly introduced, and most often from the United Kingdom. This does not exclude the possibility that some or all of the same species were already resident in North American soils before plant introductions began, and may have spread from regions south of those apparently devoid of earthworms, with or without assistance from humans, while at the same time being introduced and spread with plants and soil from Europe.

The origins and means of spread of non-lumbricid peregrine earthworms are not usually so easily traced as are those of the lumbricids, but there is good evidence that some have also accompanied plant introductions that accompanied human migrations and trade.

Six peregrine species of *"Pheretima"* s.l. were recorded by Ljungström (1972) along the eastern coast and in the north-east of Transvaal, South Africa. Indonesian, Indian, Arabian and Chinese traders and settlers have visited or lived in this region for at least the last 1500 years. The original geographical limits of *"Pheretima"* s.l. were considered by Easton (1979) to include eastern and southeastern Asia, the archipelagoes to the east as far as New Caledonia and Vanuatu and the Cape York Peninsula of Australia. It seems very likely that the southeast African pheretimas were introduced by people from that region, probably with soil that was brought with plants.

Few of the isolated volcanic and coral islands of the tropical region of the Pacific basin have been systematically explored for earthworms, but it is apparent from those that are well known that peregrine species of *"Pheretima"* s.l., together with *Dichogaster* spp., *Ocnerodrilus occidentalis* and *Pontoscolex corethrurus* are widely distributed over the region. Gates (1959) concluded that the six species of *"Pheretima"* s.l. that he found in collections from Rennell Island, in the southern Solomons, were probably introduced by humans from the larger land masses to the west. Lee (1969, 1981) similarly concluded that most, if not all, of the 15 species known from the main Solomon Islands group and of the 21 species known from Vanuatu have probably been introduced by humans.

For the Vanuatu earthworms Lee (1981) adopted the terms applied to Pacific islands snails by Solem (1959), who distinguished "coral island taxa", which he believed to have been introduced and dispersed by non-Europeans before Europeans entered the region, "tropical tramps", which he believed to have been introduced by Europeans, and endemic species. At least seven and perhaps all 18 of the species of *"Pheretima"* s.l. were regarded by Lee (1981) as coral island taxa, three species, *D. bolaui, O. occidentalis* and *P. corethrurus*, as tropical tramps, while it could not be convincingly demonstrated that any species was endemic. People have lived in Vanuatu for at least 3000 years (Shutler 1970, Brookfield and Hart 1971). They brought with them food and other plants important to their culture. The present inhabitants are Melanesians, closely related to the people of New Guinea and the Solomon Islands, and it may be assumed that their progenitors came from that region, but there is also a history of repeated migrations within and from outside the southwest Pacific, with continued dispersal and introduction of plants. It is the custom of the people, when visiting away from their village, to bring food with them. I have seen a wide range of invertebrates, up to the size of large centipedes, in baskets of taro, sweet potatoes, and other vegetables carried by visitors from island to island within the Solomons group, and many invertebrates must have been and must continue to be dispersed in this way.

B. Establishment and Rate of Spread of Populations

There is no documented case of direct competition between previously established and newly introduced earthworms. The few examples that have been studied have shown that common peregrine earthworms have occupied niches that previously had no earthworms or from which the previous earthworm population had been eliminated as a result of changed land use.

Peregrine lumbricids have moved into soils in northern United States and Canada that were devoid of earthworms and probably had had no earthworms since pre-glacial times (p. 159). In northeast Iceland, where no earthworms were present, and where it seems likely that they had not penetrated since the land was exposed by retreating ice, Bengtson et al. (1979) introduced two of the most widespread of lumbricid peregrines, *Aporrectodea caliginosa* and *Lumbricus rubellus*, into the soil in a hay field. Sixty groups, each of two or four adult individuals, were confined in soil in net bags, which were buried and were recovered after 14 months. Live adults were recovered from 82% of the bags; they included few *L. rubellus* and it was concluded that this species would soon have died out. *Aporrectodea caliginosa* had high adult survival and cocoon production rates and appeared to have become firmly established; Bengtson et al. suggested that the superior ability of *A. caliginosa* compared with that of *L. rubellus* to successfully colonize new areas is at least partly due to its relatively longer total and reproductive life span.

In a very different environment, on the small (c. 3000 ha) island of Raoul (29°15′S, 177°52′W) in the southwest Pacific, it was recorded (Benham 1905) that in 1904 earthworms were very rare; they included an acanthodrilid (*Rhododrilus kermadecensis*) and a few specimens of "*Allolobophora*", that were found only after much searching. From collections made ten years later, Benham (1915) recorded the peregrine lumbricids, *A. caliginosa, Eisenia fetida* and *Dendrodrilus rubidus* (*Bimastos constrictus*), but did not comment on their distribution or abundance. Further collections made in 1949 showed that *A. caliginosa* and *E. fetida* were widespread and numerous in the litter and A horizons of forest soils throughout the island (Lee 1953). In not more than 45 years these two species had spread all over the island, almost certainly into soils that had no earthworms or very few earthworms previously. Their initial introduction is presumed to be by New Zealanders, who attempted to settle the island at various times in the nineteenth and early twentieth centuries, but were discouraged and left because of volcanic eruptions and frequent earthquakes.

Much of the landscape in temperate regions of the world has been cleared of its original vegetation and developed as grazing and crop lands. Miller et al. (1955) and Lee (1961) recorded the sequence of changes in earthworm communities that accompanied the cutting and burning of an area of shrubland (*Leptospermum* dominant) and establishment of pasture grasses in New Zealand. Under the original vegetation there were three acanthodrilid species, *Eodrilus pallidus*, a litter inhabitant, *Maoridrilus ruber*, a topsoil species, and *Octochaetus multiporus*, a subsoil species. The immediate effect of fire was to destroy the litter species in patches where the litter actually burned, but otherwise to make little difference. This was followed for c. three months by a resurgence of the litter species, then a rapid decline of litter and topsoil species so that 10 months after the fire they could no longer be found. A period of c. four years followed when there were no topsoil earthworms; *A. caliginosa* and *L. rubellus* then became established in the upper soil horizons, apparently invading from surrounding areas, and *O. multiporus* continued to live in the subsoil.

A similar series of changes was recorded by Lee (1959, 1961) in a large pasture development area near Rotorua, New Zealand. The original vegetation of shrubs, bracken and grasses on andosols from rhyolitic volcanic ash, had a small population of the acanthodrilid *Rhododrilus similis* in the upper horizon of the soils. Sampling of a series of development sites showed that *R. similis* quickly disappeared after the vegetation was crushed, chopped and incorporated into the soil with giant discs, fertilizer was applied and grass-clover pastures were planted. In pastures up to c. three years old no earthworms could be found, but as the pasture developed and the "humus buildup" at the surface became deeper, lumbricids began to appear. First to appear, after four to five years, was *Octolasion cyaneum*, in low populations of c. $25/m^2$; in progressively older pastures *L. rubellus* was found with *O. cyaneum*, followed by *A. caliginosa* and *A. longa*. Finally, in pastures seven to ten years old, populations of *A. caliginosa* and *A. longa* were found, with a minor component of *L. rubellus*, up to c. $250-500/m^2$.

Rates of increase in population and of spread from points of introduction have been investigated where common peregrine earthworms have been deliberately introduced to areas where they were formerly absent.

In newly developed polders, where reclamation was completed in 1957 and grass was sown in 1964, van Rhee (1969b) introduced *A. caliginosa* and *Allolobophora chlorotica* into experimental plots during 1964-1967, and followed their rate of spread and population growth. The earthworms were introduced at eight points in a field 50 m × 60 m, and in three years had spread throughout the field. *Allolobophora caliginosa* spread at least 6 m/y from points of introduction, while *A. chlorotica* spread c. 4 m/y. A total of 4664 *A. caliginosa* were introduced to the field in twelve separate lots, between June 1964 and April 1967, while a total of 2588 *A. chlorotica* were introduced in twelve separate lots, between June 1964 and April 1966. Population estimates in October 1967 showed that the *A. caliginosa* population had increased by a factor of 72-fold and the *A. chlorotica* population 25-fold; estimates in April 1968 showed an increase relative to the numbers introduced of 86-fold and 47-fold respectively. *Aporrectodea caliginosa* had a rapid initial increase followed by a slowing down, as was also shown by van Rhee (1967), while *A. chlorotica* increased in numbers more slowly, but more consistently. Hoogerkamp *et al.* (1983) have continued to follow the progressive spread of earthworms in these polder soils; mean annual dispersal distances have been c. 9 m for *A. caliginosa*. They have also observed rate of spread of *L. terrestris* and found it to average c. 4 m/y.

In New Zealand, Stockdill (1966, 1982) has introduced earthworms, predominantly *A. caliginosa*, to the soils of pastures that were devoid of earthworms, and has followed their spread for long periods. The earthworms were introduced in turfs, placed on the ground surface at 10 m spacings, and initially occupied 0.07% of the total area. Spread was very slow for four to five years, with only 4.5% of the total area occupied after 4.75 years, but there was then a rapid spread, up to c. 10 m/y, with 76.9% of the area occupied after 6.75 years, and the whole area occupied at c. seven years from initial introduction (Fig. 21).

C. Special Characteristics of Peregrine Species

Many peregrine species have physiological and behavioural characteristics that distinguish them from the majority of earthworms and that are important to their success as colonists. These may be summarized as follows.

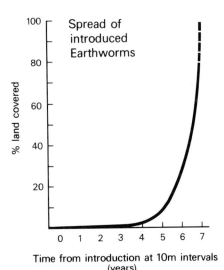

Fig. 21 Spread of lumbricid earthworms (predominantly *Aporrectodea caliginosa*) introduced to a New Zealand pasture, i.e., percentage of land area occupied, for seven years after introduction of earthworms in turfs placed on the ground surface at 10 m centres. (Diagram provided by S. M. J. Stockdill).

1. Parthenogenesis

Asexual reproduction, by simple fission and regeneration of portions of the body, has long been known in some aquatic families of Oligochaeta. The "higher" Oligochaeta, including earthworms, are hermaphroditic. On the basis of studies of the classical text book earthworm, *Lumbricus terrestris*, and of the disposition of reproductive organs of other species, it was long accepted that all earthworms reproduce only after copulation and mutual interchange of sperm, which is stored in the spermathecae until subsequent fertilization of ova. Gavrilov (1939) studied reproduction in the peregrine lumbricid *Eiseniella tetraedra* and found that, though biparental sexual reproduction occurs, it is the exception, and that uniparental parthogenesis is the usual method of reproduction. The capability for uniparental reproduction has obvious selective advantage for species that are accidentally introduced to a new environment.

In further studies of a variety of earthworms in Argentina, Gavrilov (1948) found that parthenogenetic reproduction occurred in the peregrine megascolecids *Amynthas hilgendorfi* and *Eukerria* sp., the Criodrilid *Criodrilus lacuum*, and the peregrine lumbricids *Eisenia fetida*, *Dendrobaena octaedra* and *Eiseniella tetraedra*, but could not be demonstrated in some other megascolecids, a eudrilid, and the peregrine lumbricids *Lumbricus terrestris* and *L. rubellus*.

Following on from Gavrilov's investigations, it has been shown that many earthworms may be capable of parthenogenetic reproduction, which may be facultative or obligate, or may occur in some populations of a species but not in others. Reynolds (1974) listed a total of 38 North American earthworm species that he believed to reproduce parthenogenetically. They include: nearly all of the peregrine lumbricids, with the notable exception of *Lumbricus castaneus*, *L. rubellus*, *L. terrestris*, *Aporrectodea longa* and *Allolobophora chlorotica*, which are among the most widespread species; many of the most widespread of the *"Pheretima"* s.l. group of peregrines, but not including *Metaphire californica* and *Amynthas hawayanus*; *Microscolex dubius*, *M. phosphoreus*, *Dichogaster bolaui*, *Ocnerodrilus occidentalis* and *Pontoscolex corethrurus*. Other lists of species considered to be parthenogenetic may be found in Ljungström (1970), Reynolds (1973), Jordan *et al.* (1976) and Reynolds and Reinecke (1976). The accuracy of these lists of supposedly parthenogenetic species can not be assumed. No account has been taken of the possibility of biparental reproduction by the transfer of sperm between individuals in spermatophores (Bouché 1975d, and p. 67).

Omodeo (1952, 1955) examined the incidence of polyploidy in Lumbricidae and found 13 polyploid races among 29 species investigated, with polyploid races in all lumbricid genera except *Lumbricus*. Various degrees of polyploidy were found in some species, e.g., in *Aporrectodea rosea* races with chromosome numbers 3n, 4n, 6n and 10n occur. In nearly all polyploid races parthenogenesis is obligatory, with abortive spermatogenesis due to inefficiency of the mitotic mechanism. For odd-polyploid races no maturation of reduced gametes is possible, while Omodeo considered that the remaining parthenogenetic polyploid mutants may have been derived from obligate parthenogenetic triploids. The parthenogenetic polyploids include some of the most widespread peregrine species, e.g., *A. rosea, A. caliginosa,* and Omodeo noted that parthenogenetic forms appeared to be more widespread than sexually reproducing races of the same species.

Jaenike *et al.* (1982) studied a number of clones of parthenogenetic *Dendrodrilus rubidus*, collected from northeastern United States, and found them to be pentaploid. They noted that pentaploid populations of *D. rubidus* are not known from Europe, which might be taken to indicate that the species is not necessarily introduced and is native to North America. It may also be taken as evidence that a polyploid mutant parthenogenetic race has an advantage over diploid forms of the same species in its dispersal after introduction, as was noted by Omodeo (see previous paragraph).

The incidence of polyploidy in earthworm families other than Lumbricidae is unknown, but Omodeo (1952) concluded, from cytological phenomena known at gametogenesis, that polyploidy is probably common in other families. Its relationship to parthenogenesis and to the capability of introduced earthworms to spread into new environments deserves further study.

2. Association with Man

It is well established that the most successful peregrine species are those that live in man-modified habitats. Their dispersal and establishment in alien environments are discussed on pp. 157–163.

3. Tolerance to Variability in Physical and Chemical Environment

Much evidence is presented in Chapter 1 that peregrine species, especially the most widespread of lumbricids, have wider ranges of tolerance, and often greater capability for adaptation, to high and low temperatures, relative humidity levels, pH, and concentrations of electrolytes in the soil solution, than do the majority of earthworms.

9
Altitudinal Zonation

On mountain sides in New Zealand, Lee (1959) found a general tendency towards reduction in diversity and abundance of the megascolecid populations with increasing altitude. The trend is illustrated in Figure 22, which is based on data from mountains up to c. 2500 m altitude in the North Island.

At 24 sites, between 910 m and 2160 m altitude on the slopes of Mt Kosciusko in southeastern Australia (Table 16), Wood (1974) found a total of ten species of Megascolecidae. There were from two to five, most commonly four species per site, with no apparent pattern related to vegetation or soils. Maximum abundance and biomass were found in the alpine humus soils of the subalpine and lower alpine zones, with minimum abundance in the podzolic soils of the lower montane zone. This is directly opposite to the New Zealand results, described in the previous paragraph. The podzolic soils at lower altitudes are subject to prolonged drought and high temperatures during summer, have low organic carbon contents (mean 2.7%; range, 1.4%-4.4% at 0-4 cm) and are coarse textured, with mean coarse sand content 58% (range 50%-67%) at 0-4 cm depth and 61.5% (range 55%-69%) at 12-16 cm. In contrast, the alpine humus soils of higher altitudes are less drought prone and summer temperatures are moderate; they have high organic carbon contents (mean, 5.5%; range, 3.7%-10.0% at 0-4 cm), are fine textured, with mean coarse sand content 38% (range 26%-52%) at 0-4 cm depth and 42% (range 29%-58%) at 12-16 cm. Wood (1974) calculated correlations between biomass of topsoil-inhabiting and subsoil-inhabiting earthworms at 18 of his sites and topsoil (0-4 cm) and subsoil (12-16 cm) contents of moisture, coarse sand and organic carbon. For topsoil species there was a high positive correlation between biomass and soil moisture ($r = 0.954$, $P < 0.001$), a positive correlation with organic

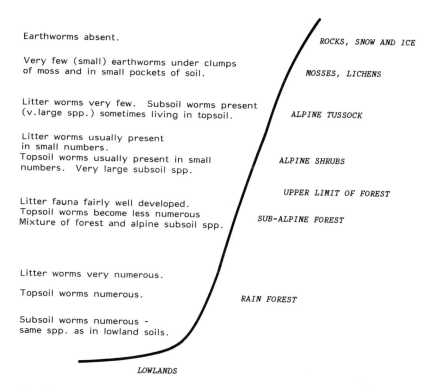

Fig. 22 Altitudinal zonation of major ecological groupings of earthworms (Megascolecoidea) on a typical mountainside in the North Island of New Zealand. (From Lee 1959).

carbon content (r = 0.630, P < 0.05), and a negative correlation with coarse sand content (r = −0.648, P < 0.05), all soil measurements for 0-4 cm samples. For subsoil species there was a correlation of biomass only with organic carbon content at 12-16 cm (r = 0.653, P < 0.05). It is apparent that earthworms in the environment of the Mt Kosciusko region would be least stressed in the high altitude alpine humus soils. In the New Zealand mountains, where severe drought is not a problem, rainfall is high and temperatures are low, there is a tendency for accumulation of surface peat or mor layers, with low pH, thereby producing a progressively less favourable soil environment with increasing altitude.

Earthworms were collected by Collins (1980) on a ridge of Gunung (Mount) Mulu, Sarawak, at 11 sites, between 130 m altitude in the lowlands

Table 16. Earthworms (Megascolecidae) from an altitudinal gradient on the slopes of Mt Kosciusko, southeastern Australia, with notes on soils and vegetation. (Data from Wood 1974).

Vegetation and Soils		Altitude (m)	Abundance (no./m²)	Biomass (g/m² fresh weight)	Total	Topsoil spp.	Subsoil spp.	Intermediate spp.
ALPINE ZONE	Tall Alpine Herbfield Fjaeldmark on Alpine Humus Soils	2230						
		2160	40.8	33.3	4	2	1	1
		2140	—	—	5	2	2	1
		2130	79.2	54.1	4	2	2	0
		2100	—	—	4	2	2	0
		2040	41.6	17.6	4	2	2	0
		1980	8.0	1.2	3	1	2	0
		1950	134.8	59.4	4	2	1	1
		1890	80.0	81.8	5	2	2	1
SUB-ALPINE ZONE	Sub-Alpine Woodland on Alpine Humus Soils	1840	15.2	5.7	2	2	0	0
		1790	—	—	3	1	2	0
		1770	105.6	35.7	5	2	2	1
		1670	60.0	26.3	4	2	1	1
MONTANE ZONE	*Wet Sclerophyll Forest on Transitional Alpine Humus Soils	1520	{ 76.0* / 26.4	27.6* / 25.5	5* / 3	2 / 2	2 / 1	1 / 0
		1400	—	—	3	1	2	0
		1360	34.4*	12.3*	4*	2	2	0
	Dry Sclerophyll Forest on Podzolic Soils	1340	{ 12.8 / 32.8	17.8 / 6.0	4 / 4	2 / 2	2 / 2	0 / 0
		1290	32.0	11.9	4	2	2	0
		1220	53.6*	47.9*	4*	2	2	0
		1210	—	—	2	2	0	0
		1140	7.2	10.2	3	1	2	0
		940	13.8	10.2	3	1	2	0
Savanna Woodland on Podzolic Soils		910	—	—	4	2	2	0

and the summit of the mountain at 2371 m (Table 17). Unlike the New Zealand mountains, the tropical Gunung Mulu is forested from base to summit. The earthworms included Moniligastridae, which were numerous at the lowest site but were otherwise found in small numbers at sites up to 1650 m altitude, and Megascolecidae, which were numerous, but with no regular trend in populations, at sites up to 1930 m, present in small numbers at 2090 m and absent only from samples at the summit.

In the southwest Pacific Solomon Islands and Vanuatu, where the mountains rise to $c.$ 2400 m and $c.$ 1900 m respectively and are forest covered, Lee (1969, 1975a, b, 1981) found little evidence of altitudinal zonation of earthworms. Annual rainfall is so high (up to 4000 mm or more), with little seasonality, that soils are constantly saturated and there is often peat or mor formation at high altitudes, with litter pH often < 4. Most earthworms are found in fallen logs, moss on the ground surface or on tree trunks, in accumulated organic matter in the axils of tree branches or leaves of Pandanaceae, or associated with epiphytes, and earthworms are found in the soil only in exceptionally well-drained sites. The earthworms are predominantly pantropical peregrine species (p. 160), associated with humans and probably only recently introduced. Their distribution, both inter-island and intra-island, bears no apparent relationship to geographical or altitudinal barriers.

Altitudinal zonation of Lumbricidae in Europe is apparently similar to that observed by Lee (1959) in New Zealand Megascolecidae. Sergienko (1969) identified earthworms and estimated their biomass along a gradient from 740 m to 1800 m altitude at Chernogora in the Carpathian Mountains, USSR. Five species were found at 740 m, four at 1000 m, three at 1300 m, and only one at 1600–1800 m. Biomass (fresh weight) was 16.5 g/m^2 at 1000 m and 2.2 g/m^2 at 1800 m. Gruia (1969) found that in Romania the biomass of lumbricids in uncultivated soils similarly fell steadily in a transect from plains to hilly land to mountains, with a maximum biomass of $c.$ 60 g/m^2 in plains soils, $c.$ 50 g/m^2 in hill soils and $c.$ 40 g/m^2 in mountain soils, and the decline in biomass was independent of differences in soil type.

The examples discussed above demonstrate that there is no general relationship between diversity, abundance or biomass of earthworms and altitudinal gradients. For any particular gradient, differences along the gradient in temperature, seasonality of rainfall, soil moisture, soil texture, vegetation, and thus the food supply of the earthworms, and also the origin of the earthworms, whether native and long established or recently introduced by humans, their ecological diversity and plasticity, and probably many other factors, affect the outcome directly and over-ride any simple relationships with altitudinal differences.

Table 17. Abundance and biomass of earthworms (Moniligastridae and Megascolecidae) from an altitudinal gradient on the western ridge of Gunung Mulu, Sarawak. (Data from Collins 1980).

Vegetation and Soils	Altitude (m)	Moniligastridae		Megascolecicae		Total Earthworms	
		Abundance (no./m²)	Biomass (g/m² fresh wt)	Abundance (no./m²)	Biomass (g/m² fresh wt)	Abundance (no./m²)	Biomass (g/m² fresh wt)
UPPER MONTANE FOREST (SUMMIT FACIES) ON ORGANIC SOILS	2370	0	0	0	0	0	0
	2090	0	0	2	0.18	2	0.18
UPPER MONTANE FOREST (SHORT FACIES) ON ORGANIC AND GLEY SOILS	1930	0	0	24	2.08	24	2.08
	1860	0	0	47	1.78	47	1.78
	1650	5	0.05	103	2.64	108	2.69
UPPER MONTANE FOREST (TALL FACIES) ON ORGANIC AND GLEY SOILS	1310	5	0.87	60	1.19	65	2.05
	1130	14	0.42	41	2.66	55	3.07
LOWER MONTANE FOREST ON REGOSOLS AND GLEY SOILS	800	15	0.21	55	1.14	70	1.35
	500	10	0.11	62	0.62	72	0.73
MIXED DIPTEROCARP LOWLAND FOREST ON PODZOLIC SOILS AND REGOSOLS	220	3	0.04	34	0.69	37	0.73
	130	51	0.21	41	0.90	92	1.11

Part II
Relationships with Soils and Land Use

10
Physical Effects on Soils

The physical effects of earthworms on soils result from excavation of burrows and production of casts. Casts consist of mixed inorganic and organic materials from the soil that are voided after passing through the intestine. Geophagous species ingest and void large quantities of inorganic soil particles, while detritivorous species also ingest soil when constructing their burrows. The ecological significance of burrows is discussed elsewhere (p. 13); this section deals with burrows and casts only in terms of their physical effects on soils. The converse, the effect of soil conditions on earthworms, is discussed elsewhere (p. 19).

I. Casts

Earthworm casts consist of excreted masses of soil, mixed with residues of comminuted and digested plant residues. Many species of earthworms, especially those that make predominantly horizontal burrows, selecting organic matter from within the soil matrix as their food, deposit their casts in their own burrows or in other sub-surface spaces. However, many species live in predominantly vertical burrows that open at the ground surface, feeding on plant debris collected from the surface adjacent to the surface openings of their burrows and depositing their casts at the surface. Casts that are deposited beneath the soil surface contribute to pedogenesis, but those deposited on the surface are probably more significant in terms of soil profile development (Bouché 1981, and p. 244) and of soil structure.

A. Physical Form and Composition

There are two basic forms of surface casts:
(a) ovoidal, or sub-spherical to spherical pellets, ranging in size from < 1 mm to > 1 cm in diameter, varying with the size of the species that excretes them;
(b) paste-like slurries that form generally rounded but less regular shapes. Composite casts, consisting of aggregated masses of the two basic forms, are common.

Darwin (1881) was aware of the widespread geographical distribution of earthworms that cast on the ground surface and he described many forms of casts that had been observed in England, France, India, Ceylon (Sri Lanka), Burma, United States, South America and Australia. Subsequent workers have found surface-casting species throughout the world whose casts are of the two basic types and composites of these, among species from all the families of earthworms.

Miyasaka (1959) reported that casts in Japan were generally granular and > 1 mm in diameter; the casts of some European lumbricids are sometimes granular, while others are less regularly shaped usually within the range c. 2–10 mm in diameter. Darwin (1881) described how lumbricids make a small pile of casts over the burrow entrance and then add more by forcing them up from below, apparently so as not to expose their tails to predators on the surface. Madge (1969) examined the casts of two Nigerian eudrilids at Ibadan. He found that *Eudrilus eugeniae*, which casts mainly in open unshaded grassland, deposits small heaps, 3–5 cm in diameter and usually > 3 cm high, of fine granular pellets, while *Hyperiodrilus africanus*, which lives in the same general area, casts mainly in shaded grassland and builds up over the burrow entrance a pipe-like composite cast, c. 2.5–8 cm high and 1–2 cm in diameter, with a vertical hole running through it but closed at the top. Similar casts produced by *Hippopera nigeriae* (a glossoscolecid) described by Nye (1955), also from Ibadan, were up to 6 cm high and 1.5 cm in diameter (Fig. 23). Darwin (1881) was sent castings from Nice in southern France that were up to > 8 cm in height and 2.5 cm in diameter; he also described similar casts from India, one of them 15 cm high and 3.75 cm in diameter. Darwin quoted Perrier, who considered the earthworms from Nice to be Indian and south-east Asian megascolecids, which must have been introduced with plants from these regions, but M. B. Bouché (pers. comm.) has found similar casts, produced by *Aporrectodea* sp. (Lumbricidae) in the region of Nice.

Very large composite casts made in the same way but of irregular shape, have been described by Darwin (1881), weighing up to nearly 90 g. Lee (1967) described casts up to 5 cm high and 5 cm wide, and weighing 25–45 g

10. *Physical Effects on Soils* 175

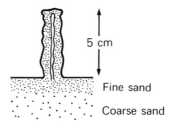

Fig. 23 Pipe-like cast of *Hippopera nigeriae* (Glossoscolecidae) found in secondary forest and tall grassland in southern Nigeria. (After Nye 1955).

(Fig. 24a), produced in large numbers by *Metapheretima jocchana* in the savanna lands of the lower Sepik valley, New Guinea, and Gates (1961) reported similar casts that weighed up to 1.6 kg from Burma. A very large cast commonly produced by an unidentified megascolecid in northern Queensland rainforests is illustrated in Figure 24b.

Particle size analyses of casts show that, in general, earthworms preferentially ingest the smaller particle size fractions, and that casts usually contain higher proportions of clay and silt, and less sand than surrounding soil (e.g., Nye 1955; Watanabe 1975; Sharpley and Syers 1976). The size of particles ingested is related to the size of the earthworms. Lee (1967) found that casts of the very large (length *c.* 500-600 mm, diameter 9-10 mm) New Guinea earthworm *Metapheretima jocchana* were derived mainly from the 0-25 cm layer of the soil. The casts contained 21%, 10%, 45%, and 23% of clay, silt, fine sand and coarse sand respectively and only 3% gravel ($>$ 2mm particle size), while soil samples from 0-25 cm contained 15% of gravel and a gravel-free sample contained 18%, 9%, 47%, and 25% respectively of clay, silt, fine sand and coarse sand. Once the gravel in the field sample was removed (and this had apparently happened, due to particle size selection by the earthworms) there was no significant difference in texture between casts and the 0-25 cm soil. Because earthworms select particle sizes and because they preferentially select organic debris when they are feeding, it is not possible to identify for certain the soil horizon from which soil in casts has come by normal methods of soil chemical or physical analysis, unless organic matter is first removed from the casts and soil samples, and some other method is used to assist in identifying the horizon of origin. Lee (1967) used the proportions of clay minerals in the casts, compared with those in soil samples, to confirm that the casts of *Metapheretima jocchana* were derived from the 0-25 cm horizon. The same methods were used by Lee and Wood (1971a) to identify the source of soil used by Australian termites to build their mounds. Bolton and Phillipson (1976) compared the maximum

176 Relationships with Soils and Land Use

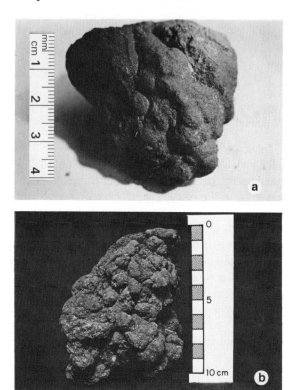

Fig. 24 Very large composite casts of two tropical megascolecid earthworms: (a) cast of *Metapheretima jocchana* from tall grasslands and secondary forest of the middle Sepik Valley, New Guinea, up to 5 cm high and 5 cm wide, and weighing 25-45 g; (b) cast of unidentified megascolecid from north Queensland tropical rainforest, c. 10 cm high and 8 cm wide, weighing c. 200 g.

size of soil mineral particles in the posterior gut of *Aporrectodea rosea, A. caliginosa* and *Octolasion cyaneum* and found them to be c. 100 μm, 200 μm and 500 μm respectively, and these dimensions are related to the relative diameters of the three species.

Comparison of particle sizes in casts and in surrounding soils have led some authors to conclude that earthworms can break mineral particles into smaller sizes by grinding, probably in the gizzard (e.g., Bassalik 1913; Blancke and Giesecke 1924; Meyer 1943; Evans 1948; Teotia *et al.* 1950; Shrikhande and Pathak 1951; Joshi and Kelkar 1952; all quoted in Edwards and Lofty 1977). It is unlikely that physical comminution of this kind can have much significance, except for very weathered mineral particles, as the pressures available are very small and the residence time of soil in the gizzard is very short (p. 6). This is not to deny the significance of comminution of organic matter and the physical rupturing of organic-mineral particle complexes, which is probably a very important contribution of earthworms to the processes of organic matter decomposition in soils.

B. Stability of Casts as Soil Aggregates

The adhesion of mineral and organic particles to form soil aggregates is a most important physical characteristic of soils. The shape and physical packing of aggregates influence, among other things, aeration, infiltration of water, water holding capacity, surface area, and thus availability of channels through which roots may grow, the availability of sites for nutrient uptake by plants, microbial and fungal activity, movement of microfauna, the proportion of aerobic compared with anaerobic sites and volumes, and, through these things, the fertility of the soil. Earthworm casts are soil aggregates, and, perhaps most importantly, as they are broken down by physical processes they are replaced by further casting by earthworms. The quantity of cast aggregates as a proportion of total aggregates (p. 179), their stability and the stability of fragments into which they disintegrate are important factors in determining soil structure.

Tests of aggregate stability, either resistance to water drops or resistance to sieving under water, have shown that earthworm casts are often more stable but sometimes less stable than other soil aggregates. Their stability depends to a considerable extent on the concentration and type of organic matter that they contain and also on microbial activity in them after they are produced.

Jeanson (1960a) measured the proportion of aggregates resistant to sieving under water in soil (from the B horizon of an alluvial soil with 25% clay, 3% C, pH 7.5) in which *Lumbricus terrestris* had lived for two months, comparing surface casts and undisturbed soil from samples to which ground lucerne or ground calcium carbonate had been added at the beginning of the experiment, with soil without addition of these substances. She showed a significant increase of $c.$ 40% ($P < 0.01$) in the number of stable aggregates where ground lucerne had been added and a significant decrease of $c.$ 16% ($P < 0.10$) where no ground lucerne was added. Other authors (e.g., Swaby 1950; Ponomareva 1950, 1953; Dobrovol'skii and Titkova 1960; van Rhee and Nathans 1961) have shown an association between the stability of casts and their organic matter content. Swaby (1950) suggested that increased stability resulted from the production of gums by bacteria in the soil as it passed through the earthworm gut and that the production of these gums was promoted by the presence in ingested soil of organic materials that favoured the multiplication of bacteria. Others have suggested that soil particles are cemented by calcium humate derived from the interaction of ingested organic matter and calcite excreted by the calciferous glands (Meyer 1943), or by mechanical reinforcement of soil particles by fibrous plant residues in ingested food (Ponomareva 1953). It has also been suggested that the casts become stabilized after they are excreted, by gums that result from microbial digestion of their organic components (Waksman and Martin 1939) or by the binding effect of fungal hyphae. Arthur (1965) noted that the bacterial population of the earthworm gut is some 1000 times that of surrounding soil (Parle 1963a) and that many bacteria produce levans and dextrans, polysaccharide molecules that are known to bind quartz and clay particles. He also pointed out that the quantity of polysaccharide necessary to produce good aggregation is probably higher than is attained in the field. He noted that casts retain their stability even when treated with acidified hydrogen peroxide, which would oxidise polysaccharide gums and calcium humate.

Parle (1963b) found that, in casts of *Aporrectodea longa*, the length of fungal hyphae and the number of yeast cells increased rapidly from the time a cast was produced, while bacteria and actinomycetes, though initially numerous, did not increase significantly. He demonstrated that the length of fungal hyphae per unit weight of cast increased by about 200% over a period of 15 days after a cast was produced, and then declined rapidly, approaching the initial level again after 25 days. Stability of the casts, measured by the sodium saturation method of Emerson (1954) declined slightly over the first five days, then increased to reach a maximum about 50% higher than the initial level after 15 days, coinciding with maximum hyphal length, and then declined to about the initial level after 25 days.

The evidence for adhesion of soil particles due to some kind of bacterial or fungal activity after the deposition of the casts is circumstantial, but fairly convincing. The increase in stability where organic matter concentration in casts is relatively high, and consequently a substrate exists for bacterial and fungal growth, is consistent. It seems likely that the production of gums (perhaps polysaccharides) or the binding effect of fungal hyphae may be significant factors in promoting an increase in aggregate stability in earthworm casts.

Ruschmann (1953b) proposed that the stability of cast aggregates may result from antagonistic relations between actinomycetes and aerobic grampositive bacteria in casts. Actinomycetes isolated from casts suppressed the aerobic bacteria that effect the destruction of soil aggregates, and Ruschmann considered this to be perhaps the most important factor contributing to the beneficial effects of earthworms on soil fertility.

Earthworm casts may sometimes accumulate so rapidly that the spongy layer they form on the soil surface is not strong enough to support the weight of grazing animals. Lee (1959) records that in New Zealand, for several years up to about 1940, very large earthworm (lumbricid) populations were found in slightly raised patches on fertile pasture lands in the Hauraki Plains area. The surface of these patches was a deep layer of earthworm casts; the hoofs of grazing animals sank into the ground, burying the crowns of pasture grasses and resulting in the death of the plants. There was a serious management problem for several years but the strength of the turf layer apparently increased, or the earthworm population decreased, and there was no longer a problem.

C. Quantities of Casts and Rates of Production

It has been estimated by various authors (e.g., Hopp 1946; Ponomareva 1950; van de Westeringh 1972) that up to 50% of the aggregates in surface layers of the soil are recognizable as earthworm casts, and in mull-type forest soils in Europe Kubiena (1953) and Dobrovol'skii and Titkova (1960) found that practically all the aggregates of the A_1 horizon were earthworm casts or residues of them. Similarly, Lavelle (1978) found that the top 15–20 cm of soils in wooded savanna lands of the Ivory Coast consisted almost entirely of casts of Megascolecidae and Eudrilidae.

Not all earthworms cast at the soil surface; most species that deposit casts do so in their burrows or in other soil spaces and it is not easy to assess the quantity of subsurface casting or its significance in aggregate formation and soil structure. Of the 37 recorded species of Lumbricidae in Britain, it has been claimed that only *Aporrectodea longa, A. caliginosa* and to a lesser extent *Lumbricus terrestris*, deposit casts on the soil surface (Madge 1969; Edwards and Lofty 1977). There is apparently some variability in the casting behaviour of lumbricid species; Jefferson (1958) reported that *Aporrectodea caliginosa* produced large numbers of casts in experiments on turf at Nottingham, (Great Britain), but not at Bingley, about 120 km distant. Bolton and Phillipson (1976) recorded that *Aporrectodea rosea* deposited casts on the surface during the initial few days of excavation of its burrow system and thereafter deposited them below the surface, while Thomson and Davies (1974) found this species to be a prolific surface caster in Canada. Nielsen and Hole (1964) recorded *A. caliginosa* casting at the surface in Wisconsin forests, while Sharpley and Syers (1976, 1977) found *A. caliginosa* casting prolifically at the soil surface in New Zealand pastures. Species that usually deposit casts in subsurface soil spaces will cast at the soil surface in compact soils, and this may account for observed differences in casting behaviour of species in different circumstances.

180 Relationships with Soils and Land Use

Surface-casting species are well known among the megascolecoid, glossoscolecoid and eudriloid earthworms of the world, although the majority of species, as in the Lumbricidae, deposit most, if not all, of their casts beneath the soil surface. For example, Lavelle (1978) records that in the savanna lands of Lamto, Ivory Coast, six species, four Megascolecidae and two Eudrilidae, out of a total of 18 species, deposit at least part of their casts on the soil surface. In New Zealand, the very widely distributed *Octochaetus multiporus* deposits most of its casts in its burrows, which are very extensive, but also deposits some casts at the soil surface around burrow entrances. Examples from Africa may be found in Nye (1955), Madge (1969), Ljungström and Reinecke (1969) and Reinecke and Ryke (1970b), from India and Pakistan in Roy (1957), Dubash and Ganti (1963), Khan et al. (1967), Petal et al. (1977) and Dash and Patra (1977) and from Japan in Watanabe (1975).

Darwin (1881) recorded many examples of cinders, paving stones, flints, sand and other materials placed on the soil surface in English grasslands that sank below the ground at rates of c. 0.25–0.5 mm/y due to burial beneath earthworm casts, which he called "vegetable mould" (Fig. 25). He measured the depth of burial of foundations and floors of ancient ruins. At Stonehenge he found that fallen monumental stones had sunk c. 130–140 mm relative to the general level of the landscape and noted that cast material also formed a sloping fringe rising to a thickness of c. 100–110 mm against the edges of the Stonehenge stones (Fig. 26). The Stonehenge stones, in Darwin's words, had "fallen at a remote but unknown period".

His observations accurately illustrate two important points concerning the depth of accumulation and subsequent erosion of surface-deposited earthworm casts. First, assuming that the rates of burial of recently deposited stones, etc., which Darwin had measured were maintained, the fallen Stonehenge stones should have been more deeply buried, since they would have required no more than c. 250–500 years to have been buried to the observed depth, and would then have fallen within well-recorded historical time. There is a limit to the depth of burial that is possible, because the most active casting species are active only in superficial soil horizons, mainly in the top 10–20 cm. The soil of this superficial layer is worked and reworked; over a period of years it will all pass through earthworms over and over again, so that the soil is stirred, and mixed with organic debris from the surface. The action of earthworms in this respect contrasts strongly with that of termites, especially those that build mounds, since many termites bring soil, commonly from depths of 50–100 cm and sometimes from many metres, to the surface, thus inverting soil profiles (see Lee and Wood 1971b). Secondly, the sloping layer of soil against the edges of the stones illustrates how the level of cast material on the general ground surface is reduced by

10. *Physical Effects on Soils* 181

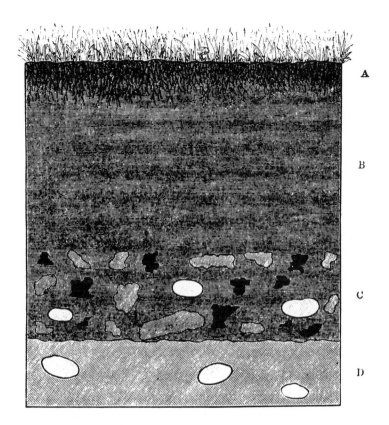

Fig. 25 Illustration showing the build-up of "vegetable mould" and consequent burial of a layer of burnt marl and cinders that were spread on the ground surface 15 years previously when the land was cleared and grass was planted. A = turf; B = "vegetable mould" without any stones; C = "vegetable mould" with fragments of burnt marl, cinders and quartz pebbles, originally at the surface; D = subsoil of black peaty sand with quartz pebbles. The combined depth of layers A and B was c. 80 mm. (From Darwin 1881).

raindrop impact and trampling by animals, so that the cast aggregates collapse, while the cast layer is protected to some extent from these processes against the vertical edges of the stones.

The quantity of surface casts produced by earthworms in a variety of environments is summarized in Table 18. Cast production is usually not continuous throughout the whole year. In the cool temperate climates of Europe and Japan casting ceases in the cold winter months, when earthworms retreat deep into the soil, and for some species there is another period

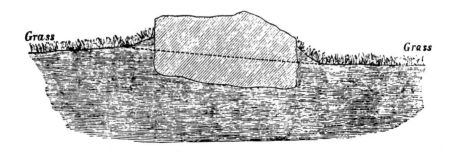

Fig. 26 Illustration showing a section through a fallen Druidical stone at Stonehenge, England, showing depth of burial due to earthworms and accumulation of a sloping fringe of "vegetable mould" around the edges of the stone. The width of the stone illustrated is c. 75 cm. (From Darwin 1881).

of little or no cast production in the summer. In the warmer temperate regions of New Zealand and Australia lumbricids cease casting through the hot dry summer months. Barley (1959c) found that earthworms were active for only about six months of the year in the mediterranean climate of South Australia. In tropical regions with wet and dry seasons, casting is generally restricted to the wet season. Graff (1971) estimated that the A_h horizon of the soil, from which casts were derived, weighed c. 100 kg/m², so that earthworm cast production represented the overturning of c. 25% of the A_h horizon each year.

II. Burrows

Earthworm burrows are of two basic types. There are those that are the permanent "home" of an individual earthworm, opening to the surface, where food is gathered, and maintained by removal to the surface as casts of material washed into them, and those that are temporary, in the sense that an earthworm makes a burrow as it moves within the soil from one place to another to find temperatures or moisture conditions that suit it best, or as it ingests soil and uses its included organic components as food. The mechanics of burrow construction are discussed on p. 7. The pressures exerted to compact the burrow walls depend on the hydrostatic pressure that can be generated in the fluid-filled coelomic cavity and that generated by contraction and stiffening of the muscles of the body wall. In *Lumbricus terrestris* the maximum pressure that can be exerted is c. 18 kPa (Seymour

Table 18. Annual rate of surface cast production by earthworms in various regions of the world.

Location	Vegetation	Taxa	Earthworms Biomass g/m²	Frequency no./m²	Weights of casts kg/m²	Period of cast production	Depth of casts produced at soil surface (mm)	Reference
England	Pastures	Lumbricidae	—	—	0.75–1.60	Spring–Autumn	2–5	Darwin 1881
France	Pasture	Lumbricidae	165.4	—	7.0	Spring–Autumn	—	Bouché 1982
Germany	Pasture	Lumbricidae	96.5	—	25.75	Spring–Autumn	—	Graff 1971
Russia	Grass ley	Lumbricidae	—	179	5.20	Spring–Autumn	—	Ponomareva 1953
Germany	Oak wood	Lumbricidae	—	—	5.80	Spring–Autumn	—	Kollmannsperger 1934
Germany	Beech wood	Lumbricidae	—	—	6.80	Spring–Autumn	—	Kollmannsperger 1934
New Zealand	Pasture	Aporrectodea caliginosa	—	—	2.5–3.0	Autumn–Spring	—	Sharpley and Syers 1976, 1977
Australia	Pasture	A. caliginosa	62–78	460–625	0.25	Autumn–Spring	0.18	Barley 1959c
Nigeria	Grassland	Hyperiodrilus africanus	—	—	17.58	Wet season (6 months)	15–20	Madge 1969
Nigeria	Secondary forest, former farmland	Eudrilus eugeniae Hippopera nigeriae	—	—	5.12	Wet season (6 months)	2–4	Nye 1955
Ivory Coast	i Grass savanna	Eudrilidae Megascolecidae	39.7	202	18.9 2.8	All year	—	
	ii Shrub savanna	Eudrilidae Megascolecidae	54.5	350	20.7 7.1	All year	—	Lavelle 1978
	iii Shrub savanna (protected from fire)	Eudrilidae Megascolecidae	35.9	400	23.6 3.0	All year	—	
South Africa	Grassland	Microchaetus modestus dominant	—	61	2.85	Summer–Autumn	—	Reinecke and Ryke 1970b
South Africa	i Grassland ii. Kommetjies	Microchaetus spp. dominant	96	74	5.0 27.0	Summer–Autumn	—	Ljungström and Reinecke 1969
Cameroon	Mountain savanna	Not determined	—	—	21.0	—	—	Kollmannsperger 1956b
Sudan Gezira	Grassland	Not determined	—	—	26.8	—	—	Beaugé 1912
India	Pasture	Eutyphoeus waltoni Pheretima posthuma	—	—	1.3	June–January (wet season)	—	Roy 1957
Japan	Pasture	Pheretima hupeiensis Eisenia japonica	7.6	30	3.8	Spring–Autumn	3.1	Watanabe 1975

1978) and although the effects of such a pressure on soil compaction must vary with the structure and texture of the soil in individual cases, it cannot in general have much effect on the bulk density of soils.

In terms of the geometry of spaces between soil particles, earthworm burrows have some special characteristics. The form and biological functions of burrows are discussed on pp. 13–15; their physical effects on soils result from the following:

(a) they are constructed channels rather than interstitial spaces between aggregates;
(b) they are continuous in vertical and horizontal planes, often penetrating from the surface to depths of tens or hundreds of centimetres and/or horizontally for similar distances;
(c) their diameters and their lengths place them among the largest of soil pores;
(d) their construction involves little significant compression of burrow walls, but deposition of illuviated materials on their walls may result in physical changes in soils;
(e) unlike other voids in the soil, they are lined by the earthworms with a protein-rich mucus that serves to lubricate the animals' passage through the soil and this often results in at least temporary stabilization of the burrow linings;
(f) earthworms that inhabit permanent burrows may maintain their integrity for years, removing to the surface soil particles that are washed into them; they may persist long after they are abandoned by the earthworms that constructed them;
(g) in soils with large earthworm populations the burrows have frequent intersections;
(h) they provide a pathway for the movement of surface water and large particles from the surface to deeper layers and ready access for plant roots to penetrate the soil.

The physical effects of burrows on soils derive largely from their influence on pore space, and thus on water infiltration and aeration in the soil, and to some extent on modification of soil around the walls of burrows. These effects are discussed below; a layer of casts on the soil surface also has important effects on porosity and discussion of this subject is included with that on the effects of burrows (see following Section).

A. Porosity

Earthworms affect the pore space in soils by burrowing and by depositing their casts as loosely packed aggregates on the soil surface. Hoeksema and

10. Physical Effects on Soils   185

Jongerius (1959) concluded from studies of soil structure in Dutch orchards that in soils lacking earthworms, pore space was determined by the packing of mineral particles and that about 30%-40% of the total soil volume consisted of air-filled or water-filled spaces. Similar soils with earthworms (biomass frequently as high as 200 g/m^2) had a total of 60%-70% pore space, about 40% water-filled and 20%-30% air-filled. At a depth of 25 cm in orchard soils with high earthworm populations they found 200 earthworm burrows per square metre of surface area and concluded that most of the air-filled spaces in soils of good moisture holding capacity were earthworm burrows. The pore spaces measured by Hoeksema and Jongerius (1959) indicate a 75%-100% increase in pore space due to earthworms. Satchell (1967) estimated that up to two-thirds of the air-filled pore space in soils may be earthworm burrows.

The diameter of the burrows varies with the dimensions of the earthworms but is generally in the range of 1-> 10 mm, which places them among the largest of soil pores. Since the basic food source of earthworms is plant litter, their burrows are frequently close to the surface but many species, especially the lumbricids that inhabit much of the area of temperate grasslands and croplands, live in semi-permanent burrow systems that penetrate to about 20-50 cm into the soil. The deepest burrow reported for a lumbricid was that of a giant species (*Scherotheca* sp.) at 5.20 m in a mediterranean karstic pasture land (Bouché 1971); burrows of megascolecid and other non-lumbricid species have been found at similar depths (Stephenson 1930; Lee 1959; Gates 1972).

Data of Wollny (1890), showing an increase in soil-air volume due to earthworms from 8% to 30% of total soil volume, have been extensively quoted in later accounts of the effects of earthworms on porosity. Wollny enclosed six large earthworms in a 5 cm diameter soil column for two months but the resulting increase in soil porosity can hardly be considered to be applicable to field situations. Hopp (1973) examined three sites in the USA, under crops and pastures and found that in the spring there was a positive correlation between earthworm populations and the proportion of large pores (Fig. 27). Nordström and Rundgren (1974), who studied earthworms at 20 forest, meadow and pasture sites in Sweden, could show no correlation between porosity and total abundance or biomass of earthworms. Porosity depends, among other things, on soil texture, so that it would be difficult to show porosity changes due to earthworms in soils of very different textures, but Nordström and Rundgren (1974) could show no correlations when soils of similar textures were compared. They concluded that there may be circumstances where a significant correlation exists but that it would be difficult to distinguish the effects of earthworms on pore space from those of roots.

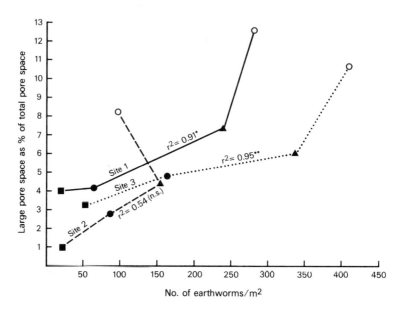

Fig. 27 Relationship between earthworm populations and large pore space, in spring, under varying agronomic practices. (Data from Hopp 1973.) ■ = bare ground, following corn; ● = young wheat, following corn; ▲ = pasture, following wheat; ○ = continuous pasture. *r significant ($P < 0.02$); **r significant ($P < 0.01$); n.s. = r not significant.

Edwards and Lofty (1977), concluded that a realistic estimate might be that earthworm burrows constitute about 5% of total soil volume, quoting the work of Stöckli (1928) and of Teotia et al. (1950), who estimated increases in porosity of 3%–5% due to the burrows in several localities.

More recent work of Douglas et al. (1980) included a study of the continuity of "transmission" pores (i.e., those > 50 μm in diameter) from A to B horizons of a clay soil (c. 50% clay), comparing ploughed with direct-drilled treatments. The mean value of hydraulic conductivity (K) at the interface of the A and B horizons for direct-drilled soil was found to be twice that of ploughed soil. In soil cores from direct-drilled soils that included the A–B horizon transition, continuous, vertical, cylindrical earthworm channels, 1–8 mm in diameter were visible and when these channels were artificially blocked at each end K decreased by 80%. Although these channels were shown to be irregularly spaced and to occupy only a small proportion of the total pore space, the experiments suggested that they provide an enhanced link for transmission of water and gases between the A and B horizons in direct-drilled soils. Total porosity and volume of transmission pores was shown to be higher in cultivated than in direct-drilled soils, but earthworm burrows, in the direct-drilled soil, were considered to be particularly significant because of their continuity from the surface through the A-B horizon transition.

Numbers of earthworm burrows per unit area have been counted by several authors (Table 19). Considerable differences are evident with differing cultural practices. Van de Westeringh (1972) illustrated the effects of heavy use of fungicides in orchards, a practice which greatly reduces earthworm populations in the upper soil layers. Hopp (1973) applied lime to experimental plots, adjusting the pH to 5.5 in one case and to 7.0 in another; the earthworm population at the lower pH was 46/m^2, while at the higher pH the population was 145/m^2. Ehlers (1975) compared corn fields that had been subjected to normal tillage practices with similar fields where there had been no tillage for four years and showed a great increase in numbers of burrows where zero-tillage methods were practised.

Table 19. Number of earthworm burrows counted on horizontal surfaces exposed at various depths in soil.

Site data	Depth (cm)	No. of burrows per m²	Reference
Orchard soils, Netherlands	25	200	Hoeksema and Jongerius 1959
Pasture soil, France	Surface	800	Bouché 1971
Orchard soils, Netherlands	10	3[a]; 27[b]	van de Westeringh 1972
	20	2[a]; 22[b]	
	30	7[a]; 17[b]	
	40	11[a]; 16[b]	
	50	25[a]; 33[b]	
	60	23[a]; 18[b]	
	70	15[a]; 7[b]	
	80	7[a]; 1[b]	
Cropland, USA	7.5	237[c]; 624[d]	Hopp 1973
Corn fields, soil from loess, Germany	2	27[e]; 117[f]	Ehlers 1975
	20	79[e]; 141[f]	
	30	187[e]; 305[f]	
	60	348[e]; 363[f]	

[a]soils treated with copper oxychloride and/or DNOC (4,6-dinitro-o-cresol) fungicides; [b]untreated soils; [c]pH adjusted to 5.5; [d]pH adjusted to 7.0; [e]normal tillage practices, [f]zero-tillage for 4 years.

B. Water Infiltration and Water-Holding Capacity

The porosity of soils is increased by earthworm burrows but because of their relatively large diameter burrows can conduct water only when it is in a tension-free state. They must be regarded as an additional drainage system, a very significant one, within the soil matrix that operates during periods of heavy rain or when soils are irrigated.

Barley (1959c) found that, in a permanent pasture near Adelaide where there was a relatively low population of lumbricids, the recognizable earthworm burrows were too wide to conduct water at the tensions normal during infiltration. He concluded that earthworms had little significance in promoting water movement through soils under the local (mediterranean) climatic regime. In the B horizon earthworm burrows occupied too small a space to contribute much to gaseous diffusion, while in the A horizon root channels were far more numerous and probably more significant in modifying pore space.

In more humid climatic zones porosity due to earthworm activity does affect infiltration and gaseous diffusion. Van Rhee (1969a) demonstrated its effect on water available for plant growth in Dutch orchard soils. He measured the soil water content at wilting point (pF 4.2) and at field capacity (pF 2.2) in three plots with and three plots without earthworms, recording means of soil samples from 20 cm and 40 cm depth at pF 2.2 and of 0–40 cm samples at pF 4.2. The differences in moisture content, which represent the water available to plants, were 37.3% and 26.5% on a weight per volume basis for the plots with and without earthworms respectively, showing that the presence of earthworms resulted in more water being available for plant growth. The bulk density of the soils with and without earthworms differed by less than 4%. He concluded that the presence of earthworms resulted not so much in an increase in total soil pores as in a redistribution of pore spaces, with a higher percentage of large pores in which water is less strongly bound by capillary forces, as compared with soils lacking earthworms. Van de Westeringh (1972) confirmed these findings in Dutch orchard soils.

Stockdill (1966) introduced lumbricid earthworms (*Aporrectodea caliginosa* and *Lumbricus rubellus*) to New Zealand pastures that had been developed in isolated areas that were not accessible to invasion by lumbricids. Five years after the introduction there were 1150 lumbricids per square metre (1040 *A. caliginosa*, 110 *L. rubellus*) where introductions had been made, compared with no lumbricids and a very small population of megascolecids elsewhere. Moisture content at field capacity for 0–30 cm depth was 17% higher in the pasture with lumbricids than in the pasture without lumbricids. Moisture available to plants (i.e., difference between moisture content at field capacity and at wilting point) at 0–30 cm depth was consistently higher in soils with lumbricids than in soils without lumbricids (Table 20). He found that macroporosity decreased markedly, from 44.9% to 21.5% at 0–10 cm and from 29.3% to 22.7% over 0–30 cm depth, with an accompanying increase in bulk density from 0.68 g/cm^3 to 0.86 g/cm^3 at 0–10 cm, but with no change over 0–30 cm depth where lumbricids had been introduced compared with soils that lacked lumbricids. He attributed these changes to the observed degradation and incorporation of a porous pasture root mat by the earthworms.

The rate of infiltration of water, when water is supplied in large quantities, is also greatly affected by the presence of earthworms. Slater and Hopp (1947), Hopp and Slater (1948), Teotia et al. (1950), Guild (1952, 1955), Scharpenseel and Gewehr (1960), Stockdill (1966), and Carter et al. (1982) found that water entered the soil at two to ten times the rate in soils with earthworms than it did in soils lacking earthworms. Bouché (1971), in a French pasture soil with 800 worm burrows per square metre, poured 100 L of water down one earthworm burrow without its overflowing; he showed that the burrows intersect and that the water could penetrate through a network of burrows.

Ehlers (1975) investigated the effects of normal tillage practices compared with four years of zero-tillage in German corn fields. The number of earthworm burrows in zero-tilled soil was much greater than in tilled soil. He measured the maximum cumulative water intake and found that in zero-tilled soils the infiltration rate due to earthworm burrows alone during a five minute period was 6.15 L/m^2 (0.12 mm water per min), while in tilled soil the infiltration rate over five minutes was 1.11 L/m^2 (0.02 mm water per min).

Table 20. Available moisture in soils (0–30 cm) under pasture plots with and without lumbricid earthworms at Hindon, New Zealand. (Data from Stockdill 1966).

Sampling time	Available moisture (cm)	
	Pasture with lumbricids (1150/m^2)	Pasture without lumbricids
June	8.5	7.9
August	8.4	6.6
November	3.4	2.5
February	6.4	5.6

Stockdill (1966), at his sites in New Zealand (see previous two paragraphs) used double-ring infiltrometers to measure infiltration rates in two ways. He measured five-hour intakes, i.e., the amount of water taken into the soil in the first five hours of initial, or "dry" runs, and a "basic rate", i.e., the intake in the third hour of a second, or "wet" run. Results showed increases from 29 cm/h to 61 cm/h (110% increase) in the five-hour intakes and from 1.4 cm/h to 2.6 cm/h (86% increase) in the basic rate when plots with introduced lumbricids were compared with plots lacking lumbricids. Hoogerkamp et al. (1983) measured very large increases in water infiltration rates in pasture soils of Dutch polders 8–10 years after introduction of lumbricid earthworms. Infiltration capacity in 24-hour periods in plots with earthworms increased to c. 120–140 times that in plots without earthworms.

Wilkinson (1975) measured infiltration rates on soils in shifting cultivation areas in northern Nigerian savanna under three crop rotation systems, where three years of cropping (cotton, sorghum, ground nuts) followed two, three, or six years of fallow under Gamba grass (*Andropogon gayanus*). Observations on the sites showed that the highest infiltration rates (up to 13–14 cm water per hour) were found under fallow, where earthworms were very numerous and where the surface soil consisted almost entirely of earthworm casts, while the lowest rates were found where earthworms were few or absent. There was a significant relationship ($P < 0.001$) between infiltration rates, measured annually, at the end of the growing season, and years of successive fallow. One year of cultivation and cropping drastically reduced infiltration to about 18% of that at the end of six years fallow and there was no significant change through the three-year cropping cycle (Fig. 28). Lal (1974) compared ploughed with zero-tillage plots under maize crops in Nigeria and found that after two years of cropping there were about 100 earthworm casts per square metre on the soil surface in ploughed plots, compared with up to $2400/m^2$ in zero-tillage plots, and that water infiltration rates were 21 cm/h and 36 cm/h respectively.

Tisdall (1978) examined the effect on water infiltration rates and macro-porosity of soil in irrigated orchards in Victoria, Australia. In one peach orchard, straw was mixed into the soil at the rate of 68 t/ha and thereafter a mulch of straw and sheep dung at the rate of 5.5 t/ha was added annually to the surface. Irrigation practice was designed to keep soil moisture wetter than 30 kPa suction during the growing season. After three years infiltration, macro-porosity and earthworm population in this orchard were compared with those of 158 sites in local orchards that had not been treated similarly. In the treated orchard the earthworm population (predominantly *Aporrectodea caliginosa, A. longa, A. rosea, Lumbricus rubellus*) was about $2000/m^2$, compared with $150/m^2$ average for the other sites. Infiltration rate, measured with ring infiltrometers, in the treated orchard was at the rate of

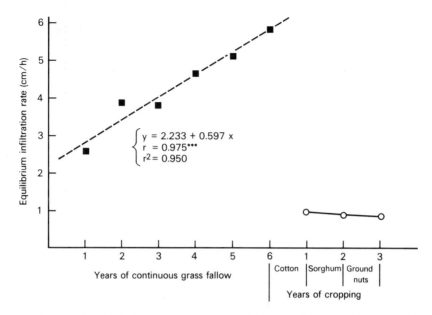

Fig. 28 Equilibrium infiltration rates (cm/h) in nine-year rotations (six years of Gamba grass fallow followed by three years of crops) in northern Nigerian savanna lands. (Data from Wilkinson 1975). ***$P < 0.001$.

50 mm in 1 min, compared with 50 mm in 83 min average for the other sites. Macro-porosity (% air-filled porosity at 4 kPa suction) was measured by draining undisturbed cores 76 mm diameter and 63 mm deep on a ceramic plate and was 19% for the treated orchard compared with 5% average for the other sites. This experiment has important implications in the context of Australian soils, which generally have low amounts of surface litter and are subject to seasonal desiccation. They also frequently lack lumbricid earthworms; introduction of suitable lumbricids, or perhaps other non-lumbricid species, together with irrigation and mulching, should produce drastic improvements in the availability of water for crop production.

Prolonged heavy use of the organophosphorus insecticide phorate, which is particularly toxic to earthworms (see Table 29), on pastures in England resulted in elimination of earthworms and changes in soil physical properties (Clements 1982). After three years' application of phorate (3.3 kg/ha a.i. every month) infiltration rates were c. 7% and hydraulic conductivity was c. 4%-7% of those in untreated plots; soil bulk density increased by up to 17%, and change in bulk density was reflected in a 50%-100% increase in penetrometer readings.

Conventional models of water infiltration into the soil do not allow for the effect of surface-opening holes of the dimensions of earthworm burrows on infiltration and surface runoff rates. W. M. Edwards *et al.* (1979) provided a numerical model that shows that the number of holes per unit area, the diameter and depth of the holes, greatly influence infiltration and surface runoff. Flow through large pores occurs only during or immediately after periods of heavy rainfall or heavy irrigation, but Edwards *et al.* (1979) proposed that such flow, through earthworm burrows or other animal burrows could account for many reports in the literature of rapid rates of movement of water (and fertilizers) into the soil, such as those included in the examples above. Edwards *et al.* (1979) considered the case of a level field with a regularly spaced pattern of vertical holes and the infiltration of water into a small cylindrical column around one hole: the column is open at the surface and is large enough in diameter that free surface water can move under gravity into the hole, and is then able to infiltrate radially away from the hole. They assumed that the soil and its initial water content were uniform with depth, that air could escape freely from the bottom of the hole, that runoff could occur only when the central hole was full and that the rainfall rate was constant and was greater than the sum of vertical infiltration through the soil surface and lateral infiltration through the walls of the hole, so that the hole would remain full of water for the duration of the rainfall. Beginning from the Richards equation (Richards 1931), which describes unsaturated water flow in uniform soil material under laboratory conditions, they derived an equation for the infiltration that would occur through the walls of the hole in horizontal layers of spacing Δz and into concentric rings of spacing Δr. From the solution of this equation they derived a model for total infiltration into the soil in the presence of a vertical hole. Details of the model may be found in the paper of W. M. Edwards *et al.* (1979).

Taking a silt loam derived from loess that had not been ploughed for several years and whose hydraulic properties were known, Edwards *et al.* (1979) used their model to calculate infiltration and runoff rates (by difference) for columns of 50 cm diameter and 100 cm depth, with or without a central hole, 60 cm deep and either 5 mm or 10 mm in diameter. Assuming an initial pressure head of the column of -1000 cm, rainfall at a steady rate of 10 cm/h for 30 min and calculating moisture tensions at radial and vertical intervals of 5 cm, the differences between columns with and without a central hole are illustrated in Figure 29. The total amount of water applied to the surface in their example would be 9813 cm^3. For the column without a hole about 5700 cm^3 would infiltrate in 30 min, while the remaining 4100 cm^3 (42%) would run off. For a column with a central hole of 5 mm diameter, infiltration in 30 min would be about 6900 cm^3 and

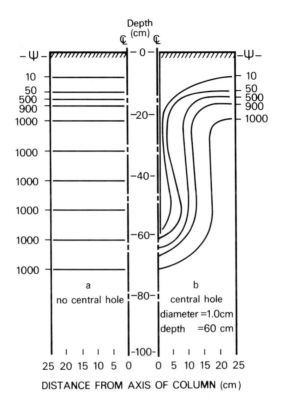

Fig. 29 Calculated tension distribution in 50 cm diameter soil columns with no central hole (a) and with a central hole 1 cm diameter, 60 cm deep (b) at time = 0.5 h after rainfall at rate of 10 cm/h. Ψ = matric potential in cm of water. (Reproduced from Soil Science Society of America Journal, Volume 43, 1979, pages 851–856, by permission of the Soil Science Society of America).

runoff about 2900 cm^3 (29%), while for a column with a 10 mm diameter hole infiltration in 30 min would be about 7670 cm^3 and runoff about 2140 cm^3 (22%). Further calculations showed that the depth of the hole very significantly affects the infiltration rate; for a column of 50 cm diameter and 100 cm depth, cumulative runoff for a 30 min period was calculated to be 3450 cm^3 (35%) in the presence of a 10 mm diameter hole 30 cm deep, compared with 2140 cm^3 (22%) for a 60 cm deep hole of the same diameter. It can be seen from Table 19 that the number of earthworm burrows per unit area of soil surface may greatly exceed that assumed in these calculations (5/m^2), although if they were very numerous they would usually be of less diameter than 5-10 mm, which was a diameter based on burrows of *Lumbricus terrestris*. The effect of closer spacing of burrows was illustrated by Edwards *et al.* (1979) who further calculated infiltration and runoff for burrows 5 mm in diameter, 60 cm deep, at the centre of a column of soil 30 cm in diameter (i.e. *c.* 14 burrows per square metre), receiving the same rainfall of 10 cm/h for 30 min; the total amount of water falling on a column of this size would be 3533 cm^3, and they calculated that only 95 cm^3 (2.7%) would run off while the remainder would infiltrate into the soil.

In a recent series of papers, Germann and Beven (1981a, b, c) have adopted an experimental approach using large blocks of undisturbed soil to investigate the effect on water flow of the complex system of macropores that exists in soils in the field. The minimum dimensions of macropores were taken to be those with a capillary potential of -1.0 cm; this corresponds to a pore diameter of *c.* 3 mm and would include the burrows of most earthworms. The soil water properties of both macropores and micropores were determined on the same samples, without interruption between the two modes of water flow. Experimental methods are described in Germann and Beven (1981a). From the experimental results Germann and Beven (1981b) developed a model to predict the relative importance of micropores and macropores in soils of different permeabilities; they concluded that the presence of macropores generally increases infiltration rates and that their effects on infiltration are greater in soils with intermediate hydraulic conductivities than in soils with very low or very high hydraulic conductivities. In their third paper, Germann and Beven (1981c) compared their theoretical model with some published results from field measurements of infiltration, including those of Ehlers (1975) (this Section), and earlier data of Burger for Swiss soils. They found a high level of agreement between the field measurements and theoretical values, and suggested that a useful first approximation to describe flow through macropores might take the form

$$Q_{ma} = Q^* e^2_{ma}$$

where Q_{ma} is the volume flux density and e_{ma} is the porosity of the macropore system of the soil, while Q^* is a variable coefficient reflecting the hydraulic connectivity, geometrical structure, tortuosity and roughness of the macropores. They calculated a series of values of Q^* for Ehlers' data and concluded that earthworm burrows give high values of Q^* and so may be a particularly significant class of macropores in their effects on water flow in soil.

Infiltration of surface water through earthworm burrows may not always be beneficial. There are reports (Smith and Peterson 1982) that surface water, infiltrating into the wide (*c.* 25 mm diameter) and extensive burrows of the giant Australian earthworm *Megascolides australis* during storms, has resulted in landslides and in the failure of the earth walls of small water storage dams.

The relationships between earthworms and soil macroporosity have important implications for the erodibility of soils, tillage practices, fertilizer and biocide penetration into soils and land use practices, which are discussed later in this book.

C. Aeration

It is unlikely that earthworm burrows are often an important factor in soil aeration but they may sometimes contribute significantly.

Kretzschmar (1978) has provided the most comprehensive data on the numbers, dimensions and distribution in the soil of earthworm burrows. In a pasture at Cîteaux, near Dijon, France, he counted and measured burrows to a depth of 2 m. The earthworm (Lumbricidae) biomass (Bouché 1975c) was 150 g/m^2, including about 5% épigées (litter-inhabiting species), 81% anéciques (soil-inhabiting, surface-feeding or near surface-feeding species, with vertical or near-vertical burrows), and 13% endogées (deep soil-inhabiting and feeding species). The total length of burrows was 380 m/m^2 of ground area, with a volume of about 5 L/m^2, and a surface area of about 4.9 m^2/m^2 ground area; 27% of the burrows were in the 0–20 cm horizon, 39% at 20–40 cm, 22% at 40–60 cm, and 13% at > 60 cm depth. Most of the burrows (72%) were sub-vertical, i.e., inclined at > 60° to the horizontal.

With a few assumptions, it is possible to calculate from these data the significance of the earthworm burrows in aeration of the soil*. Assuming the bulk density of the soil is 1.4 and the specific gravity of the soil material is 2.7, the total porosity would be 0.48 cm^3/cm^3 soil. Aeration is considered to be insufficient for maximum plant growth when > 90% of the pore space is water-filled, i.e., when < 0.048 cm^3/cm^3 is air-filled. The mean diameter of earthworm burrows (from Kretzschmar's [1978] data) is about 4 mm, and the mean cross-sectional area about 0.125 cm^2. Pores of this size would not remain filled with water unless the soil was flooded, and would otherwise be air-filled. From Kretzschmar's data, the volume of earthworm burrows would be about 0.006 cm^3/cm^3 at 0–20 cm, 0.009 cm^3/cm^3 at 20–40 cm, and 0.005 cm^3/cm^3 at 40–60 cm depth. These volumes represent 13.3%, 18.8% and 10.4% respectively of the minimum required air-filled pore space, and

---

*I am grateful to Dr W. W. Emerson for providing the additional information for these calculations to be made.

about 1%, 2% and 1% respectively of the total pore space in each of the three soil layers. The biomass (150 g/m^2) of earthworms at the site is large by comparison with most recorded earthworm populations, but is exceeded in some cases (see Table 7); its contribution to aeration accounts for c. 10%-20% of the minimum necessary for maximum plant growth. Kretzschmar (1978) excluded the 0-6 cm layer of the soil from his measurements and also excluded burrows < 2 mm in diameter.

These calculations represent possible effects on aeration at a specific sampling time. A more detailed analysis of data from the same site (Kretzschmar 1982) revealed much seasonal variation in numbers, length and diameter of burrows, and in their vertical distribution within the soil profile. Burrows ≤ 2 mm diameter were most common in spring, coinciding with the presence of maximum numbers of juvenile earthworms. Burrows of 2-3 mm diameter were principally those of the deep-dwelling endogées and of juvenile anéciques; endogées are less affected by summer drought than earthworms that live close to the surface and they continued burrowing in deep soil layers through at least part of the summer, when other earthworms were not only inactive but in many cases had sealed off the entrances of their burrows. Burrows 3-4 mm diameter included some of endogées and some of anéciques; seasonal variations in their abundance were not so clear as were those in the smaller diameter classes. Large species, which make burrows > 4 mm in diameter, were predominantly anéciques, surface feeders that make sub-vertical burrows (e.g., *Lumbricus terrestris*); these earthworms are inactive through the summer, and their burrows were proportionately most important in autumn, when they become active and re-establish their burrow systems. Total burrow length, to 1.2 m depth in the soil, varied through a period of 1.5 years from 142 m/m^2 to 888 m/m^2 surface area, with corresponding variation in burrow volume from 1.3 L/m^2 to 9.16 L/m^2. The frequency of burrows correlated fairly closely with soil moisture content (positively) and with soil temperature at 20 cm depth (negatively).

The effects of earthworm burrows on aeration, or on water infiltration and drainage, are apparently not constant at any one site, but should be recognized as important where earthworms are numerous and probably significant even where earthworm populations are low.

D. Physical Modification of Soil around Earthworm Burrows

The mechanics of burrows construction are described on p. 7. The pressures exerted on the soil are small and their effects, in terms of compaction of the surrounding soil cannot be great, although they must vary with the structure and texture of the soil. There is little quantitative information on this subject. Dexter (1978) investigated the ability of *Aporrectodea caliginosa* to burrow in an artificially compacted soil. He took A-horizon soil from Urrbrae loam, near Adelaide, consisting of 51% sand, 32% silt, and 17% clay, remoulded it into blocks and then exposed the blocks in the laboratory to *A. caliginosa* for ten days. The blocks were then impregnated with wax, cut into 5 mm thick sections, and the total length of burrows (L) was determined. He measured the resistance of the blocks to a micropenetrometer (S) and showed that there was no significant relationship ($r^2 = 0.2$) between L and S over the range S = 0.3–3 MPa, i.e., the length of burrows made by the earthworms was independent of the resistance (strength) of the soil; the range of strength measured was enormously larger than the pressures attained in the coelomic fluid which he estimated as about 0.45 kPa. Although this pressure is much lower than the estimate of Seymour (p. 8), his conclusion that the principal method of burrowing in this soil must be by ingestion and casting of soil was justified, in view of the very much larger pressures necessary to force a micropenetrometer into the soil. Taking clods of soil from the field, with earthworm burrows almost certainly formed by *A. caliginosa* and using the method of Greacen *et al.* (1967) for measurement of soil density, Dexter also showed that there was no difference between soil density or structure in the immediate vicinity of earthworm burrows when compared with soil remote from burrows.

It seems likely from the data above that the pressures exerted by earthworms have no significant effects on compaction of the soil. However, illuviation from surface horizons and deposition of clay and dissolved materials on the walls of existing burrows, and, in some cases, substances excreted by the earthworms do influence the physical characteristics of the surrounding soil. Jeanson (1964) studied thin sections of burrows of *Lumbricus terrestris* and *Aporrectodea icterica* constructed in the laboratory in a loam from the textural B horizon of a soil derived from loess. The burrows had a continuous lining, about one-tenth of the burrow diameter, of oriented clay particles, and, outside this layer, three concentric zones of humic material, calcium carbonate and iron oxides. Using an electron probe Jeanson (1971) further showed that burrows of *Lumbricus terrestris*, which had been kept in the laboratory for eight months, had concentrations of iron and manganese on the burrow walls, with the iron concentration at the burrow surface one and a half times that at 100 μm and three times that at 200 μm from the surface, the difference becoming indistinguishable at 600 μm from the surface. Calcium was concentrated in rounded "corpuscles" of calcite, a few micrometers in diameter, unlike the iron and manganese, which were evenly distributed.

She had previously found (Jeanson 1961) that the nature of burrow linings may vary with the kind of food available. When *L. terrestris* was kept in soil columns in the laboratory, and given ground lucerne as food, a blackish layer about 1 mm thick formed on the burrow walls. Microscopic examination of this layer showed that it consisted of a mixture of soil mineral particles and humic material; she concluded that this material was excreted by the earthworms and was spread on the surface of the burrow by the earthworms. Outside the blackish biogenic layer was a reddish ferric-iron rich layer 2-3 mm thick, which she concluded was physicochemical in origin, probably due to oxidation of ferrous-iron compounds in the oxygen-rich environment adjacent to the burrow. Dietz and Bottner (1981) put ^{14}C-labelled litter on the soil surface of grasslands in southern France, and used autoradiography of thin soil sections to follow the movement of the ^{14}C-labelled material down the soil profile. Most of the ^{14}C-labelled decomposition products detected in the soil over a period of about a year were carried in by drainage water, while a small proportion was mixed, with buried litter and earthworm faeces, into the 0-10 cm soil layer. Some ^{14}C-labelled material, which lined earthworm burrows to considerable depths, was considered to be included in mucus excreted by the earthworms, but it was concluded that most mixing was superficial, taking place through feeding on surface litter and deposition of casts at or near the surface.

Earthworm burrows may last for very long periods, continuing to accumulate illuviated materials. In the South Island of New Zealand I have seen large numbers of recognizable burrows of *Octochaetus multiporus*, heavily coated with humic and ferric-iron layers, in the B horizons of palaeosols formed from loess during Pleistocene interglacial periods. These burrows have been subsequently deeply buried under later loess layers, in which soil profiles have developed, and *O. multiporus* may still be found in the B horizon. The region 2 mm thick immediately surrounding earthworm burrows has been designated the "drilosphere" by Bouché (1975b) and is discussed further (p. 208) as it influences microbial activity in soils.

Richards (1978) distinguishes three types of mucus-producing gland cells in the epidermis of lumbricids: one type produces a mucopolysaccharide-protein–lipid complex that provides a lubricant for the earthworm's movement through the soil; a second type produces a carboxylated acid mucus of low viscosity, which is trapped by small projections of the cuticle to form a moist surface film for respiratory exchange; the third type, which is found only in small numbers, is protein-rich and secretions from these cells may affect the viscosity and water-retention capabilities of the substances produced by the other gland cells. Needham (1957) estimated that about 50% of all nitrogen excreted by *Lumbricus terrestris* and *Eisenia fetida* was excreted as a component of the mucus. The quantities of mucus involved are large, and much of this material is deposited on burrow walls, sticking soil particles together to make a coherent burrow lining. Agarwal *et al.* (1958), in India, reported that some *Allolobophora* spp. excrete a "waxy" fluid that cements soil particles and results in a cloddy structureless soil where the earthworms occur in large numbers. It seems most likely that this is an extreme case of sticking together of soil particles by secreted mucus but it illustrates the possible significance of the less extreme effects in most soils of mucus secretion by earthworms.

11
Chemical Effects on Soils

Earthworms affect the chemical composition of soils and the distribution of plant nutrients in them in the following ways:
(a) they take large quantities of partially decomposed plant litter and other organic detritus from the soil surface and remove it to subsurface layers; much of it is ingested, macerated, and passed through the digestive system, together with large quantities of inorganic soil constituents. Some species feed below the soil surface, on dead plant roots or on organic materials in various stages of decay, originally derived from the surface;
(b) faecal material (casts), which consists of finely divided and thoroughly mixed organic and inorganic materials, is deposited on the soil surface or in their own burrows, or in other spaces below the soil surface, resulting in the formation of organic matter enriched (humified) soil horizons;
(c) they derive their energy from the metabolism of carbon compounds in the ingested organic materials, but their digestive processes do not much affect the ingested plant material during transit through the alimentary canal;
(d) waste products of their metabolism are redistributed in the soil in faeces (casts), urine and glandular secretions, while high mortality rates contribute much dead earthworm tissue to the soil: plant nutrients derived from metabolic and decay processes are recycled in the plant/soil ecosystem;
(e) soil microorganisms are ingested together with litter and inorganic soil material. The viability and dispersal of microorganisms are affected by passage through the gut and so too are microbial populations and subsequent decomposition processes in casts;

(f) the thin layer of soil immediately surrounding earthworm burrows is affected by the excretion of nitrogenous wastes and secretion of mucus, and also by the relatively easy access of air, due to the direct opening of burrows to the soil surface;
(g) they affect the accumulation and distribution of heavy metal and radioactive elements in the soil and their more general distribution in ecosystems.

In order to understand these effects it is necessary to have some knowledge of aspects of the behaviour, relationships with microorganisms, and physiology, especially of excretion, of earthworms.

I. Contribution to Organic Matter Incorporation and Decomposition

As it affects the composition of their casts, and therefore to a considerable extent their effects on the chemical composition and the cycling of nutrients in soils, earthworms may be divided into three groups. The groups are discussed more fully on pp. 102–129, but for the present purpose they may be described as follows:
(a) litter-inhabiting, arboreal, or log-inhabiting species that do not make burrows in the soil, épigées of Bouché (1972), which ingest much organic matter and some inorganic material. These species are almost entirely confined to forest ecosystems. Their faeces consist of finely divided but little digested organic matter and their effects on the redistribution and cycling of plant nutrients are little understood. They include representatives of all the major taxonomic groups of earthworms and are very widely distributed in the forests of the world;
(b) surface-feeding species that construct and live in permanent burrows or systems of burrows in the soil, anéciques of Bouché (1972), which ingest organic matter, mostly plant litter, and varying quantities of inorganic soil materials. These species are common and widely distributed. They include the majority of species of Lumbricidae and representatives of all other taxonomic groups of earthworms. Their faeces (casts) consist of organically enriched soil material; many species deposit their casts at the soil surface, while some deposit them in their own burrow systems or in other subsurface soil spaces. They have profound effects on the incorporation of organic matter into the soil, the distribution and cycling of plant nutrients, and on pedogenetic processes;
(c) subsurface-feeding (geophagous) species that construct and live in burrow systems that are continuously extended, endogées of Bouché (1972), which may feed non-selectively on soil, but commonly selectively

ingest organic matter, mixed with large quantities of inorganic material from the soil. These species are widely distributed and include some species of all the major taxonomic groups of earthworms. Their faeces (casts) are usually enriched in organic matter relative to the surrounding soil, and are deposited mainly in their burrow systems or in other soil spaces, but occasionally at the surface, and though they must to some extent affect the distribution of organic matter and the cycling of plant nutrients, their more important effects are probably on pedogenesis.

Casts differ chemically from undisturbed soil primarily because they contain more organic matter than the mean level of the soil. The quantity of food ingested and the nature of the organic compounds excreted and otherwise derived from earthworms will be discussed here. It is important to know what quantities of organic matter are removed from the soil surface, what kinds are to be found in the excreted casts, and how the casts differ from the surrounding soil.

In soils with considerable populations of earthworms the general mass of the "undisturbed" soil, at least to a depth of 20-30 cm, consists of former systems of burrows and casts. Comparisons made between currently recognizable earthworm-modified material and surrounding soil represent the current state of flux and disequilibrium of organic matter and plant nutrients between a system of compartments within the soil whose spatial relationships constantly change. The whole soil is affected, but not uniformly at any one time.

A. Quantities of Organic Matter Incorporated in Soil

Earthworms play a key role in the removal from the surface of plant litter, dung, and other organic material and its incorporation into the soil. Satchell (1967) summarized evidence from a number of laboratory and field trials with Lumbricidae, and concluded that where earthworms flourish the amount of organic material they consume is limited by the amount available rather than by the earthworms' capacity to ingest it. The data of Franz and Leitenberger (1948) for a population of *Lumbricus rubellus*, van Rhee (1963) for six lumbricid species, and Needham (1957) for *L. terrestris* indicate that lumbricid earthworms consume about 27 mg dry weight of litter per gram wet weight of earthworms per day, and that in optimum conditions they probably consume considerably more than this. On the basis of a consumption of 27 mg g^{-1} d^{-1}, Satchell (1967) calculated that a population of *L. terrestris* he studied in a mixed forest in northwest England had the capability to consume the total annual litter fall (*c.* 300 g/m^2) in about three months.

Raw (1962) placed apple leaves under netting on the ground surface at the rate of 200 g/m² in an orchard clean cultivated between the trees, where there was a biomass of 168 g/m² of *L. terrestris*. The leaves were removed by *L. terrestris* at a rate of up to 20 g m^{-2} d^{-1} and after two months 99.5% of the leaves had been taken. In another orchard, under grass, where the biomass of *L. terrestris* was only about 53 g/m², Raw found that the same rate of removal (20 g m^{-2} d^{-1}) was maintained, despite the presence of only about one third the biomass of *L. terrestris*. Rapid removal of plant litter from the soil surface in a variety of habitats by lumbricid earthworms has been demonstrated by many other authors, including Heath (1964) in England, Heungens (1969) in Holland, Zajonc (1971a) in Czeckoslovakia, Tiurin (1937), Perel' and Sokolov (1964), Perel' and Karpechevsky (1971), and Kudrjaseva (1976) in Russia, Maldague (1970) in Canada, Nielsen and Hole (1964), Vimmerstedt and Finney (1973), and Malone and Reichle (1973) in the United States, Waters (1951) in New Zealand, and many others.

Removal of litter from the soil surface by earthworms does not imply that all the litter is immediately ingested. Some species, especially *Lumbricus terrestris*, pull leaves or parts of leaves into their burrows and leave them there, sometimes eating them when they are partially decomposed, e.g., Nielsen and Hole (1964) noted leaves lining the burrows of *L. terrestris* to a depth of 11 cm.

Stout and Goh (1980) have traced the distribution in litter and soil horizons of the "bomb" radiocarbon, the sudden additional pulses of atmospheric ^{13}C and ^{14}C that resulted from the indiscriminate explosion of atomic bombs by several military powers from 1945 onwards. Comparing two New Zealand grassland soils, one with and one without lumbricid earthworms, they showed that surface litter and soil to 10 cm depth was greatly enriched in ^{14}C in soil without earthworms, but in soil with earthworms, where there was no surface litter accumulation, the ^{14}C had been thoroughly mixed into the soil to *c.* 18 cm depth. Closely similar results were obtained in forest soil at Brogden's Belt, Oxford (England).

Dietz and Bottner (1981) put litter, labelled with ^{14}C, of the grasses *Lolium perenne* and *Brachypodium phoenicoides* on the surface of undisturbed soil cores, with and without *Aporrectodea caliginosa meridionalis* and measured the rate of movement of the ^{14}C-labelled material into the soil. The half life (time for disappearance from the surface of half the ^{14}C-labelled material) was 5.7 weeks for *L. perenne* litter in the presence of earthworms and was 20 weeks and 15 weeks respectively for *L. perenne* and *B. phoenicoides* litter in the absence of earthworms.

Little is known of litter consumption rates in earthworms other than Lumbricidae. At Lamto, Ivory Coast, Lavelle (1978) estimated that mixed populations of megascolecid and eudrilid earthworms ingest about 30% of the quantity of litter decomposed each year in grass savanna and about 27% of that decomposed in shrub savanna. Sugi and Tanaka (1978b) studied a population of six species of *Pheretima* and one species of *Allolobophora* in a warm temperate evergreen oak forest on Kyushu Island, Japan, and calculated from studies of populations, their phenology and bioenergetics that they ingested 1071.4 g m^{-2} y^{-1} of litter from the soil surface; this is 1.4 times the estimated annual litter fall of 752.3 g/m^2 (Nishioka and Kirita 1978) at this site and they concluded that the earthworms must also ingest and re-ingest the available organic matter. At a nearby site in the same forest, where the earthworm population was lower, Sugi and Tanaka (1978b) estimated that earthworms ingested 51.6% of the annual litter fall. A rather different mode of litter incorporation was described by Madge (1965) for *Hyperiodrilus africanus* (Eudrilidae) in a forest in Nigeria. During the wet season the vertical columnar casts, 3–8 cm high and 1–3 cm in diameter, of this species are pushed up through the litter layer; these casts subsequently collapse and disintegrate, burying the litter.

Apart from litter, the major source of earthworm food that is incorporated into the soil is the dung of ruminants; it is actively sought and buried and is also carried into earthworm burrows with drainage water. The potential ability to bury dung, as with plant litter, may greatly exceed the available supply. Guild (1955) fed laboratory colonies of *Aporrectodea longa*, *A. caliginosa* and *Lumbricus rubellus* on cow dung and calculated that a field population of adults (25/m^2) of these species removed dung from the soil surface at a rate equivalent to about 1700–2000 g m^{-2} y^{-1}, and allowing for consumption by juveniles the total consumption would be potentially about 2500–3000 g m^{-2} y^{-1}. Satchell (1967) calculated that given a daily dung production per cow of about 24 kg and a density of 0.8/ha, the annual dung production would be about 600–750 g/m^2, which is about one quarter of the estimated potential dung consumption by a rather small population of lumbricid earthworms. The consumption rate (dry weight per gram body weight) calculated by Guild (1955) is equivalent to about 40 mg g^{-1} d^{-1}; Barley (1959b) showed that in the laboratory *A. caliginosa*, in Australia, with sheep dung as its only food source, consumed about 80 mg g^{-1} d^{-1}.

Holter (1979) compared the efficiency of cattle dung removal by earthworms (predominantly *A. longa*) and the dung-feeding beetle *Aphodius rufipes* (Scarabaeidae) in a Danish pasture and concluded that, in two years when rainfall was sufficient to maintain their activity, *A. longa* accounted for about 50% of dung disappearance, compared wth 14%–22% due to *A. rufipes*, but in a year when there was a drought and earthworms were relatively inactive, disappearance of dung from the surface was very slow. In further studies, Holter (1983) compared the relative efficiency of dung burial by two different associations of lumbricid species in Danish pastures. Under and in some dung pats were aggregations of *Aporrectodea longa* and *Lumbricus terrestris*, while others were characterized by aggregations of *L. castaneus, L. festivus* and *L. rubellus*. Rates of dung burial were the same, irrespective of species present, for the same biomass of earthworms during a 24-day study; for 1000 "earthworm-g-days" *c.* 66 g of dung was removed, with a high correlation between earthworm biomass and rate of dung removal ($r^2 = 0.992$). There was good evidence that the presence of *A. rufipes* larvae and adults in dung in some way stimulates the feeding activity of earthworms (especially *L. castaneus*) on dung, but the stimulatory mechanism is not known.

Martin and Charles (1979) found that in New Zealand earthworms were the principal agents of dung removal (especially cattle dung, but also sheep dung) in pastures. The species most concerned were *Lumbricus castaneus, L. rubellus, L. terrestris* and *Aporrectodea caliginosa*. The proportions of the various species in total populations vary through the year. A high proportion of *A. caliginosa* survives the summer and this species is dominant, feeding often on old dung, at the beginning of winter; late in winter the older dung has gone and the area of ground surface covered by dung is reduced to about 25% of the summer maximum, and at this stage the dominant dung-feeding earthworms are *L. castaneus* and *L. rubellus*. Martin and Charles (1979) found that numbers and biomass of earthworms under dung were usually much greater than between dung pats.

B. Nature of Organic Matter in Casts

Not all of the plant litter, dung, and other organic material that earthworms remove from the soil surface is ingested by them and of the amount ingested very little is digested (p. 136). The ingested organic matter is macerated, mixed with ingested inorganic soil material, passed through the gut and excreted as casts, with most of it little changed chemically, but finely ground physically. The greater surface area of organic matter thus exposed to microbial activity must facilitate its further decomposition. For example, Powlson (1980) compared the respiratory activity (carbon dioxide output) of soils that had been mechanically ground with unground controls; grinding more than doubled the carbon dioxide output of an arable and a grassland soil. Powlson attributed the increased carbon dioxide production to the increase in surface area of organic matter, due to the maceration of only about 0.5% of the total organic matter content of the soil. Similar effects, resulting from compressing and shearing soils, were reported by Rovira and Greacen (1957).

Piearce (1972a) examined casts of *Aporrectodea caliginosa*, a geophagous species, and *Lumbricus rubellus*, a litter feeder, from a pasture soil. Casts of both species contained much amorphous humus; *A. caliginosa* casts contained only small amounts of raw humus (fragments with regularly arranged intact plant cells) and much mineral material, while those of *L. rubellus* had large amounts of raw humus and little mineral material. The proportions of the various components in casts closely resembled those found by Piearce in the gut of the same species (p. 19) and illustrate the low efficiency of degradation of plant tissue in transit through the gut of earthworms.

Czerwinski *et al.* (1974) found that the casts of Lumbricidae in pasture soils in Poland were enriched with strongly decomposed organic substances (comparable to Piearce's [1972a, 1978] amorphous humus) whose C : N ratio resembled that of humic materials.

C. Microbiological Decomposition in Casts

A clear distinction should be made between the populations of organisms reported to have been isolated from earthworm casts and microbial populations in transit through the earthworm gut. Earthworm casts contain finely divided organic debris, little digested by the earthworms, mixed with inorganic soil materials, living microbial cells, nitrogenous and other excretory products of the earthworms' metabolism. Casts have a high moisture content, and though conditions within casts when they are excreted are probably largely anaerobic they are usually deposited in an aerobic situation and provide an extraordinarily favourable microenvironment for a wide range of decomposer organisms. Nowak (1975) proposed that the most important effect of earthworms on soils may be the stimulation of microbial activity that occurs in casts, because this results in the rapid transformation of soluble nitrogen into microbial protein and consequently prevents its rapid loss to lower soil horizons by leaching.

Many investigations have shown increased populations of microorganisms in casts compared with surrounding soil (e.g., Teotia *et al.* 1950; Kollmannsperger 1952, 1956a; Ruschmann 1953; Schultz and Felber 1956).

Parle (1963b) followed the development and respiratory activity of microbial populations in casts of *Aporrectodea longa* at intervals up to 45 days from the time they were produced. Counts and direct measurement showed that the fungal population increased for up to 20 days; length of fungal hyphae increased steadily to about three to five times that in fresh casts after 15 days, and then declined slowly, but after 45 days was still much greater than in fresh casts. Numbers of yeast cells also increased over a 20 day period; total bacterial cells in fresh casts were some 100 times as high as in soil samples and did not change significantly over 20 days. Respiratory activity of casts declined continuously over a 45 day period, which would not be expected from the pattern of change in microbial numbers, and this was interpreted by Parle as indicating that bacteria and actinomycetes probably formed resting stages as casts aged (c.f. Schultz and Felber [1956] who considered that most bacteria in casts were present as spores).

The most significant chemical changes in casts compared with undisturbed soil relate to the accelerated decomposition of organic matter and to changes in the forms of nitrogen compounds, increasing the proportion of nitrogen that is readily available to plants.

Barley and Jennings (1959) maintained *Aporrectodea caliginosa* in jars of soil for 50 days, measuring oxygen consumption (i.e., decomposition) rates and cumulative production of exchangeable nitrate and ammonium ions, compared with those of soil without earthworms. Over a 30-day period the calculated oxygen consumption due to the earthworms' own metabolism would have been about 200 μL O_2/g culture medium, while the observed difference due to earthworms was 410 μL O_2/g medium. The additional 210 μL O_2/g consumed represented about 10% of total oxygen consumption and was attributed to stimulation of microbial decomposition by the earthworms. Over the full 50-day period there was an increase in exchangeable nitrate and ammonium ions in the soil with or without earthworms, but the increase was > 20% greater in cultures with than in cultures without earthworms; total exchangeable nitrate and ammonium ions after 50 days was about 6% greater with earthworms present than without earthworms.

Parle (1963b) similarly showed an increase in nitrate and ammonium ions in casts of *Aporrectodea longa* relative to surrounding soil; about 96% of this nitrogen was in the ammonium form and < 4% in the nitrate form in fresh casts, but over a period of 20 days the proportions were changed by microorganisms to about 65% in the ammonium and 35% in the nitrate form.

Loquet (1978) calculated rates of organic matter decomposition in lumbricid casts, compared with undisturbed soil, using the coefficient of mineralization of Dommergues (1968)

$$(C \text{ converted to } CO_2/\text{Total C}) \times 100$$

Calculated coefficients for casts were 3.3, 1.5, 2.0 for samples taken in autumn, spring and the following autumn respectively, compared with 2.5, 0.9, 0.6 for corresponding samples of the 0–6 cm horizon of soil, an increase in the casts of 32%, 67% and 233% respectively.

208 *Relationships with Soils and Land Use*

Jeanson (1960b) and Czerwinski *et al.* (1974) attributed much of the formation of humus in soils and its association with mineral soil material to form clay-humus complexes to mixing of organic with inorganic materials in the earthworm gut and to subsequent microbial activity in earthworm casts.

D. Microbial Decomposition around Earthworm Burrows

Bouché (1975b) defined a zone 2 mm in thickness around the walls of earthworm burrows, which he named the drilosphere. In a permanent pasture at Cîteaux, in France, with an earthworm population of 12 species (*Aporrectodea longa* and *A. caliginosa* dominant), Bhatnagar (1975) compared populations of nitrogen-fixing and denitrifying bacteria in the drilosphere with those of the remainder of the soil. He found that about 40% of the total aerobic nitrogen-fixers, 13% of the anaerobic nitrogen-fixers and 16% of the denitrifiers were in the drilosphere, with maximum populations at 20–40 cm depth in the soil. Bouché (1977a) and Kretzschmar (1978), who also worked on pasture at Cîteaux, measured the length of earthworm burrows as 380–400 m/m^2 of ground surface, with a volume of 5 L/m^2 and a surface area of about 5 m^2/m^2 ground surface. Loquet (1978) quoted unpublished data of Kretzschmar that the surface area of burrows varied through the year from 1.6–12 m^2/m^2 of ground surface. Bhatnagar (1975) attributed the preponderance of aerobic nitrogen-fixers in the drilosphere to the rich energy source available from concentrations of organic matter deposited in burrows by earthworms and to the ready access of air through the burrows. Soils with large populations of earthworms have low C : N ratios and Bhatnagar attributed this to the stimulation of aerobic nitrogen-fixers and consequent net increase in nitrogen in the drilosphere. This conclusion is at variance with the results of experiments of Nye (1955) which showed that when earthworms are kept in pots and all the nitrogen in the system is accounted for, there is no increase in total nitrogen in the system. Further investigation is necessary to resolve the apparent contradiction.

Loquet (1978) measured carbon dioxide output and invertase, urease and dehydrogenase activity in the drilosphere of the Cîteaux soil and found it to vary greatly through the year, sometimes higher and sometimes lower than in surrounding soil, but with no regular pattern. In New Zealand, pot experiments reported by Ross and Cairns (1982) in which rye grass was grown on a subsoil in the presence or absence of *Aporrectodea caliginosa* showed that the activities of invertase, amylase and phosphatase were stimulated by the presence of the earthworms. Plant growth was stimulated by the presence of *A. caliginosa* and it was concluded that the observed increase in enzyme activity was probably due to an increase in plant material available for decomposition rather than directly to the presence of the earthworms. However, the evidence of increased microbial activity in the drilosphere (Bhatnagar 1975, see previous paragraph) indicates that increases in biochemical reactivity of soils may be directly affected by substrates for microbial activity that are produced by earthworms.

E. Carbon, Nitrogen and C : N Ratio of Casts

Results of some analyses of carbon and nitrogen content and C : N ratios of casts, with corresponding analyses of surrounding soils are listed in Table 21. With a few exceptions, the carbon content of casts is about 1.5-2 times that of the soil, while the nitrogen content is about 1.2-1.7 times that of soil, so that C : N ratios of casts are generally a little higher than those of soil. This is in accordance with the inability of earthworms to digest most of the plant material they ingest, and the slight increase in C : N ratio perhaps reflects their efficiency as producers of protein (Chapter 16) and consequent demand for nitrogen, as well as nitrogen losses from the general body surface in mucoproteins and from the nephridia in urine (p. 213).

Seasonal variations in the carbon and nitrogen content of casts were recorded by Syers *et al.* (1979) in a New Zealand pasture soil where the dominant surface casting species was *Lumbricus rubellus*. The total nitrogen content of the casts varied little through the April-October (autumn-winter-spring) period when most surface casting occurred; total oxidizable carbon content increased during April-May, when litter production was at its maximum, and was at a minimum in August-October, when surface litter was at its minimum. *Lumbricus rubellus* was apparently obliged to ingest more inorganic soil material to obtain adequate organic material during the latter period, with a consequent reduction in the organic matter content of casts, a corresponding decrease in carbon content, but apparently little change in nitrogen content; the C : N ratio of casts reached a maximum of about 11.7 at the end of May and a minimum of about 10.0 in the middle of October.

The quantities of carbon and nitrogen excreted in casts may be very large. Syers *et al.* (1979) estimated total cast production on the soil surface in a New Zealand pasture as 33 t ha^{-1} y^{-1} (3.3 kg m^{-2} y^{-1}); with a mean carbon content of 4.3% and mean nitrogen content of 0.39% (Table 21), this represents an annual output of 142 g/m^2 of carbon and 13 g/m^2 of nitrogen in casts. The total nitrogen in surface casts represented 73% of that of the litter removed from the soil surface (610 g m^{-2} y^{-1}), demonstrating the low

Table 21. Organic carbon and nitrogen content and C:N ratios of earthworm casts and surrounding soil, from various parts of the world.

Earthworms	Habitat	Locality	Organic C% Casts	Organic C% Soil	Nitrogen % Casts	Nitrogen % Soil	C:N Ratio Casts	C:N Ratio Soil	Reference
LUMBRICIDAE									
Lumbricus terrestris									
Lumbricus spp.	Old pasture	Germany	8.6	3.9	0.54	0.35	16.0	11.2	Graff 1971
Aporrectodea caliginosa									
Aporrectodea caliginosa	Grazed pasture	Poland	4.8	2.7	0.48	0.28	10.0	9.6	Czerwinski et al. 1974
Aporrectodea rosea									
Octolasion lacteum									
Dendrobaena octaedra	Grazed pasture (former sheep fold)		4.8	2.6	0.46	0.31	10.4	8.3	
Lumbricus rubellus									
Lumbricus rubellus	Pasture	New Zealand	4.3	3.5	0.39	0.23	11.0	15.2	Syers et al. 1979
Lumbricus terrestris	Pot experiment — arable soil	Germany	1.3	0.8	0.13	0.10	10.0	8.8	Aldag and Graff 1975b
MEGASCOLECIDAE									
Amynthas hupeiensis	Pasture	Japan	2.5	1.6	0.24	0.20	10.8	8.1	Watanabe 1975
Metapheretima jocchana	Natural grassland	New Guinea	3.0	2.3	0.19	0.14	15.8	15.4	Lee 1967
Millsonia anomala	Natural grassland	Ivory Coast	5.8	1.9–8.2	0.57	0.17–0.57	10.2	10.8–15.3	Lavelle 1978
GLOSSOSCOLECIDAE									
Glyphidrilus sp.									
Alma stuhlmanni	Grass swamps	Uganda	2.9	1.7	0.28	0.18	10.4	9.4	Wasawo and Visser 1959
Alma emini			2.1	0.05	0.22	0.01	9.5	5.0	

assimilation efficiency of ingested organic matter by *L. rubellus* and indicating that most of the ingested organic nitrogen passes through the gut without change. Graff (1971) measured a total annual cast production of 25.75 kg/m² in a pasture in Germany, representing about 25% of the total weight of the A_h horizon of the soil. The carbon content of the annual output of casts was 2137 g/m², or about 42% of the total carbon content of the casts and the A_h horizon, while the nitrogen in casts totalled 134 g/m², about 34% of the total nitrogen in casts and the A_h horizon. Czerwinski *et al.* (1974) found that, in Poland, annual cast production in a grazed pasture was 0.78 kg/m² and in a pasture used the previous year as a sheep-fold, and consequently with much added sheep dung, it was 3.53 kg/m². The casts contained respectively 75% and 85% more organic carbon than undisturbed soil, representing an enrichment of the surface soil layer in the pasture of 28.4 g/m² of organic matter, including 16.5 g/m² of carbon and 1.57 g/m² of nitrogen and in the former sheep-fold an enrichment of 145.1 g/m² of organic matter, including 78.4 g/m² of carbon and 5.29 g/m² of nitrogen.

There is a considerable increase in polysaccharide content of casts relative to undisturbed soil. For example, Bhandari *et al.* (1967) found that 36.6% of the acid-hydrolysable fraction of organic carbon in casts of *Eutyphoeus waltoni* in an Indian pasture was in the form of polysaccharides, while only 7.8% of that in undisturbed soil occurred as polysaccharides. This may be little more than a reflection of the abundance of undigested cellulose from plant tissue in the casts. Parle (1963b) found six to seven times as much polysaccharide in casts as in undisturbed soil. The stability of casts has been attributed to polysaccharide gums, perhaps produced by bacteria (e.g., Swaby 1950) but Parle (1963b) found no correlation between polysaccharide content and cast stability in casts up to 30 days from their initial production.

II. Nitrogen Transformations Due to Earthworms

The influence of earthworms on soil nitrogen and nitrogen cycling was reviewed by Lee (1983) and is summarized here. Nitrogenous products of earthworm metabolism are returned to the soil in four ways:
(a) in casts: most of the nitrogen in casts is in plant tissue that has passed through the gut with little chemical change, but some nitrogenous metabolic wastes are also included;
(b) in urine: most nitrogenous waste products of metabolism are excreted in urine;
(c) in mucoproteins: these are secreted onto the body surface to lubricate the earthworms' movement through the soil and to maintain the moist surface layer essential for respiratory exchange;

(d) in dead earthworm tissue: earthworm tissue is about 60%-70% (dry weight) protein and has a nitrogen content of about 12%.

B. Nitrogen in Casts

The concentration of nitrogen available for plant growth is low in casts and probably has little influence on plant growth, as is illustrated by the following examples.

Barley and Jennings (1959), in South Australia, kept *Aporrectodea caliginosa* at 15°C in pots containing 500 g of Urrbrae fine sandy loam (red-brown earth), supplied with finely ground clover litter mixed into the soil as food, and measured exchangeable ammonium, nitrate and other soluble nitrogen contents of casts from pots with earthworms compared with those of soil from similar pots without earthworms. Casts contained < 0.2% total exchangeable and soluble nitrogen; analyses of casts and soil collected over a period of 315 hours showed that the total nitrogen ingested by the earthworms (on wet weight basis) was 32 μg g^{-1} h^{-1}, including 2.3 μg of exchangeable and soluble nitrogen, while the excretion rate of exchangeable and soluble nitrogen in casts (on a wet weight basis) was 4.2 μg g^{-1} h^{-1}, an increase of 1.9 μg (83%) compared with the control. Selection in favour of organic matter compared with mineral soil material by earthworms when feeding increases the total organic matter content of casts compared with that of the soil; further analyses showed that the total ammonium and nitrate N accumulated in pots with earthworms over a 50-day period was 8 mg/g wet weight earthworms. The biomass of earthworms was 3 g/kg soil; estimates of earthworm biomass (wet weight) in Urrbrae fine sandy loam under pasture were 62-78 g/m^2 and assuming that they would be active in the top 15 cm of the soil, which has a bulk density of about 1.5 (E. L. Greacen, pers. comm.), this represents a biomass of 280-350 mg/kg soil. In the mediterranean climate of the Adelaide region Barley (1959a) found that lumbricid earthworms are active for about 150 days of the year and it may thus be calculated that their casts would contribute only about 2.2-2.8 mg m^{-2} y^{-1} (22-28 kg ha^{-1} y^{-1}) of additional available nitrogen to the soil.

Aldag and Graff (1975b) similarly kept *Lumbricus terrestris* in pots, supplied food in the form of finely ground cattle dung, and measured the nitrogen content of casts compared with that of the soil. Total nitrogen in casts was 40% higher than in surrounding soil and differences were most marked in the soluble and exchangeable fractions, NO$_3^-$ and NO$_2^-$-N (0.28 μg/g casts, 0.01 μg/g soil) and NH$_4^+$-N (0.7 μg/g casts, 0.3 μg/g soil). If the figures for soluble and exchangeable nitrogen are applied to known rates of cast production of European lumbricids, which are generally in the range of

5-7 kg m^{-2} y^{-1} (see Table 18), additional inputs of available nitrogen due to earthworm casts would be about 3.5–5 mg m^{-2} y^{-1}, or 35–50 g ha^{-1} y^{-1}. These amounts are very similar to those calculated from the data of Barley and Jennings (1959) and in neither case are they significant in terms of the requirements of plants.

It is likely that the casts of the meronephridial species would be more enriched with exchangeable nitrogen than those of holonephridial species, since the urine produced by the septal nephridia is discharged into the gut and must be voided with the casts. Dash and Patra (1977, 1979) studied a mixed population of *Lampito mauritii*, a meronephridial megascolecid species, and an unidentified ocnerodrilid, a holonephridial species, in a grassland at Orissa, India. Both species were active in the top 25 cm of soil; the mean monthly biomass of the two species was 30.25 g/m^2, with *L. mauritii* contributing about 90% of the biomass and presumably of the casts. Casts contained 0.47% nitrogen, compared with 0.35% in the surrounding soil, which is in the same range as has been reported for a variety of other species (Table 21). Dash and Patra (1979) found an increase in nitrogen in casts, equivalent to 9.29 g m^{-2} y^{-1} and some of this probably resulted from excretion of urine into the gut, but they did not distinguish between the forms of nitrogen present. It is likely that most of it was in non-exchangeable forms, in plant tissue.

No data are available on the concentration of exchangeable forms of nitrogen in casts of meronephridial species. Meronephridial earthworms, especially those of the pheretimoid group of genera (Sims and Easton 1972; Easton 1979), are very widely distributed in tropical and southern temperate regions of the earth and there is a need for research on the chemical composition of their casts and their influence on nutrient cycling.

B. Nitrogen in Urine and Mucoproteins

In holonephridial species, which include the majority of earthworms (all Lumbricoidea, Biwadriloidea, Sparganophiloidea, Almoidea and some Megascolecoidea), urine is excreted directly to the exterior through serially arranged nephridiopores and most of it enters the drilosphere, the thin layer of soil immediatley surrounding earthworm burrows (p. 208). Mucoproteins are secreted from the body surface and must also find their way into the drilosphere.

Laverack (1963) estimated that about half the total nitrogen metabolized and voided is excreted in urine, mainly as ammonia and urea, with a small proportion as allantoin and uric acid. Experiments of Tillinghast (1967), in which excretion from the gut in casts was temporarily prevented, indicated that in active, feeding *Lumbricus terrestris* urea was excreted in urine, but that most of the ammonia was excreted into the gut and must be voided in casts. This finding is not easily reconciled with the very low NH_4^+-N content of casts recorded by Barley and Jennings (1959) and by Aldag and Graff (1975b) (p. 212), though Parle (1963a) found that freshly voided casts contained about 300 µg NH_4^+-N/g dry weight. These anomalous findings may result from differences in the general soil environment of the availability of microbial urease, with the consequence that NH_4^+-N in casts may be more rapidly converted to urea in some soils than in others.

The relative proportions of ammonia and urea excreted are variable between species and within the same species, in response to changes in nutritional level, temperature, soil texture, and water availability (Needham 1957; Laverack 1963; Tillinghast *et al.* 1969; Khalaf El-Duweini and Ghabbour 1971). Bishop and Campbell (1965) showed that in *Lumbricus terrestris*, when starved, there is an increase in production of the urea-cycle enzymes and a corresponding decline in ammonotelism until the animal becomes entirely ureotelic. Florkin (1969) considered that this change probably results from problems of osmoregulation in the starving animal, related to the amount of metabolic water that is available to be lost from the body with the urine. Tillinghast (1967) suggested that ammonia excreted into the gut may exchange with other cations and thus provide a mechanism to conserve cations such as Na^+, essential for the control of blood pH, and that it may also act as a buffer and account for the uniformity of pH (6.4–6.6) along the length of the intestine of *L. terrestris*.

The rate of excretion of urine by earthworms in natural conditions is not known. Wolf (1940) found that *L. terrestris* kept immersed in water produced urine at a rate of about 60% of its body weight per day, while Bahl (1947) calculated that *Metaphire posthuma*, also in water, produced about 45%–50% of its body weight per day. It is unlikely that such large quantities could be produced in soil, as it would be difficult for earthworms to replace water at such a rate.

Needham (1957) measured nitrogenous components of excretion in the holonephridial species *Lumbricus terrestris*, *Eisenia fetida* and *Aporrectodea caliginosa* collected from soils in England. He kept them under water in small flasks at a constant temperature of 23°C, and measured the amounts of nitrogen compounds accumulated in the water. He distinguished between fasting and feeding earthworms and examined the effect of variation in the volume and frequency of renewal of the water, i.e. the effect of varying concentration of accumulated waste products in the medium, on the quantities and proportions of the nitrogenous components produced. The method suffers from the disadvantage that nitrogen excreted by all pathways accumulates in the water and for feeding animals there would be some proportion of faecal nitrogen. The possibility of nitrogen contamination directly from food was reduced by feeding the earthworms on elm leaves, which contain almost no water-soluble nitrogen compounds. Some of Needham's data are summarized in Table 22. He converted all his measurements to specific outputs, i.e., μg (nitrogen) g (wet body weight)$^{-1}$ d^{-1}. The residual nitrogen, not accounted for by the AA fraction (amino-N + NH_4^+-N) and the urea fraction in Table 22 was not identified, but was considered to be largely in mucoprotein. From Table 22 it may be seen that the specific output of nitrogen of *E. fetida* is considerably higher than that of *L. terrestris*, while that of *A. caliginosa* is much lower than that of either of the other species. Animals fed on elm leaves produced less nitrogen than fasting animals; *L. terrestris*, when fed on filter paper, which has little or no nutritional value, produced little less than when fasting, but food intake was found to be the major factor affecting nitrogen excretion. Changes in water volume or in frequency of renewal of the water used as the medium had little effect on the pattern of ammonia or urea excretion. The proportion of ammonia to urea was much higher in *E. fetida* than in *L. terrestris* and *A. caliginosa*. Needham (1957) also measured the titratable acid content of the daily urine output of *L. terrestris* and *E. fetida*; he found that acid production is higher in feeding than in non-feeding individuals of both species, but that *E. fetida* always had a higher output of acid than *L. terrestris*. Increased acid produced by *L. terrestris* when feeding is apparently neutralized by an accompanying increase in ammonia production (Table 22) but *E. fetida* does not produce more ammonia when feeding (Table 22).

Needham's results may be compared with those of Khalaf El-Duweini and Ghabbour (1971), who examined the nitrogenous fractions of urine produced by *A. caliginosa* and *Metaphire californica*, collected from soils in the much hotter and drier climate of Cairo. *Metaphire californica* is a meronephridial species, a representative of the large number of mainly tropical species that make up the "pheretima" group of genera (Megascolecidae). *Metaphire californica* has two nephridial systems, (a) a segmentally arranged system of meronephridial tubules associated with the septa that filters the coelomic fluid and discharges into the intestine, and (b) an integumentary system isolated from the coelom and discharging to the exterior (Bahl 1919, 1947). The septal system produces urine, while the main function of the integumentary system is apparently to rid the body of excess water, or to maintain a moist film on the body surface, and Khalaf El-Duweini and Ghabbour distinguished its product as "perspiration". Khalaf El-Duweini and Ghabbour starved the earthworms for five days, until the gut was empty, and kept their specimens in unsaturated air at room temperature (19°C–22°C), conditions that might more closely resemble the

Table 22. Specific rates and proportions of components of nitrogenous excretion in *Lumbricus terrestris*, *Eisenia fetida*, and *Aporrectodea caliginosa* from England. Specific rates are given as μg/g wet weight of earthworm per day. (Data from Needham 1957).

Earthworm	Total nitrogen excreted μg g⁻¹ d⁻¹	AAᵃ + Urea N Fractions μg g⁻¹ d⁻¹	% of Total Nitrogen	Ammonia + Urea N Fraction μg g⁻¹ d⁻¹	% of AAᵃ + urea N fractions	Ammonia N Fraction μg g⁻¹ d⁻¹	% of total ammonia + urea fraction	Urea N Fraction μg g⁻¹ d⁻¹	% of total ammonia + urea fraction
Lumbricus terrestris									
Fasting	331.4	159.2	48	159.1	100	30.0	23	129.1	77
Feeding on filter paper	316.4	155.5	49	155.4	100	29.4	19	126.0	81
Feeding on elm leaves	268.8	94.8	35	94.8	100	54.0	57	40.8	43
Eisenia fetida									
Fasting	569.0	345.5	61	307.5	89	234.4	76	73.1	24
Feeding on filter paper	318.9	169.8	53	130.3	77	80.1	61	50.2	39
Feeding on elm leaves	401.8	204.8	51	153.1	75	104.8	68	48.3	32
Aporrectodea caliginosa									
Fasting	133.3	97.2	73	—	—	—	—	83.2	—
Feeding on elm leaves	87.5	57.3	65	—	—	—	—	19.9	—

ᵃAA fraction = Amino nitrogen + ammonia nitrogen.

natural environment than the water filled flasks used by Needham. They measured ammonia and urea output by the two species; in Table 23 their data have been recalculated in the same units as used by Needham (1957), µg (nitrogen) g (wet weight)$^{-1}$ d^{-1}.

The specific rate of total nitrogen excretion for fasting Egyptian *A. caliginosa* is not known; the total ammonia plus urea nitrogen (127 µg g^{-1} d^{-1}) is higher, but comparable with the ammonia plus urea nitrogen figure for fasting English *A caliginosa* (97.2 µg g^{-1} d^{-1}). The total ammonia plus urea nitrogen (391 µg g^{-1} d^{-1}) for fasting *M. californica* is comparable with the ammonia plus urea figure for fasting English *E. fetida* (307.5 µg g^{-1} d^{-1}). Khalaf El-Duweini and Ghabbour (1971) compared their results for *M. californica* with results of Bahl (1947) for *Metaphire posthuma*, kept in water, as in Needham's experiments; ammonia concentrations for *M. californica* in unsaturated conditions were about twice and urea concentrations about ten times the concentrations measured by Bahl for *M. posthuma* in water. Khalaf El-Duweini and Ghabbour concluded that earthworms taken from normal soil conditions and kept in water produce increased quantities of more dilute urine, with an accompanying shift from ureotelism to ammonotelism; this is in accordance with Florkin's (1969) interpretation of the opposite process of a shift from ammonotelism to ureotelism in conditions of water shortage (this Section). Tillinghast *et al.* (1969) further confirmed that *L. terrestris* kept in water at 23°C increased their ammonia output and decreased their urea output as the volume of water was increased, but did not do so when kept at 8°C. The general similarity of patterns of ammonia, urea and mucoprotein output by the same species in different environments and the reasonable coherence of the range of patterns in widely different species indicate that these patterns are probably similar in all earthworms and vary similarly with experimental conditions.

Assuming an average of nitrogen in ammonia and urea of *c.* 200 µg g^{-1} d^{-1} and that earthworms are active in most parts of the world for only about six months of the year, their input of nitrogen to the soil in these forms would be in the range < 1 to *c.* 100 kg ha^{-1} y^{-1}, commonly *c.* 18-50 kg ha^{-1} y^{-1} for lumbricid populations and *c.* 18-35 kg ha^{-1} y^{-1} for non-lumbricid populations (Lee 1983).

C. Nitrogen Derived from Dead Earthworm Tissue

Production in earthworm populations is discussed in Chapter 6. In general, annual biomass production is about two to seven times mean annual biomass, and assuming a reasonably steady state in mean biomass from year to year, the annual mortality must also be about two to seven times the mean annual biomass production.

Table 23. Specific rates and proportions of the ammonia and urea fractions of nitrogenous excretion in fasting specimens of *Aporrectodea caliginosa* and *Metaphire californica*, from Cairo, Egypt. Specific rates are given as μg/g (wet weight) of earthworm per day. (Data after Khalaf El-Duweini and Ghabbour 1971).

Earthworm	Total ammonia + urea nitrogen μg g^{-1} d^{-1}	Ammonia nitrogen fraction		Urea nitrogen fraction	
		Nitrogen μg g^{-1} d^{-1}	% of total ammonia + urea nitrogen	Nitrogen μg g^{-1} d^{-1}	% of total ammonia + urea nitrogen
Aporrectodea caliginosa	127	14	11	113	89
Metaphire californica	391	73	19	318	81

Most of the nitrogen in earthworm tissue is incorporated into proteins; the protein content of earthworms is of the order of 60%-80% of the dry weight of the tissue; e.g., 60%-61% for *Eisenia fetida* (Sabine 1978); 64.7% for *Aporrectodea rosea*, 68.2% for *A. longa*, 71.1% for *A. caliginosa*, 74.0% for *Lumbricus castaneus*, 77.3% for *A. icterica* (Bouché 1981); 62.0%-71.5% for *L. terrestris* (Lawrence and Millar 1945). Barley (1961) estimated the nitrogen content of lumbricid earthworms as about 12% of the dry weight and Bouché (1981) found that it ranged from 10.3% to 12.4% in five lumbricid species; Satchell (1963) found that *Lumbricus terrestris* in an English woodland contained about 1.75% nitrogen on a wet weight basis, while Dash and Patra (1977, 1979) reported a nitrogen content of 3% of wet weight for a population of *Lampito mauritii* and an unidentified ocnerodrilid from India. Dead earthworms decay rapidly; Satchell (1967) added dead *Lumbricus terrestris* to soil in pots and found that after two to three weeks, when the earthworms had entirely disintegrated, the nitrogen derived from them was present in the soil as nitrate (25% of the total), ammonia (45%), soluble organic compounds (3%), with the remaining 27% probably held in the form of undecomposed residues (e.g., setae, cuticle) and additional microbial protein.

The mean biomass (dry weight) of lumbricid earthworms in northern and southern temperate regions ranges from c. 5 g/m^2 in boreal forests to c. 45 g/m^2 in the most densely populated of New Zealand pasture soils. Inputs of nitrogen from dead tissue must range from c. 1-2.5 g m^{-2} y^{-1} to c. 9-22.5 g m^{-2} y^{-1}, or between extreme limits of < 10 g m^{-2} y^{-1} to c. 225 kg ha^{-1} y^{-1} (Lee 1983).

D. Significance in Ecosystems of Nitrogen Input from Urine, Mucus and Dead Tissue

Satchell (1963) estimated the nitrogen inputs of a population of *Lumbricus terrestris* in an English mixed oak-ash woodland. From biomass and production data he calculated that the annual nitrogen input from dead earthworm tissue was 6-7 g/m^2. Nitrogen inputs from excretion of urine and secretion of mucus, based on Needham's (1957) measurements and corrected for soil temperatures, were nitrogen 3.3 g/m^2, giving a total of about 10 g m^{-2} y^{-1} (100 kg ha^{-1} y^{-1}). The input of nitrogen from leaf fall in European woodlands is about 30-70 kg/ha from the trees, plus an undetermined but lesser input from the ground flora and the total must be of the same order as the estimated output of *L. terrestris*. Other sources of nitrogen that enter the soil include decaying roots, the dung of vertebrates and invertebrates, dead bodies of animals other than earthworms and fixation of atmospheric nitrogen by free-living and symbiotic microorganisms. Observed requirements of the *L. terrestris* population indicate that nitrogen must circulate rapidly through decomposer organisms, so that it may be used over and over again during a year; there may be intense competition for nitrogen among the soil fauna.

Bouché (1978) estimated the annual nitrogen turnover of lumbricid earthworm populations in French forests, using as a basis the data of Satchell (1963) and additional data from Satchell (1970) and Lakhani and Satchell (1970). The calculated nitrogen turnover was 6.5 mg/m^2 in mixed forest, 6.0 mg/m^2 in oak forest, 3.0 mg/m^2 in beech forest and 1.3 mg/m^2 in spruce forest, indicating that earthworms play an important part in nitrogen cycling in European forest ecosystems.

Keogh (1979) estimated nitrogen turnover of a population of *Aporrectodea caliginosa* in a grazed New Zealand pasture, using the data of Needham (1957) for nitrogen excretion by this species, and calculating nitrogen input from dead tissue on the basis of Satchell's (1963) measurements of the nitrogen contents of *Lumbricus terrestris*. For a pasture with a dry matter production of herbage of 15 000 kg ha^{-1} y^{-1}, about 600 kg ha^{-1} y^{-1} of nitrogen is available to be cycled from organic matter to "mineral" forms of nitrogen in the soil. Keogh calculated that earthworms were responsible for processing about 109-147 kg ha^{-1} y^{-1} of nitrogen, or about 20% of the total. Nowak (1975) estimated that in Polish pastures, 3.4%-17.6% of the total nitrogen input was processed by earthworms.

The quantities of nitrogen that may be cycled through earthworms appear to be extraordinarily large compared with the amounts generally recognized as being available for plant growth. It should be realized that much of this nitrogen is in a form readily assimilated by soil microorganisms and that soil microorganisms are an important component of the diet of earthworms (p. 23). It seems likely that there is rapid recycling of nitrogen through earthworms, microbial tissue and back to earthworms, on a time scale much shorter than is reflected in the annual cycles generally studied in relation to plant growth. Some of this nitrogen, in readily available forms, would become available to plants, as in the traditional concept of "the nitrogen cycle"; there are minor cycles within the broad traditional cycle and nitrogen may be recycled many times each year through earthworms and other decomposer organisms (p. 144).

E. Fixation of Atmospheric Nitrogen

Nitrogen-fixing microorganisms are common, together with other microorganisms, in the gut contents of earthworms and other soil-inhabiting animals (e.g., Bocock 1963; Raw 1967; Newell 1967; Satchell 1967; Breznak et al. 1973; Edwards 1974; Harding and Studdart 1974; French et al. 1976). Their presence could indicate that there is a true symbiotic relationship that promotes nitrogen fixation, but it seems more likely that they are simply in transit through the gut with ingesta and that the presence of nitrogen-fixing microorganisms has no special significance.

Citernesi et al. (1977) isolated and identified nitrogen-fixing bacteria from the gut contents of a variety of soil animals, including the earthworm *Eisenia fetida*. Samples from *E. fetida* yielded strains of the common free-living anaerobic nitrogen-fixers *Clostridium beijerinckii*, *C. butyricum* and *C. paraputrificum*, and it was demonstrated by the ethylene reduction method that nitrogenase activity apparently due to these organisms, was detectable in living animals. *In vitro* cultures of the bacteria showed that the strains isolated did fix nitrogen, but had lower nitrogenase activity than strains of the same bacteria isolated from sources other than the gut of soil animals. The authors were not able to determine whether the association of N-fixing bacteria and earthworms was commensal or fortuitous.

Kaplan and Hartenstein (1977) also tested a range of soil invertebrates, including five species of earthworms, *Eisenia fetida*, *Octolasion tyrtaeum*, *Lumbricus terrestris*, *Lumbricus* sp., and *Allolobophora chlorotica*, for their ability to fix nitrogen by using the acetylene reduction method, and for nitrate reductase activity. The animals were collected at three times in the year (spring, summer, autumn) and immediately tested; the nitrogen fixation tests were done on living animals at 15°C or 23°C and assays for nitrate reductase *in vitro* at 25°C. There was no evidence of nitrogen fixation or reduction of nitrate to nitrite in any of the earthworms or in any of the other soil animals tested (three species of isopods, two species of millipedes, one species of snail, and three species of slugs).

There is some conflict between the results of Citernisi et al. (1977) and those of Kaplan and Hartenstein (1977). The latter authors, while they were not able to show nitrogen fixation, conceded the possibility that in other conditions and in other soils symbiotic relationships might exist between nitrogen-fixing organisms and earthworms or other groups of soil animals included in their tests. It is generally true that in nitrogen-rich environments nitrogen-fixing organisms, though they may be present, are inactive and that they actively fix nitrogen only in conditions of low nitrogen availability (see, e.g., discussion in Lee 1983). Further investigation is necessary to resolve the possibility that nitrogen may be fixed by microorganisms in the gut of earthworms exposed to nitrogen-poor substrates.

III. Phosphorus

Only very small quantities of phosphorus (P) are excreted by earthworms in liquid wastes (Bahl 1947) but considerable quantities are ingested with organic matter, passed through the intestine and excreted in casts. As with nitrogen, much of the P in casts is contained in undigested plant tissue but concentrations of total and available P in casts, though lower than in plant litter, are generally higher than are found in the underlying soil. Lunt and Jacobson (1944) found 150 μg/g available P in casts compared with 21 μg/g in surface soil (0–15 cm) in United States croplands; Graff (1971) found 137 μg/g in casts and 22 μg/g in surface soil (0–10 cm) in a German pasture; Czerwinski et al. (1974) found 380 μg/g and 290 μg/g in casts compared with 35 μg/g and 70 μg/g respectively in surface soils (0–10 cm) in Polish pastures; there are many other records of increased available P in casts of a wide variety of earthworms; e.g., Puh (1941) in China; Kozlovskaya and Zhdannikova (1957), Atlavinyte and Vanagas (1973) in Russia; Enwezor and Moore (1966), Lal (1974), in Nigeria; Gupta and Sakal (1967), Petal et al. (1977) in India; Zajonc (1971a) in Czeckoslovakia.

Increases are commonly of the order of five to ten times the available P in casts compared with that in surface soil.

The most significant studies of P availability in earthworm casts have been made in New Zealand. Sharpley and Syers (1976) cleared all casts from the soil surface at sampling sites in a pasture and after 14 hours collected the fresh casts (attributed to *Aporrectodea caliginosa* but perhaps some may also have been from *Lumbricus rubellus*) which were frozen until analysed, and at the same time took soil samples from the same sites. Water-dispersed samples showed 20% fine materials (< 4 μm) in casts compared with 6% in 0–5 cm soil samples. Inorganic and organic extractable P fractions were concentrated particularly in the < 4 μm fraction of casts, which contained 971 μg/g inorganic and 1370 μg/g organic P, compared with 427 μg/g and 731 μg/g respectively in the < 4 μm fraction of 0–5 cm soil. Concentrations of extractable P were also high in the 4–20 μm fraction (22%) of casts, with 583 μg/g inorganic P and 864 μg/g organic P, compared with the 4–20 μm fraction (18%) of 0–5 cm soil samples which had 265 μg/g and 451 μg/g respectively. Coarser particle size fractions from casts had P contents sometimes higher and sometimes lower than those of corresponding particle size fractions of 0–5 cm soil samples. The rate of release of inorganic P from casts to 0.1 mol/L NaCl solution was initially about four times that from soil from 0–5 cm depth and sequential extractions showed that this ratio was maintained throughout the three following days. The high rate of release to solution of inorganic P indicates that the additional P in casts is loosely bound and this was further confirmed by measurements of the amounts of isotopically exchangeable P in casts, which were similarly higher than those of soil from 0-5 cm. Barley (1961) considered that increase in the pH of casts relative to soil was responsible for higher "solubility" of nutrients in casts; Sharpley and Syers (1976) found that the pH of casts was 6.22, compared with 5.86 for soil (0-5 cm) and since pH has little effect on P sorption by soils over the pH range 4-6.5, they concluded that the greater release of P from casts was due to a shift in the P sorption isotherm relative to that in undisturbed soil.

Sharpley and Syers (1976, 1977) further investigated seasonal variability in the P content of casts and its release. Cast production was confined to the period from April to September (autumn-early spring; c. 150 d/y) and totalled about 30 t ha^{-1} y^{-1} (3 kg m^{-2} y^{-1}), with a total accumulation in the casts of 9 kg/ha (0.9 g/m^2) inorganic P and 13 kg/ha (1.3 g/m^2) organic P. The total P content of casts was fairly constant throughout the cast production period but the inorganic P component decreased from a maximum in April-May to a minimum in August, while the organic P component showed an opposite trend. The mechanism of release of the inorganic P from casts is probably mainly due to enhanced microbial activity in the casts; the decline in rate of release of the inorganic P between May and August was related to declining soil temperature and was attributed by Sharpley and Syers to a resultant decline in microbial and phosphatase activity in the casts. About 90% of the additional inorganic P relative to that in underlying soil was water-extractable.

The increased availability to plants of P in earthworm casts is not entirely due to enhanced microbial and phosphatase activity. Mackay *et al.* (1982, 1983) showed an increase in availability of P from ground phosphate rock and from superphosphate applied to the soil surface where *Lumbricus rubellus* was present. A 32% increase in Bray-extractable soil P relative to that in soil without earthworms was noted at 70 days after addition of ground rock phosphate; the increased availability was attributed to the intimate mixing of ingested phosphate rock particles with soil in earthworm casts and to movement of particles from the surface down into earthworm burrows with infiltrating rain water. Ryden *et al.* (1973) recognized four major phases in which inorganic forms of phosphate may exist in soils; (a) water-soluble compounds in the soil solution, (b) physically sorbed compounds on the surfaces of soil particles, (c) strongly chemically bonded compounds derived from the physically sorbed forms, and (d) sparingly soluble compounds precipitated in forms such as calcium phosphate. Sharpley and Syers (1977) concluded that most of the additional P shown to be present in casts must be held in physically sorbed rather than chemically stabilized forms and would consequently be readily available to plants.

IV. pH and Exchangeable Cations

The pH of earthworm casts is generally closer to neutrality than that of the soils from which they are derived. Edwards and Lofty (1977) summarized data of Salisbury (1925) for 18 sites with soil pH ranging from 5.1 to 7.3, which showed that in more acid soils pH of casts was about 0.3-0.6 higher than the soil, while in soils close to pH 7 casts were little different in pH and in some cases had a lightly lower pH than the soil. Most soils that support high numbers of earthworms have pH < 7, while soils with pH considerably higher than 7 are mostly semi-arid or arid and have few earthworms. A few examples of pH changes in casts relative to soils are included in Table 24. The increase in pH of casts derived from acidic soils may result from excretion of ammonia into the intestine (p. 213) or to the production of calcium carbonate in calciferous glands and its release into the intestine (see next Section).

Casts have higher concentrations of exchangeable calcium, magnesium and potassium than the underlying soil. Some examples, for a variety of earthworms and habitats, are summarized in Table 24; the increases in available cations are related to the higher content of plant tissue in casts than in soil. Concentration of readily available nutrients (cations, phosphorus, nitrogen) in casts where surface casting species are numerous provides a favourable environment for seed germination and plant growth.

A. Calciferous Glands

Many species of earthworms have prominent paired lateral oesophageal pouches or glands arranged serially in consecutive oesophageal segments or extending through several consecutive segments. In some species calciferous glands are associated with the oesophageal pouches. The lining of calciferous glands is folded to form lamellae, which consist of two similar layers of secretory epithelium separated by a blood sinus; calcigenous cells of the epithelium secrete amorphous calcium carbonate which crystallizes in the lumen of the oesophageal pouch to form spheroliths of calcite, which are excreted into the oesophagus (Jamieson 1981).

Calciferous glands are not peculiar to any major taxonomic groups of earthworms and even within genera may be present in some species and absent in others. Their physiological significance has been debated for more than 100 years and is still not clearly understood, though it seems likely that they have a number of functions whose significance may vary with changing environmental stresses.

Table 24. pH and exchangeable cations of earthworm casts compared with surface soil layers.

Earthworm taxa	Location	Vegetation	Casts = A Soil = B (depth in cm)	pH	Ca	Mg	K	Reference
Lumbricidae	Poland	Pasture	A B(0–10)	6.3 5.4	3400 1880	420 240	460 90	Czerwinski et al. 1974
Lumbricidae		Pasture	A B(0–10)	6.4 6.0	2700 2200	330 250	590 380	Lunt and Jacobson 1944
	USA	Cropland	A B(0–15)	7.0 6.4	2790 1990	490 160	360 32	
Hippopera nigeriae	Nigeria	Secondary forest	A B(2.5–15)	6.9 6.4	1400 500	300 160	130 180	Nye 1955
		Scrub	A B(0–5)	— —	1990 1190	160 140	220 140	Cook et al. 1980
Hyperiodrilus sp.	Nigeria	Garden land	A B(0–5)	— —	1700 720	150 120	310 90	
Glossoscolecidae (3 spp.)	Uganda	Grass swamps	A B(0–30)	— —	1380 870	310 270	770 550	Wasawo and Visser 1959
Amynthas hupeiensis	Japan	Pasture	A B(0–10)	5.6 5.2	1760 520	240 170	— —	Watanabe 1975

Exchange cations (ppm)

Laverack (1963), Nakahara and Bevelander (1969) and Jamieson (1981) have reviewed information on their structure and function. Functions attributed by various authors to calciferous glands include:
(a) neutralization of ingested humic materials (Darwin 1881);
(b) absorption of nutritious materials from the gut (Michaelsen 1895);
(c) absorption of oxygen (Combault 1907, 1909);
(d) excretion of excess calcium absorbed from the gut (M'Dowall 1926; Robertson 1936, 1941);
(e) fixation and excretion of respiratory carbon dioxide (M'Dowall 1926; Robertson 1936, 1941);
(f) regulation of acidity in nitrogenous excretion (Needham 1957);
(g) regulation of blood pH (Guardabassi 1957);
(h) pH, ionic and osmotic regulation of body fluids (Prentø 1979);
(i) excretion of excess water and regulation of body water content (van Gansen-Semal 1959; Chapron 1971).

Darwin's hypothesis has had some support from subsequent workers, and may represent one function of the glands. Arthur (1965) pointed out that calcium carbonate does not dissociate at the pH (6.4–7.0) prevailing in the gut of lumbricids and so could not be involved in neutralizing organic acids that might be ingested or be formed in the intestine. This is not necessarily so. The presence of excess calcium carbonate in the gut does not negate the possibility that some may dissociate when gut pH falls and thus may have a role in buffering the gut contents against acidity.

The theories of Michaelsen (1895) and Combault (1907, 1909) are not supported by experimental evidence.

At least some of the calcite spheroliths formed in the calciferous glands pass right through the intestine and are excreted in casts, as was illustrated by Wiececk and Messenger (1972) who found white, sand-sized spheroids of calcite in the O_1 and underlying A_1 horizons of acidic grey-brown podzols in forested areas of northern Illinois, USA. The dominant earthworm in these soils was *Lumbricus terrestris* and Wiececk and Messenger showed that calcite spheroids extracted from 14 of 15 soil samples were of the same weight (1.46 mg), external morphology, internal mineralogy, crystallinity and fabric as were spheroliths extracted from the anterior portion of the alimentary canal (containing the calciferous glands) of *L. terrestris* from the same sites. Eight soil samples from the 0–25 mm horizon had a mean of 2.55 g/m^2 of these calcite spheroids and the soil pH was 8, while another six samples had a mean of 0.83 g/m^2 at 0–25 mm depth and pH of 6.2–7.8. Ponomareva (1948) similarly found spheroidal calcite concretions, about 1 mm in diameter, in potted soils initially free of calcite, in the presence of earthworms. Arthur (1965) concluded that calcium ions are taken up into the blood from plant tissue as it passes through the gut and are utilized to absorb some of the respiratory carbon dioxide, which is voided as calcite via the calciferous glands into the intestine and excreted in casts. Kale and Krishnamoorthy (1980) similarly concluded that, in *Pontoscolex corethrurus*, calcium is concentrated in and excreted in concretions formed in the calciferous glands. Casts contained 1.3 times more calcium than surrounding soil, but had 11.8 times more soluble (ionic) calcium than did the soil.

Among British Lumbricidae Piearce (1972a) distinguished litter-feeding species (*Lumbricus* spp., *Dendrodrilus rubidus subrubicundus*, *Allolobophora eiseni*) that have active calciferous glands containing concentrations of calcite granules, from humus-feeding or soil-feeding species (*Allolobophora* spp., *Aporrectodea* spp., *Octolasion* spp., *Dendrobaena veneta*) that have apparently inactive calciferous glands. He concluded that there was some evidence that the glands fix metabolic carbon dioxide in some species but not in others and some evidence that the glands (in *L. rubellus*) are capable of excreting excess dietary calcium. He also showed that the litter-feeding *L. rubellus* absorbs more calcium from its food than does the soil-feeding *Aporrectodea caliginosa*, so that *L. rubellus* may be obliged to have a specialized physiological mechanism (in the calciferous glands) to dispose of excess calcium that is not necessary in *A. caliginosa*, simply because it has to dispose of less calcium; or, he concluded, the calciferous glands might have some other function in acid–base balance.

The significance of calciferous glands in absorbing carbon dioxide from the blood is generally agreed and this function may be particularly important when external carbon dioxide levels are high or oxygen levels are low (Prentø 1979). A more general function in pH, ionic and osmotic regulation of various body fluids also seems likely. Prentø (1979) proposed that calcium is taken up through the intestinal wall by chloragogen cells, passed to the blood, and thence to the calciferous glands, where bicarbonate ions are formed and the calcium ions are taken up and eventually excreted as calcium carbonate. Prentø proposed that this mechanism would facilitate rapid increases or decreases in concentration of ions in extracellular fluids and so might be important in regulating osmotic pressures during dehydration and rehydration. Excretion of excess water by the epithelium of the calciferous glands against the osmotic gradient may proceed simultaneously with calcium excretion or independently of calcium excretion, according to evidence of Chapron (1971). Jamieson (1981) concluded that if sufficient calcium is available the calciferous glands have the primary role of secreting calcium and secondarily excrete water, while, if calcium is in low concentration, they lose their regulatory role and continually excrete water. Jamieson proposed that this may explain the paucity of earthworms in soils subject to desiccation and low in calcium, such as podzols, but this is probably only one of a number of factors involved, since podzols are generally acidic, relatively low in other nutrients and have high C:N ratios.

It is well established that earthworm populations are at a maximum in soils of pH 5.5–7.5 and that the pH of soil in the gut of lumbricid earthworms is c. 6.4–7.0. The large amounts of soil that pass through earthworms where populations are high (Table 18) may thus be a prime factor in determining soil pH and consequently in determining the ionic balance of the soil solution and the availability of plant nutrients.

The wider significance of calciferous glands as sites for excretion of potentially toxic heavy metal ions is discussed in the next section and their general significance in ionic regulatory mechanisms is discussed on pp. 60–61.

V. Heavy Metals

In recent years there has been increasing awareness of the danger to human health, to the survival of wildlife and the general contamination of soils and to plants that result from environmental pollution by heavy metals. The term "heavy metals", as widely used in this context, is imprecise. It includes principally the more common members of the d-block transition series and the b sub-group metals of the periodic table of elements; these elements have a biological function or are toxic to some organisms (Hughes et al., 1980). The most important as environmental pollutants are lead (Pb), cadmium (Cd), mercury (Hg), zinc (Zn), copper (Cu), nickel (Ni), antimony (Sb) and bismuth (Bi) but many other elements are involved. Atmospheric contamination from industrial plants and from release of Pb in exhaust fumes of motor vehicles using leaded fuel, land-fill disposal of industrial and municipal wastes and the spreading on agricultural land of sewage sludge and slurries of animal dung as fertilizers or for disposal are among the more important sources of heavy metal pollution of the environment. A comprehensive review of atmospheric heavy metal pollution of terrestrial ecosystems may be found in Hughes et al. (1980).

Much of the atmospheric component of heavy metal pollution is deposited onto the soil surface or on plants, and thence on soil, while the contribution from land-fill waste disposal, sewage sludge and animal dung is added directly to the soil and to groundwater. Because earthworms ingest large quantities of soil and decaying plant material they are unusually susceptible to accumulation of pollutants. They are preyed on extensively by birds, some mammals and other vertebrates, so heavy metals accumulated in their tissues or contained in their intestines are readily passed on in food chains. Not all heavy metals are absorbed by earthworms from their food but those that are absorbed may accumulate, both in earthworms and their predators, if they should lack adequate biochemical or physiological mechanisms to eliminate them rapidly from their bodies.

A. Absorption and Excretion of Heavy Metals

Most attention in heavy metal contamination of the environment has been given to Pb and Cd. The mechanism of absorption of Pb by earthworms is not clearly understood; some may enter through the intestine from ingested food but McIntyre and Mills (1975) concluded that Pb also enters through the body wall when earthworms are in contact with Pb-contaminated soil colloids.

It should not be presumed that heavy metals associated with the body wall are necessarily absorbed into the tissues, as some may be absorbed by mucus on the body surface and not incorporated into the epithelium. Fleming and Richards (1981a, b) devised a technique for eluting heavy metals from the mucus layer and showed that in *Lumbricus rubellus* the mucus may act as a barrier to penetration of heavy metals. In further experiments they directly examined the distribution of Pb and Fe at the body surface of *E. fetida*, by ultramicroscopy of thin sections through the body wall, and were able to show concentrations of these heavy metals in the surface mucus with little penetration into the cuticle or epidermal cells (Fleming and Richards 1982).

Wielgus-Serafinska and Kawka (1976) investigated the distribution and levels of Pb accumulation in the organs and tissues of *Eisenia fetida* after seven days feeding in the laboratory on soils to which PbO_2 had been added at (mg/g) 0.0, 0.1, 0.2, 0.4 and 1.0. Lead was found to have accumulated in various tissues, but its location and concentration varied between the different levels of Pb added to the soil (Table 25). Total Pb concentration in the tissues showed a general upward trend with increasing soil concentrations with mean values ranging from about 4.7 μg/g fresh weight at a soil concentration of 0.1 mg/g to about 12.5 μg/g fresh weight at a soil concentration of 1 mg/g, but there was wide variation between samples and no significant difference could be demonstrated between the means of total Pb content in the earthworms for the four treatments of added Pb. Earthworms from all soils with added Pb, however, had significantly higher Pb content than controls from soil without added Pb. Wielgus-Serafinska and Kawka (1976) concluded that increasing environmental concentrations of Pb stimulate excretory mechanisms in *E. fetida* that lead to its rapid

Table 25. Location of lead in tissues of *Eisenia fetida* after seven days' feeding in soil with various concentrations of added lead. (Data from Wielgus-Serafinska and Kawka 1976).

Location in tissues	Concentration of PbO_2 added to soil (mg/g)				
	0.0	0.1	0.2	0.4	1.0
Epithelial gland cells of body wall	–	+++	++	+	+
Epithelial cells of peritoneum	–	++	+	+	+
Gland cells of intestinal mucosa	–	–	+	++	–
Giant nerve cells	–	–	–	+++	–

+++, abundant; ++, moderate; +, little accumulation; –, absent.

removal from the tissues, so that increasing environmental Pb concentrations are not paralleled by increasing concentrations in earthworms. Concentration of Pb in epithelial gland cells would facilitate its excretion with mucus and this may provide an important pathway for its excretion.

The pathways of Pb accumulation and excretion vary between species. Andersen and Laursen (1982) concluded that Pb is handled in at least three ways in *L. terrestris*:
(a) immobilization in the chloragogen cells of the gut wall;
(b) excretion through the calciferous glands; and
(c) storage and subsequent elimination through waste nodules (or brown bodies).

Brown bodies are small dark-coloured ovoid or flattened masses found in the coelomic cavity in the hindmost segments of earthworms (Stephenson 1930). They contain large cysts of *Monocystis*, setae, encysted nematodes, disintegrated amoebocytes and other waste materials; they may be excreted through the dorsal pores, but are usually too large to be voided in this way and are probably retained until the death of the individual. In *Aporrectodea* spp. and *Allolobophora* spp. the calciferous glands are not well developed in contrast to those of *L. terrestris*. In *Aporrectodea longa* fed on sewage sludge high in Pb, the total Pb content was 6 µg/g compared with 24 µg/g for *L. terrestris* in the same environment, but the Pb content of waste nodules of *A. longa* was 89 µg/g compared with 57 µg/g in *L. terrestris*. Andersen and Laursen (1982) concluded that Pb is probably more efficiently handled by *A. longa* than it is by *L. terrestris*, probably due to the superior ability of *A. longa* to utilize the waste nodule immobilization/elimination pathway.

Tara *et al.* (1979) extended the investigations of Wielgus-Serafinska and Kawka (1976) to include Pb and Zn accumulation in *Lumbricus terrestris*. Zinc was found to accumulate in the peritoneal epithelium, in nerve cells of the ventral nerve cord, and in the chloragogen cells that form the outer layer of the intestine, while Pb accumulated in epithelial glandular cells of the body wall and in the nerve cord and chloragogen cells.

Andersen and Laursen (1982) found that Cd is particularly concentrated in chloragogen cells in *L. terrestris*, where it is bound in the form of Cd-metallothioneins with small amounts deposited in waste nodules. In *A. longa* they found that the concentration of Cd was about equal in chloragogen cells and waste nodules. Zinc, Mn and Fe were shown to be excreted through the calciferous glands in *L. terrestris*, with Fe also accumulating in the gut wall, especially in juveniles. Excretion of Mn and Fe through the calciferous glands was also demonstrated by Bouché (1983) in *Aporrectodea velox*.

Ireland and Richards (1977) and Ireland (1977) found that Pb was concentrated particularly in the chloragogen cells of *Dendrodrilus rubidus* and *Lumbricus rubellus* collected from a highly contaminated mining area. Electron-dense flecks in the chloragogen cells were shown to contain Pb (Ireland and Richards 1977), but formation of these flecks was attributed to deposition as phosphate of unbound Pb during preparation of the tissue for electron microscopy. Although much Pb may be present in an unbound form, there is evidence that Pb is sequestered in intracellular chloragosomes and debris vesicles. Ireland (1977) considered that Pb was probably firmly bound within the chloragogen cells and was then excreted via the nephridia when these cells detached from the intestinal wall and degenerated in the coelomic fluid.

Laverack (1963) concluded that the chloragogen cells of earthworms are to some extent analogous with the liver of other animals, since they contain large quantities of glycogen and fat, have a high concentration of acid ATPase (adenosine triphosphatase) and appear to be involved in energy storage; on the other hand, he pointed out that they have high concentrations of ammonia and urea, which indicates that chloragogen cells are important in detoxication mechanisms. They are also apparently involved in calcium metabolism (p. 224). The proposition that significant numbers of chloragogen cells detach from the intestinal wall and disintegrate in the coelom, whence their contents are excreted, is from Cuénot (1898) and Liebmann (1946), but was challenged by Abdel-Fattah (1954) who could find no evidence for this process. It seems most likely that the chloragogen cells perform several functions including:

(a) the synthesis and storage of energy-rich materials such as glycogen, which can be made available to other tissues as required,
(b) the accumulation of ammonia and urea, which are passed on by diffusion to the coelomic fluid or the blood to be excreted through the nephridial system, and
(c) the accumulation of unwanted inorganic materials, including Ca and

heavy metals such as Pb and Cd absorbed by the gut and their immobilization in the small spheroidal chloragosomes and debris vesicles that the cells contain.

It seems likely also that when the chloragogen cells die they detach and disintegrate in the coelomic fluid and their contents are absorbed into the blood, stored in waste nodules, scavenged by amoebocytes, or excreted directly through the nephridial system or through the dorsal pores.

Earthworms apparently possess a number of mechanisms for immobilization and excretion of heavy metals. Their relative importance varies between different metals and varies between species for the same metal; their overall efficiency probably varies similarly.

The majority of heavy metal content in ingested food is not absorbed but passes through the gut and is excreted in casts. It is very likely that much of the heavy metal content of pollutants (especially Pb) is present in forms that are insoluble and unavailable for absorption by animals or plants; nevertheless it is important to know how much is absorbed and then passed on in food chains.

B. Toxicity of Heavy Metals to Earthworms

Ireland and Richards (1977) showed that glycogen levels of chloragocytes and intestinal wall cells in *Lumbricus rubellus* collected from a site with heavy Pb contamination were $< 20\%$ of those from *L. rubellus* from a control site with low lead content. The glycogen level of tissues varied inversely with their Pb content; there may be direct inhibition of carbohydrate metabolism due to Pb, or the presence of high concentrations of Pb in the cells may divert the cells' metabolism towards the processing of the Pb and away from glycogen synthesis, or differences between the nutritional status of worms from the Pb-rich habitat, which was a poor soil, and those from the more organic matter rich soil of the control might account for the differences in glycogen levels in their cells.

Copper toxicity, owing to the former long-term use of Cu-containing fungicides in orchards, was shown by van Rhee (1963, 1967) to completely eradicate earthworms where Cu levels in soil were > 80 μg/g. High Cu levels, below the lethal limit, were shown to reduce the reproductive rate of earthworms (van Rhee 1969c, 1975). Large amounts of pig wastes, containing > 1000 μg/g Cu applied to pasture soils in Holland, caused a decline in earthworm populations (van Rhee 1977b).

Experiments conducted by van Rhee (1975), in which earthworms (*Aporrectodea caliginosa*) were kept for eight and a half weeks in soil to which was added various heavy metal contaminants, showed that in relation to controls without added heavy metals:

(a) Cu at 110 µg/g did not affect body weight or mortality but may have caused a slight reduction in rate of cocoon production;
(b) Zn at 1100 µg/g resulted in a reduction to about 50% of body weight, cessation of cocoon production, inhibition of development to maturity and 22% mortality;
(c) a mixture of Hg at 10 µg/g and Co at 20 µg/g caused a slight check in increase of body weight and development to maturity and a reduction of about 65% in rate of cocoon production;
(d) a mixture of all four elements at the same rates resulted in a reduction to about 34% of body weight, cessation of development to maturity and cocoon production and 22% mortality, which was the same as that due to addition of Zn alone.

These levels of heavy metal contamination were based on levels that result from the discharge of mud containing industrial wastes from the harbour bottom at Rotterdam onto reclamation areas.

Effects of Cd, Cu, Pb, Ni and Zn on growth and cocoon production of *E. fetida* were tested by Malecki *et al.* (1982) in laboratory cultures. Survival and body weights of *E. fetida* were compared with untreated controls after four, six and eight weeks in cultures to which were added a series of concentrations of acetates, carbonates, chlorides, nitrates, oxides and sulphates of the five metals. Carbonates and oxides were less toxic than the other compounds, probably because of their relatively low solubility. Cadmium was most toxic, with significant decreases in growth rate at 100 µg/g Cd for acetate and chloride and at 50 µg/g for sulphate; minimum concentrations of the other metals that significantly retarded growth were 200 µg/g for Ni (as chloride), 100 µg/g for Cu (as nitrate), 2000 µg/g for Zn (as chloride or nitrate) and 12 000 µg/g for Pb (as acetate). Cocoon production was totally inhibited by acetates of the five metals at concentrations of 50 µg/g for Cd, 400 µg/g for Ni, 2000 µg/g for Cu and at ≥ 5000 µg/g for Zn and Pb. It is apparent that the chemical form in which toxic heavy metals are presented is an important factor in the level at which toxicity appears. Differences between the compounds tested may explain the conflicting data in the literature on concentrations that have deleterious effects on earthworms. The very high concentrations of Pb necessary to influence growth and reproduction may be due more to the generally low solubility of the Pb compounds that are found in soils and the ability of earthworms to sequester absorbed Pb than to any less toxicity of Pb compared with other heavy metals.

C. Accumulation and Concentration of Heavy Metals

Some aspects of heavy metal uptake by earthworms and of the distribution of heavy metals in earthworm tissues have been reviewed recently by Ireland (1983).

Most studies of heavy metal accumulation by earthworms have concerned Pb, Cd and Zn but there is some information on a wide range of other elements. Analyses have sometimes been of total element contents and sometimes of water-extracts or acid-extracts as an indication of elements available to plants and as some measure of the significance of earthworms in transfer of heavy metals from contaminated soil into plants and through food chains.

Ireland (1975a) investigated Pb, Zn and Ca contents of *Dendrodrilus rubidus* at two sites in Wales, one very heavily contaminated adjacent to a mine, with low Ca (1713 μg/g Pb, 1975 μg/g Zn, 332 μg/g Ca in soil) and the other less contaminated but with higher Ca (127 μg/g Pb, 172 μg/g Zn, 13 030 μg/g Ca in soil). Acetic acid-extractable Pb in the soil was about 7% of total soil Pb at the heavily contaminated site and about 2% at the less contaminated site; comparisons of Pb extracted by acetic acid from the tissues of *D. rubidus* and from the soil at the heavily contaminated site showed a concentration factor in the earthworm of 32.8 times (4160 μg/g compared with 127 μg/g in the soil), while at the less contaminated site the concentration factor for *D. rubidus* tissues was 38.5 times (100 μg/g compared with 2.6 μg/g in the soil). When contaminated earthworms from the high Pb/low Ca soil were transferred to the low Pb/high Ca soil the Pb and Ca content of their tissues fell over a five day period; when transfers were made in the opposite direction Pb and Ca in the tissues increased over a 20-day period. The ratio of concentrations of Pb:Ca always remained close to 1.0:0.9 and there was a close relationship in uptake rates of Pb and Ca. In the same experiments, at the heavily contaminated site the acetic acid-soluble Zn content of *D. rubidus* tissue was 584 μg/g and that of soil 10 μg/g, a concentration factor of 58.4 times, while at the less contaminated site the figures were respectively 114 μg/g and 18 μg/g, a concentration factor of 6.3 times.

Ireland (1975b) further extracted Pb, Zn and Ca from contaminated soil, *D. rubidus* casts and putrefied *D. rubidus* to determine their possible influence on uptake of these elements by plants. Water extracts of casts contained less Zn and Ca and slightly more Pb than soil, while water extracts of putrefied earthworms contained 53 times more Pb, 126 times more Zn and 88 times more Ca than soil. A further extraction with acetic acid yielded 155 μg/g Pb in soil, 114 μg/g in casts, 1870 μg/g in putrefied tissue, with corresponding figures (μg/g) of 5.9, 14.7 and 881 for Zn, and

197, 520 and 17 610 for Ca. Ireland concluded that *D. rubidus*, living in heavy metal contaminated soil, could increase the amounts of plant-available Zn and Ca, but not Pb by excretion in casts, while the dead tissues of the earthworm could contribute substantially to the availability to plants of all three elements (concentration factors of 12 times for Pb, 149 times for Zn and 89 times for Ca).

The significance of concentration of plant-available heavy metals in earthworm tissues is difficult to assess. Earthworms feed selectively on organic debris as well as ingesting soil and though plant roots would absorb nutrients from dead earthworm bodies and might thus be expected to absorb the available heavy metals, the total amounts involved must be very small compared with the total heavy metal contents of soils, or even with the total of available heavy metals accessible in the soil. Since there is no evidence that methods developed for estimating levels of plant-available nutrients from soils apply equally well to estimation of concentrations in earthworm tissue it is probably best to consider total heavy metal concentrations and quantities in soils, plants and earthworms not so much as an indication of toxicity to earthworms themselves, but as an indication of possible risks of their transfer through food chains and impact at higher trophic levels.

On the basis of total heavy metal content of earthworms relative to soil, the concentration factor for Pb is generally in the range 0.1–1.0, but was 2.43 for *Dendrodrilus rubidus* in very heavily contaminated acidic soil close to a mine spoil heap in Wales (Ireland 1975a). For Cd, concentration factors are generally in the range of about 11–22 in relatively unpolluted soils with soil Cd levels about 0.2–0.8 $\mu g/g$ and earthworm levels about 3–9 $\mu g/g$, but as soil Cd levels rise the concentration factor tends to fall. Martin and Coughtrey (1976) recorded concentration factors of 4.96 and 3.78 where soil Cd levels were 38 $\mu g/g$ and 29 $\mu g/g$ respectively and earthworm levels in both cases were 143.7 $\mu g/g$. For Zn, concentrations tend to increase with increasing pollution, but not proportionally, so that the concentration factor in earthworms tends to fall as soil concentrations rise; Gish and Christensen (1973) recorded concentration factors of 1.83 where the soil level was 13.4 $\mu g/g$ and the earthworm level was 24.6 $\mu g/g$, and 1.37 where the soil level was 25.1 $\mu g/g$ the and earthworm level was 34.5 $\mu g/g$. Similar results for Cd but not for Zn were recorded in Canada by Carter *et al.* (1983) who compared the heavy metal content of *L. rubellus* after 10 days' culture in uncontaminated soil with that of *L. rubellus* kept in the same soil to which sewage sludge was added at rates (grams of sludge per kilograms soil) of 5, 10, 25, 50 and 100. The mean heavy metal content of the sludge ($\mu g/g$) was Cd 99, Cu 320, Ni 67, Pb 573 and Zn 792. The Cd content of earthworm tissue increased steadily with increasing concentration of added sludge, up to 50 grams of sludge per kilogram of soil, when the Cd content of the tissue was c. 34 µg/g, or c. six times that of the sludge-soil mixture, but was not significantly higher in cultures with 100 grams of sludge per kilogram of soil. Concentrations of Zn and Cu in tissues did not increase, indicating that there was active regulation of Zn and Cu, but not of Cd. More detailed treatment of concentration ratios may be found in the review of Hughes *et al.* (1980).

Ireland (1979) compared concentrations of Pb, Cu, Cd, Zn, Mn and Ca in tissues of *Lumbricus rubellus*, *Dendrobaena veneta* and *Eiseniella tetraedra* from three soils with varying concentrations of these elements. *Lumbricus rubellus* had the highest concentrations of Zn and Mn, *D. veneta* of Cd, and *E. tetraedra* of Pb. Concentrations of Cu, Zn, and Mn in tissue appeared to be more effectively regulated than those of other metals; Cd was concentrated more by earthworms than any other of the elements tested, especially in *D. veneta* where, at one site, it was about 1.8 times the concentration in soil. Ireland and Richards (1977) recorded Pb and Zn levels in tissues of *Dendrodrilus rubidus* and *L. rubellus* from a highly contaminated site; for Pb *D. rubidus* had about twice the level of *L. rubellus*, while for Zn the situation was reversed. They considered the possibility that the differences were due to differences in food selection, but could find no evidence for this.

Rates of uptake apparently differ with differences in soils and also differ seasonally at the same site. In Denmark Andersen (1979) compared total heavy metal concentrations in earthworms from plots treated with sewage sludge, at rates equivalent to 30 t/ha dry weight, or with inorganic fertilizers, with untreated control plots. Concentration of Pb in earthworm tissue never exceeded that of the soil or soil–sewage sludge mixtures and the mean concentration factor in earthworm tissue relative to soil was 0.24. Andersen compared the Pb content with the much higher concentrations recorded in *Dendrodrilus rubidus* by Ireland (1975a). The soils at Ireland's sites were strongly acid (pH 4.2-4.5), while at Andersen's sites they were slightly acid (pH 5.7-6.0) and Andersen considered that the higher Pb content of Ireland's specimens may have resulted from differences in the Ca requirements of earthworms at the two sites.

A number of investigations have indicated a close link between Pb and Ca uptake in earthworms; Pb is apparently absorbed with the Ca and may substitute for Ca in the formation of granules in the calciferous glands (Ireland 1975a). *Dendrodrilus rubidus* has active calciferous glands and would be expected (for homeostatic requirements of its metabolism) to have a high demand for and turnover of Ca in soils of low pH. The species Andersen (1979) examined were *Aporrectodea caliginosa*, *A. rosea*, *Allolobophora chlorotica* and *Lumbricus terrestris*. The *Aporrectodea* and *Allolobophora* spp. have reduced calciferous glands and have a low demand for Ca compared with *D. rubidus*. *Lumbricus terrestris*, like *D. rubidus*, has large and active calciferous glands, but its demand for and turnover of Ca would be expected to decrease with increasing soil pH and to be much lower at Andersen's site than would that of a similar earthworm at Ireland's site. Anderson also found seasonal differences in Pb content and in Pb:Ca ratio of *L. terrestris*; in early summer, when the worms were very active, Pb content was high and paralleled Ca content but this relationship was not found in specimens collected in winter, when the worms were inactive. Ireland and Wooton (1976) measured Pb, Zn and Ca concentrations in *D. rubidus* at intervals through the year, at three sites with varying levels of soil contamination with the three elements. They found that Pb levels in tissue reflected soil levels, with highest concentrations in winter; there was a good correlation between Pb and Ca levels in tissue at sites with high soil Pb content and Ca level in tissue was more closely related to Pb level in tissue than it was to Ca level in the soil; Zn levels in tissue were not related to soil Zn levels where soil levels were high.

For three sandy soils and three loamy soils that had been treated annually for ten years with municipal waste compost contaminated with heavy metals, Ma (1982) examined body content of Cd, Zn, Pb, Cu, Ni, Mn, Fe and Cr in resident populations of *Aporrectodea caliginosa*. The experiments were conducted on small field plots sown with rye grass; some plots received compost during the experiments while others did not. Cadmium and Zn content of *A. caliginosa* were higher in sandy than in loamy soils, though the content of Cd and Zn was higher in the loamy soils; Pb was also accumulated more rapidly in sandy than in loamy soils. Rates of heavy metal accumulation generally increased with addition of compost, but in some cases there was a decrease. Concentration factors for Cd in adult worms were c. 10–30 times in loamy soils and c. 30–140 times in sandy soils, for Zn c. two to eight times and c. six to 75 times respectively, and for Pb c. 0.2–1.2 times and c. 0.3–2.6 times respectively; concentration factors for Ni, Fe, Mn and Cr were < 1, while for Cu they were sometimes > 1. The order of concentration factors in *A. caliginosa* for all soils tested was Cd > Zn > Cu ⩾ Pb > Ni > Fe ⩾ Mn > Cr. Significant negative correlations were found between concentration factors and pH for Cd, Zn and Pb but not for the other heavy metals tested; significant negative correlations also existed between concentration factors and the cation exchange capacity of the soil for all elements tested except Cr, and between concentration factors and soil organic matter content for Cu only.

It cannot be assumed that the data of Ma (1982) would necessarily be true for other species of earthworms, nor that they would apply even for the same species in all soils. The relationships illustrated with the cation exchange capacity of the soils used in the experiments indicate that the solubilities of the heavy metal ions introduced into the soil solution depend upon the preexisting physico-chemical state of the soil.

The use of leaded fuels in motor vehicles raises the possibility of heavy contamination close to roads and has led to the investigation of heavy metal levels in earthworms from such habitats. Zietek and Pytasz (1979) measured Pb and Zn concentrations in earthworms from 12 sites in Poland, each 5-10 m from the edges of main roads. For both metals they found a high correlation between concentrations in earthworm tissues and soils but correlations between Pb and Zn concentrations and traffic density were usually low and only occasionally high. Ash and Lee (1980), on the other hand, measured Pb, Cd, Cu and Fe concentrations in *Allolobophora chlorotica*, *Lumbricus terrestris*, and *L. rubellus* from roadsides in Yorkshire and from an isolated moorland site in northern Scotland and found that levels of all these metals were higher on roadsides than in the Scottish site, and concluded that pollution from vehicles was responsible for this. In contrast to the results of Andersen (1979), discussed previously, Ash and Lee (1980) found that *A. chlorotica*, which has inactive calciferous glands, had higher Pb levels than *L. terrestris*, which has large and active calciferous glands.

The manner in which vehicles are driven is known to greatly affect the size of Pb-bearing particles emitted from the exhaust, and might account for the discrepancy between the results of Zietek and Pytasz (1979) and those of Ash and Lee (1980). Hughes *et al.* (1980) point out that vehicles moving at steady cruising speeds produce Pb-bearing particles of <0.5 μm diameter, while in stop-start uban driving conditions the particles are > 5 μm in diameter. These differences must greatly affect the rate and distribution of deposition of the particles from the air; they are known to affect the chance of capture of the particles on leaf surfaces and this effect varies between species of plants, depending on the morphology of their leaf surfaces. It is also possible that the nature of the particles affects heavy metal uptake by earthworms. Investigations of Muskett and Jones (1980) indicate that, though concentrations of Pb and Cd in air and in soil are high close to roads and fall off with increasing distance from roads, there is no evidence of any detrimental effect on a wide range of macro-invertebrates.

Andersen (1979) found that total Cd was concentrated relative to soil content by factors of about 38 times in *Aporrectodea rosea*, 18 times in *A. longa*, 17 times in *A. caliginosa* and *Allolobophora chlorotica*, and 22 times in *Lumbricus terrestris*; these are mean concentration factors for four treatments, two where the soil was top-dressed with inorganic fertilizers and two where slurries of sewage wastes from different sources were added (p. 323-327). The Cd contents of the soils with added sewage sludge were higher (0.65 μg/g and 0.99 μg/g) than in those with added inorganic fertilizers (0.29 µg/g and 0.14 µg/g) but the Cd content of earthworms was consistently higher, by a factor of up to about 2.5 times, in specimens from soils with the inorganic fertilizers. Andersen (1979) concluded that some Cd in the soil that would otherwise have been available to the earthworms must be rendered unavailable by the sewage sludge. His data might alternatively be explained if Cd in sewage sludge is in forms less soluble than those included in the inorganic fertilizers used in the experiments.

Copper in high concentrations is rather toxic to earthworms but is found in their tissues at low concentrations with no apparent ill effect. Ash and Lee (1980) compared Cu concentrations in tissues of *Lumbricus terrestris* and *Allolobophora chlorotica* from urban areas and farmlands in the midlands of Britain with those in the same species collected from northern Scotland. Concentration of Cu was much lower in specimens from Scotland (0.83 µg/g for *L. terrestris*, 0.20 µg/g for *A. chlorotica*) than in those from any of the other sites. Concentrations differed between the two species at individual sites, but not consistently, sometimes one species having the higher concentration, and sometimes the other. In *L. terrestris* the highest Cu concentration was 7.5 µg/g in specimens collected beside a main road in western Yorkshire, while in *A. chlorotica* it was 8.9 µg/g in specimens collected from a roadside in Leeds. These concentrations of Cu are one to two orders of magnitude lower than those reported by van Rhee (1975) as having little if any effect on earthworms (pp. 325–327).

Bull *et al.* (1977) reported on Hg concentrations in *Lumbricus terrestris* collected within a radius of 500 m of a chlor-alkali works, compared with specimens collected 10–30 km from the works. Concentration in the earthworms was 1.29 µg/g (wet weight) close to the factory compared with 0.041 µg/g at 10–30 km, but these concentrations were only 34% and 39% respectively of those in soil from the same sites, so there was no evidence of Hg concentration in earthworms. Perhaps most interesting was that 8%–13% of the total Hg in *L. terrestris* was in the form of methyl-Hg compounds; there was no known history of use of methyl-Hg biocides in the area. Methylation of Hg is well known in aquatic ecosystems but has not previously been recorded in terrestrial ecosystems.

Gissel-Nielsen and Gissel-Nielsen (1975) found that earthworms in soil with high selenium (Se) concentrate this element up to 100 times the level in soil. In some experiments Se concentration in earthworms exceeded 40 µg/g, but with no apparent deleterious effects on the earthworms, nor on birds and small mammals that preyed on them.

D. Transfer along Food Chains

There is little quantitative data to relate heavy metal content of earthworms to uptake, retention or damage to animals that feed on them.

Ireland (1977) fed aquatic toads (*Xenopus laevis*) on Pb-contaminated earthworms (*Dendrodrilus rubidus*) and showed that much Pb accumulated in the bones, kidneys and liver of the toads. White-footed mice (*Peromyscus* sp.), similarly fed, accumulated Pb in bones but little was found in soft tissues; he concluded that toads may be more important than mice in passing Pb from earthworms to predators further along food chains.

The presence of high levels of heavy metals in earthworm tissue is not in itself adequate proof that they will be absorbed by predators. For example, Gissel-Nielsen and Gissel-Nielsen (1975) measured the Se content of blackbirds and moles, well-known earthworm predators, and found them to have Se concentrations only about half that of earthworms (*Lumbricus terrestris*) on which they were feeding. Apparently Se from the earthworms was not retained or concentrated by their predators.

E. Use of Earthworms to Monitor Heavy Metal Pollution of Soil–Plant Systems

It has been suggested that analysis of earthworms might be useful in monitoring the biological availability of heavy metals in soils. Helmke *et al.* (1979) measured the concentrations of 29 elements, including some of the heavy metals, in the tissues and in casts of *Aporrectodea caliginosa* and in soil collected from field plots in the USA that had been treated with anaerobically digested sewage sludge at rates of 15, 30 and 60 tonnes per hectare dry weight and compared them with analyses of controls from untreated plots. They concluded that analyses of the earthworm tissues could provide a useful measure of bio-availability of some elements in contaminated ecosystems, particularly Cd, Co, Hg and Zn.

It is apparent from the discussion above that there are many inconsistencies in the available data on relative concentrations of heavy metals in earthworms compared with those in pollutants, in soils and in plants. Interspecific differences in rates of uptake of individual heavy metals are apparent but the differences are themselves inconsistent between reported investigations. There are differences in rates of uptake attributed to soil pH, to interactions in the environment and in the earthworms' internal chemistry between various combinations and relative concentrations of heavy metals that occur together, to particle size of aerial pollutants, to the chemical form in which the heavy metals occur, to adsorption of heavy metals onto surfaces of clay or organic matter particles, and to selection as food by the earthworms of particular fractions of the materials available to them in the soil, which may contain relatively high or low heavy metal concentrations.

Analytical methods designed for soil and plant materials may not necessarily reflect availability of elements for absorption by earthworms, nor their availability for absorption by animals that prey on earthworms. This is not to deny that earthworms absorb and accumulate heavy metals and that accumulation may create environmental hazards, especially for animals that participate in food chains that include large quantities of earthworms. However, predators have their own detoxification and excretion mechanisms for heavy metals. It is unreasonable to expect that simple and consistent relationships will exist between levels of uptake and accumulation in all species of earthworms and concentrations in their predators or in the environment. Earthworms, as is evident from much of the data presented in this book, display a wide range of interspecific and intraspecific physiological and biochemical capabilities as well as adaptability of behaviour when exposed to environmental stress. They can not be regarded as a consistent substrate for uptake of heavy metals (nor any other kind of environmental pollutant) that may be used as a standard laboratory material for chemical analysis. Analysis of earthworm tissue might prove to be a useful and rapid method for monitoring pollution at a particular site, but a new set of standards would probably be necessary to interpret data from another site.

There is need for a better understanding of the sequence from input of pollutants to the environment, their uptake and accumulation by earthworms and their subsequent effects on food chains that include earthworms.

VI. Radioactive Elements

A. Uptake and Dispersal of Radioactive Elements

There is little evidence that earthworms absorb significant quantities of radioactive contaminants from soil. Pokarzhevsky and Krivolutsky (1975) studied the distribution of labelled strontium (^{90}Sr) between compartments of a Russian birch forest ecosystem which was deliberately contaminated with the radioisotope. Of a total of 41.3 Ci/ha (1528 GBq/ha) of ^{90}Sr they found 77.5% (32 Ci/ha) (1184 GBq/ha) in the 0-5 cm soil layer, 12.1% in the vegetation, 10.4% in the ground litter and leaf fall, and only 0.0006% (0.235 mCi/ha) (8.7 MBq/ha) shared between a wide variety of invertebrates and vertebrates. Earthworms, with a biomass of 7.5 kg/ha (dry weight) contributed only 0.37 µCi/ha (13.7 kBq/ha) of the radioactivity. The contribution of earthworms to total radioactive accumulation by invertebrates was proportionally very small; earthworm biomass was about 9.7% of the biomass of all invertebrates (77.5 kg/ha) but they contributed only 0.19% to the total accumulated radioactivity of the invertebrates (299 µCi/ha) (7.4 MBq/ha). Maldague and Couture (1971) studied the uptake of labelled iron (^{59}Fe), while Crossley et al. (1971) studied that of labelled caesium (^{137}Cs) by earthworms in experimental cultures. Although about 12% of the radioactive materials ingested was apparently retained within the body, the earthworms rapidly lost the isotopes when transferred to uncontaminated cultures, indicating that the isotopes were present in soil within the gut and were not assimilated into the tissues. Three species, *Lumbricus terrestris*, *Octolasion lacteum*, and *Aporrectodea hortensis*, gave similar results with ^{137}Cs in the experiments of Crossley et al.; the rate of loss after removal of the earthworms to uncontaminated soil was linear and was proposed as a method for measuring the rate of movement of food through the gut.

Peredel'skii et al. (1957) found that the presence of earthworms greatly increased uptake of labelled cobalt (^{60}Co) by grasses, legumes, mustard, hemp and buckwheat grown in contaminated soils. In further experiments of Peredel'skii et al. (1960a, b) maize and beans were grown in pots, with or without earthworms, and with the addition of labelled calcium (^{45}Ca) and ^{90}Sr. Uptake of the isotopes by the plants was greater if the isotopes were mixed uniformly with the soil than if they were added only to the lower layer of the soil. Presence of earthworms increased uptake of ^{90}Sr by roots of maize and beans but decreased uptake of ^{45}Ca by roots of maize and had no effect on its uptake by roots of beans. These effects relate to the distribution of roots of the plants in the pots and to the redistribution of soil and included isotopes by the activities of the earthworms.

B. Mortality Due to Radioactivity

High levels of irradiation are necessary to produce significant mortality of earthworms. Eno (1966) showed that there was no significant effect from radiation intensities up to 6 kR (1.5 C/kg) in 10 weeks following irradiation of *Lumbricus terrestris* and *Eudrilus eugeniae*. At radiation intensities of 16-64 kR (4.1-16.5 C/kg) mortality of *L. terrestris* was higher than that of *E. eugeniae*; at 256 kR (66 C/kg) both species quickly died, and at 512 kR (132 C/kg) death was immediate. Khan (1966) irradiated populations of *Eisenia fetida*, *Eutyphoeus incommodus* and *Metaphire posthuma*; treatment with gamma radiation of 3-8 kR (0.8-2.1 C/kg) resulted in increases in earthworm populations relative to controls, perhaps due to availability of food from death of more susceptible organisms, but radiation levels > 20 kR (5.2 C/kg) resulted in reduction of earthworm populations. Lofty (see Edwards and Lofty 1977) exposed *Allolobophora chlorotica* to doses (kR) of 5, 10, 25, 50, 100 and 200 gamma radiation. At 200 kR (52 C/kg) all earthworms died within a few hours; at 100 kR (26 C/kg) there was about 33% mortality within the first two days; at lower dosage rates there was little immediate effect; survival curves then ran more or less parallel for the survivors of the 100 kR (26 C/kg) dosage and all lower dosage rates for the subsequent six months.

Krivolutsky *et al.* (1972) contaminated the soil of plots in a Russian birch-pine forest with ^{90}Sr at a level of 1.8–3.4 μCi/m^2 (67 kBq/m^2). During the subsequent spring to autumn (four months) period, total soil animal populations of contaminated plots were about one-half to one-third of those in untreated control plots; earthworm numbers were badly affected, with reductions in population to as little as about one-eighth of that of control plots.

C. Use of Earthworms as Indicators of Radioactive Soil Pollution

Krivolutsky *et al.* (1982) investigated the effects of radioactive contamination of soils with products of uranium decomposition, especially ^{90}Sr, and also radium (^{226}Ra), on spiders, molluscs, insects, myriapods and earthworms, and found that over a seven-year period of exposure earthworms were the most sensitive indicators of radioactive pollution. Earthworms (Lumbricidae) declined in population and showed an increase in numbers of mucus-producing epidermal and intestinal gland cells when exposed to an increase of background radiation as low as 100 mCi/h (3.7 GBq/h) above normal levels.

As with heavy metal pollutants, it cannot be assumed that all earthworms in all soils will react similarly but their habit of ingesting large quantities of soil, which must bring them into close contact with any radioactive pollutants, probably enhances their susceptibility to radiation damage compared with most other soil-inhabiting invertebrates.

12
Earthworms and Pedogenesis

Pedogenesis and soil fertility result from a combination of *abiotic factors*, including the composition of the parent material, physical and chemical weathering, temperature regimes, hydrology and landscape stability, and *biotic factors*, especially uptake by plants of water and nutrients, photosynthesis, litter fall and decomposition of litter and dead root material.

Darwin (1881) recognized that earthworms were particularly significant contributors to the biotic components of soil processes, and his initial findings and observations have been confirmed and amplified by many subsequent workers.

I. Effects on Soil Profile Development

The physical nature, depth and relationship of the organic (O) horizon of the soil to underlying (A, B, C, etc.) horizons is a most important factor of pedogenesis, and in particular affects the structure, aeration, chemical characteristics and fertility of soils. In the dynamics of the litter layer and its eventual incorporation into deeper horizons lie the principal regulating mechanisms that form and maintain soil profiles. The disintegration, decomposition and incorporation of litter result from a combination of solution by percolating rain water, a minor component of atmospheric oxidation, but most importantly from the activities of the "decomposer industry" (Macfadyen 1963): the soil animals, fungi and microorganisms that live in and derive their energy from the litter layers. Satchell (1974) provides examples that show that > 50% of the total above-ground primary production in a range of terrestrial ecosystems is assimilated by the decomposers. Earthworms are important contributors to the "decomposer industry"; they particularly affect the disintegration and incorporation into underlying soil horizons of forest litter. Müller (1878) distinguished two major humus types in Danish beech (*Fagus*) forests: *mull*, in which the disintegrated litter was intimately mixed with mineral soil particles, and consisted largely of earthworm casts, and *mor*, in which the litter layer was sharply marked off from the underlying soil horizons, was rich in fungal mycelium and contained very few earthworm casts. The significance of mull and mor, as they relate to the feeding behaviour of earthworms, is discussed on pp. 17-18.

The difference between mull and mor is not attributable simply to the presence or absence of earthworm casts. It is related also to the chemical composition of the litter, especially whether it contains much polyphenolic material, and this is dependent on interactions between forest composition, temperature, rainfall, the concentration of some elements, especially calcium, in the underlying soil horizons, the leaching products dissolved from the litter by percolating rainfall, especially whether or not these are strongly acidic. Müller's distinction of mull and mor has stood the test of time (Satchell 1974) and is a useful concept for distinguishing forms of O and A horizons that are more or less affected by earthworm activity. Mosaics of mull and mor can be found on soils formed from uniform parent material, in patterns that follow vegetation patterns. For example, in northern New Zealand, mixed kauri (*Agathis australis*)/broadleaved forests have deep mor layers, with very few earthworms, surrounding individual kauri trees and well developed mull, with very large populations of megascolecid earthworms, under the dominant broadleaved trees (Lee 1959). Patches of deep podzolized soil underlie the mor surrounding kauri trees. Decomposing litter of kauri trees is highly lignified, contains much polyphenolic "humic" material, is strongly acidic (pH 3-4), is rich in fungal mycelium and is sharply marked off as an A_O horizon from the underlying A_1 horizon. The surrounding broadleaved trees stand on loams and clay loams with a thin superficial layer of freshly fallen litter overlying a deep granular A horizon, much of it consisting of earthworm casts, near neutral pH (5.5-6.5) and without the mycelial mats in the A_O horizon that are characteristic of the podzolic soils under kauri.

The mixing of soil horizons by earthworms results in a deepening of the humified zone of the soil profile. Nielsen and Hole (1964) recorded the thickness and distribution of litter at six forested sites with grey-brown podzolic soils in Wisconsin, USA. The litter horizons at two sites were respectively 1.6 cm and 4.5 cm deep; the first site had burrows of *Lumbricus terrestris* at 27/m^2, while the second site had none. At all sites the weight and depth of the litter layer varied inversely with the numbers of *L. terrestris* burrows. Burrows extended to depths of 1-2 m and A_1 horizon material lined them with decreasing thickness down to the C horizon. At one site all litter was removed for five years; numbers of *L. terrestris* burrows and organic matter content of the A_1 horizon fell progressively throughout that time. At another site, where B_1 material formed the surface, a new A_1 horizon began to form over a five year period, especially around the *L. terrestris* burrows, and it was estimated that a continuous A_1 horizon would be formed in 30-40 years. *Aporrectodea caliginosa* was also present, casting at the surface, but it did not remove litter into its burrows as did *L. terrestris*. Raw (1962) found that *L. terrestris*, in an English apple orchard, removed about 1.2 t/ha (dry weight) of fallen leaves from the soil surface during the winter; this was > 90% of the total autumn litter fall.

In a soil survey of South Dakota USA, Buntley and Papendick (1960) found a large area (*c.* 3400 km^2) along the eastern border of the state where soil profiles were transformed by the effects of burrowing and casting of earthworms. The western edge of the worm-modified soils corresponds approximately with the 560 mm isohyet and is not closely related to isotherms, nor to the boundaries of Pleistocene surface deposits. The worm-modified soils have a fairly thick, black A_1 horizon, grading through mixed horizons of B_2 with A_1, A_1 with B_2, B_{3ca} with B_2, B_2 with B_{3ca} horizons, or other combinations, before grading into unmixed parent material (loess or till). The authors distinguished the worm-worked soils as "vermisols"; they resemble the "thick" or "fat" chernozems of Europe, and grade fairly abruptly to the west into typical chernozems. The general fabric of the vermisols, to a depth of *c.* 90 cm, consists almost entirely of worm casts and filled worm channels in varied states of preservation, contrasting with the surrounding chernozems, whose fabric is related to that of the parent material on which they have developed. Horizon boundaries are entirely destroyed in the vermisols. The earthworms concerned were reported by Buntley and Papendick (1960) to be *Lumbricus terrestris*, but Gates (1979) casts some doubt on the validity of this identification.

In a recent series of papers Satchell (1980a, b, c, d) described investigations of a site in northeast England where birch (*Betula*) trees were planted about 30 years previously (in 1948) on an area of podzols in heather (*Calluna vulgaris*) moor land. It was originally proposed that the birch might be expected to transform the soil into a brown earth with a mull-type humus layer, perhaps in a period of 100 years, by eliminating the *Calluna* and stopping the podzolizing process, by cycling nutrients from the subsoil through the trees to the litter and by allowing the development of earthworm populations that would induce the formation of a surface mull layer, mix the soil horizons and improve soil structure, aeration and drainage. Satchell began sampling the experimental plots for earthworms in 1956, eight years after the establishment of birch, and sampled them again ten years and 27 years after planting. Only two species of earthworms, *Dendrobaena octaedra* and *Allolobophora eiseni*, were found throughout the sampling programme and there was no evidence of significant increases in populations or diversity throughout the 27 years. From all the results it was calculated that the mean earthworm population was $1.63/m^2$ in plots that had birch litter on them, while at surrounding plots under *Calluna* no earthworms could be found at the ten-year and the 27-year sampling times. On some plots, where the raw humus had been broken up and mixed into the underlying soil to promote decomposition, a new and even thicker raw humus layer, with a pH of 3.35 and a C:N ratio not significantly different from the 0–3 cm of adjacent control (*Calluna*) plots, had formed by the end of the sampling programme. Some birch plots had a little more exchangeable calcium than those under *Calluna*, but the underlying soil profiles remained as fully differentiated podzols. The original hypothesis assumed that a brown earth would form if an earthworm population that included a substantial proportion of *Lumbricus terrestris* were to become established, and Satchell showed that a minimum biomass of 100 g/m^2 of *L. terrestris* would be necessary to induce such a change. He calculated that the nitrogen requirement of such a population would be c. 3.5 g m^{-2} y^{-1}, but the total assimilable nitrogen available in the birch ecosystem would be only c. 1.33 g m^{-2} y^{-1}, so it was not possible for an adequate *L. terrestris* population to live there. He further showed that other essential requirements of *L. terrestris* could not be met, e.g., *L. terrestris* rarely occurs where soil pH is below 4.3 and the pH in the 0–3 cm layer of the soil under birch was 3.3–3.5 after 27 years and had not changed in the previous 15 years, and there was insufficient calcium available in the soil profile to satisfy the requirements of this earthworm. Satchell therefore concluded that it was unlikely that nutrients brought to the surface in birch litter could change this soil to a conditon tolerable to *L. terrestris*, and since it seems that this would be a necessary requirement for the soil to be transformed from a podzol to a brown earth it is unlikely that such a change will occur, even over a much longer period of time. In other circumstances, where plant nutrients are not limiting, earthworms are capable of inhibiting podzolization, as was shown in Germany by Wittich (1953), who found that the litter of silver fir and Norway spruce was broken down very rapidly and incorporated into the soil where there were large populations of earthworms, but formed a sharply demarcated mor layer, overlying podzols, where earthworms were few or absent.

Deliberate or accidental introduction of surface-feeding or near surface-feeding, burrowing earthworms into soils that lack them, but have no inherent limitations for earthworms, may result in dramatic changes in pedogenesis.

Tréhen and Bouché (1983) investigated the possible benefits of introducing lumbricid earthworms to the acid brown hydromorphic soils of mesophilous and xerophilous heathlands of central Brittany, France. The soils are of low natural fertility and are little used. They have small populations of lumbricids, mainly *Dendrobaena octaedra*, with some *Lumbricus rubellus* and *Eiseniella tetraedra*, and occasionally other species. Earthworms, mainly *Lumbricus terrestris*, *Aporrectodea caliginosa* and *A. giardi* were collected, in blocks of soil, from several sites in nearby pastures on other soils, and were introduced into small areas (1.5 m^2) in the heathlands that were lightly cultivated, some of which were treated with agricultural lime (55% $CaCO_3$) at the rate of 2 t/ha. After one year, populations of introduced earthworms had almost died out at introduction points that had received no lime; they had declined in numbers also where lime had been added, but the survival rate varied greatly according to the original source of the population, and was up to 82.4% for earthworms from one of the pastures.

In further investigations Tréhen and Bouché (1983) studied the effects on the soil and on the earthworms present of surface application of compost, consisting of ground domestic garbage, at rates of 2700 t/ha and 4000 t/ha, in an area of dry heathland (*Ulex europaeus*, *Erica cinerea*, *Calluna vulgaris*). There was a great increase in total earthworm population, especially of *Dendrobaena octaedra*, *Lumbricus rubellus* and *Allolobophora eiseni*, which was previously rare. The action of the augmented earthworm population was to produce a deep humified surface layer and to improve soil structure, and Tréhen and Bouché considered that soil fertility would benefit, with the added advantage of hygienic disposal of domestic garbage.

Langmaid (1964) found much modified soils where earthworms had been introduced into podzols in virgin forest sites around isolated lakes in Canada. Sites that lacked earthworms had distinct L-H litter horizons overlying the typical podzol A_e and B_{fh} horizons; these horizons were altered by earthworms to a uniform single horizon with a very abrupt smooth lower boundary. Earthworms at these sites included *Dendrobaena octaedra*, a litter-inhabiting species, *Lumbricus terrestris* and *L. festivus*, soil-inhabiting, litter-feeding species, and *Aporrectodea caliginosa*, a deep soil-inhabiting species; the soil was reworked to a greater depth where *Lumbricus* spp. were present than where they were absent.

Perhaps the best-documented example of the pedological effects of earthworm introductions was provided by Stockdill (1966, 1982), who introduced *Aporrectodea caliginosa* to the soils of an isolated upland area in southern New Zealand, where a pasture improvement programme was in progress and no lumbricid earthworms were present. Five years after earthworms were introduced a thick turf mat on the soil surface had disappeared and a deep humified A horizon had formed, while water infiltration rates and field capacity had greatly increased (pp. 188-190) and there were significant changes in fertility (pp. 264-266).

Reclaimed polder soils in Holland, which initially lack earthworms, have provided evidence of the pedological effects of introducing earthworms. Plots were established by van Rhee (1969a) in apple orchards on areas of Ijssel Lake, drained in 1957, and large numbers of *Aporrectodea caliginosa* and *Lumbricus terrestris* were introduced to them in 1966. By late 1967 *A. caliginosa* was well established while *L. terrestris* populations had declined; plots with *A. caliginosa* showed an increase in large and capillary pore space and aggregate stability and an increase of about 40% in water available for plant growth relative to plots without earthworms. Conversely, in old Dutch orchard soils that had been treated with fungicides for many years, destroying all earthworms, van de Westeringh (1972) found a deterioration of structure in the 0-40 cm horizons, with a sharp boundary between an organic mat (A_0 horizon) in the worm-free soils, contrasting with a mull-type A horizon in untreated plots where earthworms (*A. caliginosa, A. rosea, Allolobophora chlorotica, L. terrestris*) were numerous.

Further data on the effects of earthworms on polder soils of the Ijssel Lake region were reported by Hoogerkamp *et al.* (1983). From the time of entry of earthworms to previously unoccupied soil, the surface root mat of pastures is incorporated into the soil in *c.* three years; a dark-coloured surface layer (A_1 horizon) begins to form, gradually increasing in thickness to 5-8 cm in eight to nine years. The weak plate-like or moderate blocky structure of the upper 0-10 cm becomes granular, and a system of channels (earthworm burrows) 1-4 mm wide develops, extending to *c.* 50 cm depth after *c.* eight years. These physical changes are accompanied by a redistribution of carbon and nitrogen in the profile, due to the incorporation of the surface root mat, so that the carbon content of soil at 0-20 cm at one site increased from *c.* 0.18 kg/m² to *c.* 1.7 kg/m² surface area, nitrogen from *c.* 0.02 kg/m² to *c.* 0.1 kg/m², with a slight increase in C:N ratio, after eight years of earthworm activity. Conductivity of water at saturation, for air at pF 2 and oxygen diffusion at pF 2 are also significantly increased, and these changes are reflected in less mottling (gleying) of worm-inhabited areas than where worms are absent.

Introductions of earthworms as a means of incorporating surface root mats, increasing the depth of humified layers, increasing water infiltration rates, and thus affecting soil profiles have previously involved only the cosmopolitan species of Lumbricidae. Much of this work has been misguided, due to inadequate knowledge of the ecology of the earthworms introduced. In New Zealand, in the 1940s and 1950s, widespread trials of earthworm introduction resulted from observations of spectacular improvements in pasture growth that followed the unaided spread of lumbricid species on farms (pp. 264–266). The species concerned were *Aporrectodea caliginosa*, *A. longa* and *Lumbricus rubellus*. These species were transferred to areas where, in most cases, they were already common, and consequently the introductions had no effect. In exceptional cases, at sites where they were absent and where soil conditions were suitable for their survival, they were responsible for pedogenetic changes and increases in plant growth, as recorded by Stockdill (1966, 1982; see earlier) in an isolated region of the South Island. Lack of understanding of earthworm ecology has also led to the widespread introduction of *Eisenia fetida* as a "soil improver". *Eisenia fetida* inhabits dung-heaps, compost-heaps and similar concentrations of organic-rich materials. It is easily reared in very large numbers, but will not survive to become a dominant species in agricultural or pastoral lands.

The presence of Lumbricidae as the dominant group of earthworms in temperate regions of much of the world and the range of lumbricid species present are accidental consequences of European colonization, resulting from the transportation of European plants, the elimination of indigenous vegetation from vast areas and consequent elimination of most, if not all of the indigenous vegetation and its replacement with palaearctic plant species. Changes in ecological conditions consequent on land clearance resulted in elimination of the majority of the indigenous earthworms and their place was taken in temperate regions, with varying success, by the introduced European lumbricids. In some regions to which lumbricids have been introduced, where rainfall is adequate and reasonably evenly distributed throughout the year, e.g., New Zealand (see Lee 1959, 1961) lumbricids have successfully filled the vacuum that resulted from the elimination of indigenous earthworms. However, in vast areas of the major continents where rainfall is highly seasonal and much of the year is characterized by drought, lumbricids are present in pastures and crop-lands, but the species that are present are more suited to wetter climates, so their populations are small and their period of activity in the year is short, e.g., see Barley's (1959a) study of lumbricid earthworms in South Australian farmlands.

Some lumbricids that have not been widely dispersed by humans, and some earthworms other than lumbricids are much better adapted physiologically to live in seasonally dry climates. There is a good chance that non-lumbricids would exert a stronger influence on the soils of such regions. The selection of suitable species depends essentially on the earthworms' ability to withstand periods of drought by retreating deep into the soil and entering a state of quiescence, or on their ability to conserve water while remaining active. In the latter case, the morphology and physiology of the nephridia, i.e., the urine-excreting organs and processes are critical. There is a variety of types of nephridia in earthworms, and their differences are discussed on pp. 215-217. Lumbricids are all holonephridial and exonephric (except *Helodrilus antipae*). The lumbricid species that have been dispersed widely by European man are from the wetter parts of Europe; they are not well adapted to dry conditions, and are obliged to lose water at rates of 5%-20% of their body weight per day in urine. Other lumbricids, particularly those that have evolved in the seasonally dry climates of southern Europe, appear to be better adapted to drought. Some megascolecids, including the *"Pheretima"* s.l. group of genera, and species of *Lampito, Megascolex, Tonoscolex, Travoscolides* and *Woodwardiella*, have enteronephric meronephridia, and excrete their urine into the gut, where much of the water is resorbed before the nitrogenous wastes are excreted along with the casts.

Careful selection of alternative lumbricids or of species from the enteronephric meronephridial group and introduction to seasonally dry temperate regions where they are absent has not, to my knowledge, been attempted, but because it is likely that they would be active for a greater proportion of the year and/or be able to maintain higher population levels than the lumbricids that are now present, they might exert a significant and beneficial effect on pedogenesis and soil fertility. Much of the temperate regions of Australia, Africa and the Americas have long dry seasons, but these regions have neither southern European lumbricids, nor the appropriate groups of enteronephric meronephridial earthworms, which are confined mainly to Asia, although some (e.g., *Metaphire californica*) are more widespread.

II. Earthworms and Landscape Processes

A. Soil Erosion

Darwin (1881) recognized the significant role played by earthworms in soil erosion when he observed and measured the down-slope elongation of earthworm casts due to raindrop impact and surface wash in permanent grasslands. On slopes of 8°-11°30′ at Holwood Park, in England, he found that casts had flowed down-hill following rain to form smooth, narrow patches of earth 15-19 cm in length, and even on slopes of 1°-5° casts had flowed down-hill so that "fully two-thirds of the ejected earth lay below the mouths of the burrows". He measured similar down-slope movement due to surface runoff water in several other localities and also drew attention to the continued down-hill movement of particles from disintegrated casts, the accumulation of muddy material around the base of grass leaves in small hollows in otherwise level ground, and the accumulation of recognizable fragments of earthworm casts in drainage ditches transverse to the slope of the land. He reported observations in India of very large accumulations of earthworm casts being entirely removed down-slope during the monsoonal rains into streams and carried away to flood plains more than 1000 m below, and concluded that earthworms thus contribute greatly to denudation of steep land surfaces and to the building of the alluvial soils of flood plains. On very steep slopes, grazed by sheep, he measured the thickness of fine materials on the more or less parallel ledges that result from movement of the grazing animals along ridge sides, and noted that much of the fine soil material that accumulates on the ledges probably consists of disintegrated earthworm casts that have rolled or been washed down from above. He also noted that strong winds blowing down-slope, especially when accompanied by rain, result in accelerated movement of cast material down-slope and that there may also be up-slope movement when the wind blows very strongly up-slope. Strong winds during periods without rain were also considered to result in deflation and movement to leeward of material from casts that had beome desiccated, but Darwin did not have any conclusive evidence for this process. Drummond (1888), describing how termite mounds and above-ground runways were eroded by rain and wind and the debris carried away by streams, on the high plateau between Lakes Nyasa and Tanganyika, recalled that Herodotus described Egypt as "the gift of the Nile", but that ... "had he lived today he might have carried his vision farther back still and referred some of it to the labours of the humble termites in the forest slopes about Victoria Nyanza". Equally he might have attributed some of it to the labours of the humble earthworms whose casts are washed and blown from the catchment areas into the streams that feed the Nile and, indeed, most of the rivers of the world.

Sharpley and Syers (1976) and Sharpley *et al.* (1979) assessed the significance of earthworms (*Lumbricus rubellus* and *Aporrectodea caliginosa*) as a source of loss of phosphorus and nitrogen from New Zealand pasture soils, due to down-slope movement of particulate matter from their surface casts during periods of surface runoff. Poisoning of earthworms with carbaryl in permanent pasture on a 13° slope resulted in a reduction of 75% in the sedimentary load in runoff water, from 1120 kg ha^{-1} y^{-1} in control plots with earthworms to 290 kg ha^{-1} y^{-1} where the earthworms were poisoned. Losses of particulate phosphorus fertilizers in runoff water were also greatly reduced by the elimination of earthworms; however, infiltration rates in worm-free plots were only one-third of those in plots with earthworms, and there were large increases in losses of dissolved inorganic P, ammonium-N and nitrate-N in surface runoff in the worm-free plots compared with losses where earthworms were present.

Table 26. Relationships between earthworm populations, annual loss of surface soil by sheet erosion and annual surface runoff at Ithaca (USA). (Data from Hopp 1973).

Treatment for previous ten years	Earthworms mean no./m²	Annual sheet erosion (t/ha)	Surface runoff (mm)
Continuous cultivation	0	75	45
Three-year rotation	23	13	10
Continuous pasture	76	0	5
Virgin ground	200	0	7

Carbaryl is toxic to soil animals other than earthworms and some of the observed differences could have been due to reduction in other soil animal groups, but the authors' evidence indicated that lack of surface casts on the plots without earthworms was the major reason for the reduction of losses of particulate matter in the runoff water.

Hopp (1973) measured surface runoff and soil loss by sheet erosion and their relationships with earthworm populations in soils with several patterns of land use at Ithaca (USA). His observations are summarized in Table 26 and they clearly demonstrate that both erosion losses and surface runoff were greatly reduced by the presence of earthworms, and that the decreases were directly related to the number of earthworms in the soil.

B. Microrelief Features

Pickford (1926) first described crater-like mounds, about 1 m in diameter with a central bowl-like depression 30–100 cm deep, which are known as kommetjies, and are found in sparse grassland over an area of about 16 km², forming the Kommetjie Flats, near King Williams Town, in South Africa. The sides of the depressions are covered with large tower-like casts, up to about 75 mm high, of a species of *Microchaetus* (Glossoscolecidae), and Pickford (1926) considered that they were formed by the earthworms. During heavy rains the depressions collect water, and it was proposed that the earthworms were then confined to the less waterlogged soil of the mound walls, so that their casts would accumulate there and build up the mounds. Ljungström and Reinecke (1969) have also studied the kommetjies. They collected 13.5 kg of casts from 1 m², and allowing for losses of material from them, due to rain, considered that at least 27 kg m^{-2} y^{-1} of casts would be deposited by the *Microchaetus* sp. Some farmers regard the kommetjies as an aid in controlling erosion, since they hold the rainwater and allow it to infiltrate into the soil instead of running over the surface.

To the north of Lake Kyoga, in the Teso District of Uganda, are large areas of swamps, carrying seasonal grassland, which are used during the dry

254 Relationships with Soils and Land Use

season for cattle grazing. Some of the swamps have their surfaces raised into a large number of closely spaced mounds, greatly reducing the value of the land for grazing, and it was suggested by Harker (1955) that earthworms were responsible for the formation of the mounds. Wasawo and Visser (1959) found that the mounds were of two kinds, both with steep sides, either more or less columnar, 20–60 cm high and with a flat top 20–100 cm in diameter, or rectangular, 25–60 cm high, up to about 4 m long and 40–60 cm wide at the top. Eighty mounds were counted in an area of 25 m^2. Three species of glossoscolecid earthworms, *Glyphidrilus* sp., *Alma stuhlmanni* and *A. emini* in decreasing order of abundance, were found living in the mounds, very often all three species living together in individual mounds. Populations of earthworms in columnar mounds at Ochuloi Swamp, during the wet season, averaged about 300 m^2. During the wet season, when surface water just covers the mounds, the hind ends of the earthworms, which are specially adapted for respiratory exchange (pp. 50–52), protruded from their burrows above the waterlogged soil, while in the dry season the worms were found in a dormant state at depths of 40–150 cm in the soil beneath the mounds, with none in the soil beneath the intervening hollows. All three species cast at the surface during the wet season and depths of 3.5–7.0 cm (mean, 5 cm) of casts were measured at the beginning of the dry season. Vegetation on the mounds was dominated by *Cyperus haspan* (Cyperaceae), with *Echinochloa pyramidalis* (Poaceae) between mounds. Chemical analyses of fresh casts, soil from the top of mounds and soil from below the ground surface between mounds showed that casts and mound soils had more carbon, nitrogen, higher C:N ratios, and considerably more exchangeable potassium, sodium, calcium, magnesium, iron and manganese than the soil between mounds; the mound soil contained slightly less of most of the exchangeable cations than the fresh casts. Wasawo and Visser (1959) concluded that the soil forming the mounds was derived from casts, and that the mounds may be initiated as slightly higher areas between tracks made by grazing cattle and then enlarged by the casts of the earthworms, which would find in them a favourable habitat for the wet season.

Forms of microrelief were described by Haantjens (1965) from the grassy plains of the Sepik River valley in New Guinea, and their formation was also attributed to the activities of large earthworms. Haantjens distinguished two types of formation, (a) pitted soils, in which there are irregular rounded holes, 15–45 cm deep and up to 4–5 m wide, or steep-sided trenches 45–75 cm deep and up to 24 m long and 2–3 m wide, found usually in grassland but occasionally under forest, and (b) sorted stripes, which are irregular ridges and mounds, 15–25 cm high, of fine sandy texture with some fine gravel, separated by a network of coarse gravelly strips, which are found only in grassland. Concentrations of very large casts of the earthworm

Metapheretima jocchana, on the edges of the pits and on the rises between the gravel strips, led Haantjens to the conclusion that the earthworms move fine soil materials laterally out of the pits and gravel stripe areas to build up the higher areas as a place to live above the water table during the wet season, when the lower ground is flooded. Lee (1967) examined the same area as had Haantjens (1965) and collected casts (see Fig. 24a, p. 176) and soil samples from the higher ground supposedly built up by *M. jocchana*. Excavation across pits and sorted stripes (Lee 1967) showed that the earthworms and their burrows were concentrated in the soil beneath the raised ground, were very rare beneath the bottoms of the pits and were entirely absent beneath the stone stripes. Particle size analyses, chemical analyses and identification of clay minerals in samples from casts and surrounding soils indicated that the casts were derived from the 0–25 cm layer of the soil, i.e., from the soil of the raised areas, not from the soil beneath the bottoms of the pits and gravel stripes.

There is much similarity between the examples of microtopography described above and attributed to earthworms. If their formation and maintenance is due to earthworms they require soil to be collected from beneath adjacent low-lying ground and to be moved laterally and vertically upward, to be cast onto the mounds or ridges, because it is well established (pp. 251–253) that casts are disintegrated and washed downslope by heavy rain and resulting surface runoff. If soil is moved onto the higher ground during the dry season by earthworms to provide a haven during the wet season, it should be easy to find extensive burrow systems and active earthworms during the dry season under the adjacent depressed areas; but they are not found in the Ugandan swamps, nor in the New Guinea grasslands. The earthworms apparently live in the upraised areas and their surface casts, far from building up the mounds and ridges, must accelerate their destruction by providing, from the soil in the raised areas, a source of readily available sediment for removal by surface runoff. Lee (1967) proposed that the microtopographic features seen on the Sepik plains could be initiated by differential intensities of burning of the grassland, leaving small patches bare of all vegetation and allowing wind and water to remove the surface soil. Haantjens (1969) discussed the frequency of strong winds, fire, and the soil moisture regime of the soils, and concluded that Lee's proposition was an unlikely explanation of the microtopography. Whatever is the explanation, it seems most unlikely that the microtopography results from earthworm activity; in the circumstances described the raised areas provide a refuge for earthworms during wet seasons, when much of the landscape is flooded, but the effect of earthworms casting on the surface of raised areas must contribute to the eventual levelling of the microtopographic features rather than to their construction.

13
Earthworms and Plant Growth

I. Promotion of Plant Growth

A. Degradation of Pasture Root Mats

In the absence of a mechanism to remove dead roots, leaves and other plant litter from the soil surface beneath pastures and natural grasslands, a matted layer of organic debris and living roots accumulates. This root mat interferes with infiltration of water into the soil and tends to become acidic, to reduce the productivity and to change the species composition of planted pastures. Earthworms that feed on the dead plant material of the root mat, on animal dung that falls on it and make their burrows in the underlying soil, contribute to the removal and incorporation of the mat into the soil. Darwin (1881) measured rates of removal from the surface of dead plant material and commented on the significance of surface casts in assisting the process of burial. The importance of some species of earthworms in this regard is specially evident where they are introduced into land that has been developed for grazing and no suitable earthworm species have previously been present.

Waters (1951) related the biomass of earthworms (predominantly *Aporrectodea caliginosa*) to dry matter production in a rye grass/white clover pasture in New Zealand (p. 265). Degradation and incorporation by the earthworms of the dead roots that otherwise form a root mat were apparently critical requirements for high pasture production, but maintenance of a large population of earthworms was dependent on the maintenance of clover in the sward. The limiting factor for the earthworm population was probably the quantity of nitrogen, provided primarily by the symbiotic bacteria associated with clover and secondarily by return of nitrogen to the soil in sheep dung and urine.

Barley and Kleinig (1964) studied the effects of earthworms in irrigated pastures at Deniliquin in New South Wales, Australia. Four years after pasture was established they found earthworms (the megascolecids *Plutellus attenuatus, Megascolex exiguus, M. macleayi*, and a lumbricid, *Eiseniella tetraedra*) along watercourses, but none in the pastures. At this time there was a surface mat of partly decomposed litter and dead roots, with living roots, about 1 cm thick, while in older pastures, established 10–15 years previously, similar mats were 2–4 cm thick and contained up to 1200 kg/ha of nitrogen. Some lumbricid earthworms (*Aporrectodea caliginosa* and *A. rosea*) were found in isolated small areas eight years after pasture establishment. In 1955, about 175 *A. caliginosa* were introduced at each of five points, about 10 m apart, in three fields where grass/clover pasture had been established five to eight years previously. By 1963 populations of earthworms up to 300/m^2 (biomass about 160 g/m^2) including *A. caliginosa* and small numbers of *Microscolex dubius* were found in the inoculated areas, except in heavy textured soils. In 1966 and 1968 Noble et al. (1970) resampled the fields and examined pasture mats in areas inoculated and not inoculated with earthworms. Populations in 1966 were larger than in 1968. In 1966, inoculated plots had a mean earthworm population of 357/m^2 (biomass 80 g/m^2) all *A. caliginosa*, while non-inoculated plots had a mean of 23/m^2 (biomass 3 g/m^2) 38% *A. caliginosa* and 62% *M. dubius*; in 1968 corresponding figures for numbers of earthworms were 95/m^2 (biomass 50 g/m^2) all *A. caliginosa*, and 15/m^2 (biomass 4 g/m^2) 70% *A. caliginosa* and 30% *M. dubius*. Pasture mats in the inoculated areas in 1968 averaged about 7600 kg/ha and in non-inoculated areas about 72 300 kg/ha; total nitrogen in inoculated areas was 60 kg/ha in the mat, 1055 kg/ha in the 0–2.5 cm soil layer and 415 kg/ha in the 2.5–5 cm soil layer, while corresponding totals for non-inoculated areas were 1063 kg/ha in the mat, 495 kg/ha in the 0–2.5 cm layer and 271 kg/ha in the 2.5–5 cm layer. Soil bulk densities and pasture yields were not significantly different in inoculated and non-inoculated areas, total dry matter production totalling 2995 kg/ha in inoculated and 2732 kg/ha in non-inoculated plots.

Experiments conducted in New Zealand by Stockdill (1966) and others (e.g., Nielson 1951a, 1953), where lumbricid earthworms were introduced into the soils of pastures that previously lacked them, demonstrated significant pedological changes due to the earthworms (p. 248). Five years after introduction of *A. caliginosa*, Stockdill (1966, 1982) showed that pasture litter, animal dung, surface-applied fertilizers and insecticides had been incorporated into the soil, with associated reductions of organic matter content of the 0–2.5 cm horizon and corresponding increases in the 2.5–7.5

cm horizon. Incorporation of pasture root mat into the soil by earthworms is particularly well illustrated in three photographs provided by Mr Stockdill (Figs 30, 31, 32). These photographs of soil profiles were taken over a transect of c. 20 m length in a rye grass/white clover pasture, planted 23 years previously, at Hindon, New Zealand.

Similar experiments, reported by Hoogerkamp et al. (1983) on young polder soils in Holland, yielded similar results to those in New Zealand. Earthworms (*A. caliginosa*, *A. longa*, *L. terrestris* and *L. rubellus*) were introduced as mixed populations, at a density of c. $10/m^2$, soon after sowing of rye grass pastures in 1971 and 1972 on calcareous polder soils that were reclaimed in 1957. After 10 years earthworm populations reached c. $200/m^2$ around points of introduction, dispersing at a rate of c. 9 m/y for *A. caliginosa* and 4.5 m/y for *L. terrestris*. In the absence of earthworms pasture root mats were 0.5–2.5 cm thick after 10 years, with a mean weight c. 195 000 kg/ha and an abrupt boundary to the underlying mineral soil. Invasion of earthworms was followed by disappearance of the root mat over a period of c. three years, with development of a humus-stained A_1 horizon that increased in thickness to 5–8 cm in eight to nine years. It was calculated that to eliminate the root mat from these pastures within five years it would be necessary to introduce earthworms at a rate of c. 2000 per inoculation point at spacings of c. 50 m. From infrared line scanning aerial photographs, Hoogerkamp et al. were able to show that the earthworm-inoculated areas were warmer at night and cooler during the day than surrounding earthworm-free areas. The effect was particularly strong in early spring, and presumably reflected the greater biomass of grass in the worm-inoculated areas and an increase in heat exchange rates between soil and air in the absence of a pasture root mat.

In the province of Zeeland, Holland, where polders were flooded in 1944 and again in 1953, earthworm populations died out. In a seven-year-old pasture, re-established on the polder areas, van Rhee (1963) found that soils that lacked earthworms had a thick surface mat of partially decomposed grass litter and dead roots, but where earthworms had been introduced the mat was noticeably reduced; the earthworm population totalled $280/m^2$ (biomass wet weight 175 g/m^2) comprising *A. caliginosa* (72.5% of population), *A. longa* (18.5%) and *Lumbricus rubellus* (8.5%). Selective introduction of earthworm species in the field showed that, of these species, only *A. longa* was effective in reducing the pasture mat and its effects were measurable only one year after introduction. Experiments with artificial profiles (van Rhee 1963), with a layer of litter placed on top, showed that *A. longa* and *Lumbricus terrestris* dragged litter down into their burrows and cast at the surface, resulting in incorporation of the surface organic layer, while *A. caliginosa*, *A. rosea* and *Allolobophora chlorotica* removed

13. *Earthworms and Plant Growth* **259**

Fig. 30 Soil profile in 23-year-old ryegrass–white clover on a yellow-brown earth (inceptisol) at Hindon, New Zealand, developed in the absence of lumbricid earthworms. (Photo: S. M. J. Stockdill).

Fig. 31 Soil profile in 23-year-old ryegrass–white clover pasture, on the same soil and at same locality as in Fig. 30, c. two years after introduction of lumbricid earthworms. Note great reduction in surface root mat, remnants of root mat buried by cast material, general darkening (humification) of A_1 horizon. (Photo: S. M. J. Stockdill).

Fig. 32 Soil profile in 23-year-old ryegrass–white clover pasture on the same soil and at the same locality as in Fig. 30, c. ten years after introduction of lumbricid earthworms. Note absence of pasture root mat, more granular structure of soil. (Photo: S. M. J. Stockdill).

the soft tissues of the litter at the surface, leaving the harder material of the veins behind, and constructed ramifying systems of subsurface burrows in which their casts were deposited, so that they were ineffective in organic matter incorporation compared with *A. longa* and *L. terrestris*.

Differences in the efficacy of removal of surface litter by different earthworm species are illustrated by experiments of Dietz and Bottner (1981) who compared rates of incorporation of litter of rye grass and of *Brachypodium phoenicoides* in the presence and the absence of *Aporrectodea caliginosa meridionalis*. Soil from a *B. phoenicoides* grassland in southern France was separated into 10 cm layers, sieved, and repacked in cylinders 30 cm high and 5.5 cm diameter; ^{14}C-labelled litter (1 g per column) was placed on the surface and the columns were then replaced in the field. Cores were removed and the distribution of ^{14}C was traced by autoradiography at intervals for 68 weeks. In the presence of the earthworms c. 85% of the labelled ryegrass litter disappeared from the surface in 17 weeks and c. 60% of the ^{14}C had been lost, apparently in respired carbon dioxide; in columns without earthworms c. 60% of the labelled litter had disappeared. By the end of the experimental period all the labelled litter had disappeared, with or without earthworms. The distribution of ^{14}C in the soil did not differ greatly between columns with or without earthworms, with most of the labelled materials confined to the 0-5 cm layer in both cases.

The lumbricid species most commonly implicated in the degradation and burial of pasture root mats are different in southern compared with northern hemisphere pastures. The apparent anomaly could be due to misidentification in one case or the other, but this is not likely, and the observed differences probably reflect real differences in behaviour. In New Zealand and Australia *L. terrestris* is present but is sporadic, with isolated populations confined to small areas (Lee, 1959 and B. G. M. Jamieson, pers. comm.). I have found (Lee 1959) that in New Zealand *A. longa* is common only in the soils of pasture lands in seasonally dry (eastern) regions of the country. The most common and widespread earthworms of New Zealand pasture land soils are *A. caliginosa* and *L. rubellus*, with *Octolasion cyaneum* somewhat less common (Lee 1959), while in southern Australia they are *A. caliginosa*, *A. chlorotica* and *A. rosea* (Barley 1959a, Barley and Kleinig 1964). *Aporrectodea caliginosa* is most commonly found to break down pasture root mats in Australasia, with *A. chlorotica* and *A. rosea* less commonly involved (see above references and also Stockdill 1966, Noble *et al.* 1970, Martin and Charles 1979). In Europe and perhaps also in North America (Buntley and Papendick 1960, Vimmerstedt and Finney 1973) *L. terrestris* is the most important of the species involved, while *A. caliginosa* and *A. rosea* are not important (van Rhee 1963). There is evidence of intra-specific differences in tolerance of environmental stresses between populations of lumbricids in their Eurasian homelands and populations in other regions of the world where they have been introduced (pp. 163-165); such differences may be reflected in intra-specific behavioural differences in the rather different assemblages of species that make up the genetically isolated lumbricid populations that now inhabit superficially similar habitats in widely separated parts of the world.

Fertilizers, insecticides and similar materials applied as dusts or granules to the ground surface in pastures are incorporated by earthworms, together with the root mat material, into underlying soil horizons. This process has been documented in New Zealand by Stockdill (1966, 1982). He showed that where agricultural lime was applied to pastures on acidic hill soils in Otago, five years later the calcium content in the soils of plots with lumbricid earthworms (*A. caliginosa* dominant) was 3040 μg/g at 0-2.5 cm, 2080 μg/g at 2.5-7.5 cm, 1400 μg/g at 7.5-15 cm, and 360 μg/g at 15-22.5 cm depth, while in plots lacking lumbricid earthworms the corresponding calcium

262 Relationships with Soils and Land Use

Fig. 33 Layer of lime originally applied to soil surface at time of pasture establishment, six years previously, under strongly developed surface root mat, in soil without lumbricid earthworms. Hindon, New Zealand. (Photo: S. M. J. Stockdill).

contents were (μg/g) 4000, 1720, 960 and 300 (Fig. 33). Ten years after lime was applied soils with earthworms had considerably lower calcium content than those lacking earthworms, and Stockdill (1966) attributed this to the increased rate of water infiltration into the soil (p. 188), with a consequent increase in leaching losses, due to earthworm burrows. When Stockdill's experiments were in progress DDT was widely applied to the ground surface to control larval populations of the scarabaeid beetle *Costelytra zealandica*, a serious pasture pest. Two years after application of DDT (2.25 kg/ha) there were larvae of *C. zealandica* at 430/m^2 in plots lacking earthworms, while in plots that had earthworm populations of 650/m^2 there were no *C. zealandica* larvae. The DDT content of soil samples from worm-free plots was 20 μg/g at 0–1.25 cm depth, 1.8 μg/g at 1.25–2.5 cm and 0.1 μg/g at 2.5–3.75 cm, while corresponding samples from the plots that had earthworms showed DDT contents (μg/g) of 11.0, 6.4 and 0.2. Some commonly used insecticides have little or no effect on earthworms (p. 304) and the presence of an active earthworm population may often be helpful in achieving control of soil-inhabiting pests.

II. Influence on Soil Fertility and Effects on Plant Production

The chemical composition of earthworm casts is discussed in Chapter 11, and the pedogenetic effects of litter incorporation by earthworms on pp. 201–205; consequent effects on soil fertility and plant growth are of great significance. Maldague (1970) concluded, from a study of tropical and temperate forest soils, that fertility results from the maintenance of soil conditions such that decomposition processes operate at a level adequate to release plant nutrients from the litter at a rate that will sustain optimum growth. He distinguished between inherent mineralogical fertility, which derives from the chemical composition of a soil's parent material and the rates of weathering and leaching that release plant nutrients from the mineral fabric of the soil, and biological fertility, which derives from the rate of decomposition of plant litter and consequent release of its contained nutrients for recirculation in the plant–soil system. Maldague's concept applies equally to soil under grassland or any other vegetation.

Both kinds of fertility-promoting processes proceed simultaneously in all soils; at any time there is an equilibrium between them, but the balance between one and the other varies from place to place and from time to time at the same place. The causal agents of litter decomposition, and hence of nutrient cycling and biological fertility, are the soil organisms, and earthworms are among the most important contributors to the processes involved. Maldague (1970) related the rate of nutrient cycling to the rate of energy consumption in the decomposition process. Reichle (1977), while agreeing that metabolic activity as measured by energy consumption, is a key to the interpretation of rates of nutrient cycling, pointed out that the relationship is not uniform; saprovores, which include earthworms, have a low assimilation efficiency and a high production efficiency (Chapter 6), while herbivores and, even more, carnivores, have a higher assimilation efficiency and a lower production efficiency, so the relationship between energy consumption and nutrient cycling varies between these groups within the soil biota.

The influence of earthworms on plant growth is here considered in pasture, forest, cropland and orchards.

A. Pasture soils

Darwin (1881) remarked that "earthworms prepare the ground in an excellent manner for the growth of fibrous-rooted plants and for seedlings

264 *Relationships with Soils and Land Use*

of all kinds". Quantitative assessment of their contribution to plant growth in grasslands was not forthcoming until about 60 years later.

Hopp and Slater (1948) placed an unproductive clay subsoil from Maryland in barrels, seeded the soil with a pasture mixture, and tested the effect on hay yield of inoculation with earthworms, with and without added fertilizers. Hay production increased by 3.31 t/ha with earthworms and fertilizers and by 2.72 t/ha with earthworms but without fertilizers, when grass clippings were left on the soil surface during the autumn as a mulch to protect the earthworms from freezing through the winter. When no autumn mulch was applied the effect of earthworm inoculation was small. The presence of earthworms particularly stimulated the growth of clover in the pasture seed mix, and the increase in growth was attributed primarily to the influence of earthworms on soil structure, associated with increased rates of water infiltration.

Most of the published work on the effects of earthworms on pasture growth has been done in New Zealand, where pastoral industries are of great economic importance. The initial impetus for research into earthworms in New Zealand came from the observations of a farmer (Mr A. S. Ashmore) at Raetihi in the North Island, who observed that grass grew better than elsewhere on his farm in his orchard, where lumbricid earthworms were present, and that the improvement in grass growth was spreading from his orchard into surrounding fields as lumbricids spread into them. He began taking pieces of turf, containing earthworms, from his orchard, in about 1940, and introducing them into earthworm-free pastures on his farm, and the remarkable improvement that resulted in pasture quality and production attracted the attention of agricultural scientists. Nielson (1951b) set up test plots in 1949 in a field where Mr Ashmore had introduced earthworms in 1940, and where the earthworms by that time were present in large numbers through about three-quarters of the area of the field. Plots in portions of the field where lumbricids had been present for several years and where improvement in pasture quality was obvious had populations of $c.$ 1210/m^2, comprising *Aporrectodea caliginosa* (1040/m^2), *Lumbricus rubellus* (160/m^2) and *A. rosea* (10/m^2); plots along the edges of the limits of observed improvement in pasture growth had lumbricid populations $c.$ 240/m^2, comprising *A. caliginosa* (110/m^2) and *L. rubellus* (130/m^2); plots outside the limits of observed pasture improvement had *L. rubellus* only (65/m^2). The botanical composition of hay cut from the field showed increasing proportions of rye grass with increasing lumbricid earthworm populations, from 2% of total dry weight in plots outside the limits of observed pasture improvement, to 13% along the limits of observed improvement, to 42% where earthworms were present in large numbers. Native grasses of low productivity (*Agrostis tenuis* and *Danthonia*) together

13. Earthworms and Plant Growth 265

comprised respectively 76%, 55% and 19% dry weight of the production of these plots. Nielson (1951b) set up pot experiments in which worm-free undisturbed blocks of soil 38 cm square, with established pasture, were taken from seven soil types and kept for 20 months, with and without *A. caliginosa* introduced at populations equivalent to 690/m^2. At the end of the trial the total pasture production of pots with earthworms was 28%–111% higher than in pots without earthworms, and there was a high positive correlation between dry matter and *A. caliginosa* biomass (r = 0.77, r^2 = 0.59, P < 0.02).

Nielson's pioneer work was followed by that of Waters (1951) who related the biomass of earthworms (predominantly *A. caliginosa*) to dry matter production in New Zealand rye grass/white clover pastures and found a high positive correlation (r = 0.87, r^2 = 0.76) between them. In a pasture producing 8.68 t/ha dry matter, where sheep dung and urine was returned to the soil surface, the biomass of earthworms was 153.9 g/m^2 (wet weight), while in the absence of clover and without return of sheep dung and urine dry matter production was 1.63 t/ha and earthworm biomass was 48.4 g/m^2 (wet weight). Sears and Evans (1953) similarly found a strong relationship between earthworm populations (*A. caliginosa* and *L. rubellus* dominant) and pasture production. In pot trials Waters (1951) showed an increase, in the presence of earthworms, of 77% in production of rye grass alone and an increase of 113% in rye grass/white clover production, though without any increase in the clover component of the mixture.

Stockdill (1959) introduced lumbricid earthworms in 1949 into six-year-old pastures (rye grass/cocksfoot/clover) in Otago, New Zealand, where the only earthworms present at the time were a small population of the subsoil-inhabiting indigenous octochaetid *Octochaetus multiporus*. Stockdill (1959) dug in an unspecified number of individuals, contained in a square of turf taken from a pasture with abundant lumbricids, at marked points in the pasture. Two years later the lumbricids had spread to about 0.5 m around the introduced turf; three and a half years later Stockdill found lumbricids up to c. 400/m^2 at the points of introduction, 100/m^2 at a distance of 1 m and 10-20/m^2 at a distance of 2 m. After four years circular green patches about 2 m in diameter were distinguishable in early spring; earthworm casts were numerous within the green patches, declining to very few 10 m from the point of introduction. Four and a half years after introduction of the earthworms, plots of 1 m^2 were enclosed, five at earthworm introduction points and five outside the area of spread of the lumbricids. By this time the dominant lumbricid present was *Aporrectodea caliginosa*. The grass on the plots was cut, weighed and returned to the soil surface three times during the following spring; total grass production for the three cuts was equivalent to 4927 kg/ha dry weight for plots with lumbricids and 2874 kg/ha for plots without lumbricids, an increase of 2053 kg/ha (71.4%) in plots with lumbricids. Nine years after the introduction of the lumbricids the earthworms had spread over a diameter of 40 m from the points of introduction and pasture production had increased in the same proportion over the areas now occupied by the lumbricids. Stockdill has found that earthworms introduced to a pasture on a grid pattern at 10 m spacings can be expected to completely populate the soil in six to seven years (see Fig. 21, p. 163), a little less under high rainfall, more in drier areas. He has developed machinery to cut turfs from pastures with very high earthworm populations and to place the turfs at suitable spacings in introduction programmes. Duff (1958) demonstrated similar increases in pasture production; introduction of *A. caliginosa* to improved natural grasslands led to a 70% increase in spring production five years after introduction. In Holland, Hoogerkamp et al. (1983) also recorded increases due to the presence of earthworms in dry matter production of pastures, planted on polder soils, but the increases were not on the scale of those reported from New Zealand; the mean increase was c. 10% per year in six cuts from pasture after c. 10 years of earthworm activity.

Watkin and Wheeler (1966) related numbers and species composition of the earthworm population to production and species composition of highly productive New Zealand pastures. No relationship was evident until pastures were three years old, but in three-year-old pastures there was a positive correlation ($P < 0.05$) and in five-year-old pastures the correlation was more marked ($P < 0.01$) between earthworm numbers or biomass and pasture production. The dominant species under a pure grass sward was *Aporrectodea caliginosa*, while under mixed grass and clover it was *Lumbricus rubellus*.

Springett and Syers (1979) showed that roots of rye grass seedlings grown in the presence of *L. rubellus* or its casts grow upwards into casts on the soil surface (Fig. 34). They attributed this to the high available P content of the casts, which apparently results in a nutrient gradient sufficient to produce negative geotropism in the plant roots. No such effects were apparent when plants were grown in the presence of *Aporrectodea caliginosa* or its casts.

B. Croplands

Wollny (1890) conducted pot experiments that demonstrated some improvement of crop yield due to addition of earthworms, but Hopp and Slater (1949a) were probably first to quantify this effect. They grew various crops in pots and compared the effects of adding live earthworms (*A. caliginosa* and *Diplocardia* sp., or these two together with *Lumbricus*

Fig. 34 Root of ryegrass (arrow) growing upward from soil surface into a cast (probably of *Lumbricus rubellus*), apparently following a favourable nutrient gradient. From a pot experiment conducted on a silt loam from Palmerston North, New Zealand. (Photo: G. P. Mansell).

terrestris and *Allolobophora chlorotica*) or equivalent weights of dead earthworms on crop growth and yield. Addition of the dead earthworms increased yield by an average of 81% for crops of millet, lima beans, soybeans, wheat and hay, while addition of live earthworms produced an additional mean increase of 83%. Their results indicated that improved water infiltration resulting from the activity of the live earthworms may have been the most significant cause of better growth, while increases in available N might have been most significant where dead earthworms were added. it is likely that readily available sources of nitrogen would also have been provided on a continuous basis by live earthworms (pp. 211–219), so that both of the effects hypothesized by Hopp and Slater (1949a) would have operated in their presence.

Total dry matter production of grass, wheat and clover was shown by van Rhee (1965) to be increased by 287%, 111% and 877% by the presence of earthworms (predominantly *Aporrectodea longa* and *Lumbricus terrestris*) in pot experiments, and Ponomareva (1952) showed increases of 400% in crop growth due to earthworms. Atlavinyte *et al.* (1968) measured the growth of barley in pots, using six different soils and comparing the effects of

Aporrectodea caliginosa, *Lumbricus terrestris* and *L. rubellus*, one species at a time, at population densities equivalent to about 115/m^2, 230/m^2 and 460/m^2 respectively. Relative to controls without earthworms, maximum recorded increases in crop yield were 92%, 187% and 201% respectively for the three levels of earthworm populations. The effects varied from one soil type to another and increased with increasing survival rate of the earthworms, which was in the order *A. caliginosa* > *L. terrestris* > *L. rubellus*. Further experiments in the field (Atlavinyte 1974a) on a soddy podzolic loam, showed that *A. caliginosa* at a population density of 400–500/m^2 increased barley productivity by 78%–96% relative to soils without earthworms, and the effect was shown to be consistent in unfertilized soil or in soil with added straw as a fertilizer or addition of fertilizer nitrogen, phosphorus and potassium.

Edwards and Lofty (1980) measured the effects of earthworms on growth and yield of barley at a site where cereals had been direct-drilled for six years. Four treatments were applied, as follows:
(a) soil fumigated and inoculated with *Lumbricus terrestris* and *Aporrectodea longa*, which are deep-burrowing species;
(b) fumigated and inoculated with *A. caliginosa* and *Allolobophora chlorotica*, which are shallow-burrowing species;
(c) fumigated, no earthworms added; and,
(d) not fumigated, no earthworms added.
No significant differences in yield were apparent in the treatments, but there were significant increases in total weight of roots and their depth of penetration in the earthworm inoculated plots, especially in those inoculated with the deep-burrowing species. It was concluded that the increased root growth probably resulted from roots following earthworm burrows as they penetrated the soil, and that there may also be some stimulation of root growth due to enrichment in available plant nutrients of casts deposited on the burrow walls.

Very little is known of the effects of non-lumbricid earthworms on crop production. Khan (1966), in west Pakistan, found that the growth of maize on a loamy soil was enhanced by the addition of casts of *Metaphire posthuma*, and that their effect was greater than was obtained with the addition of farmyard manure.

Aldag and Graff (1975a) examined the effects of earthworms on crop quality. They measured the protein content of oats grown for 18 days in pots containing a brown podzolic loess soil in the presence of *Eisenia fetida*, with and without additions of cattle manure. In the presence of *E. fetida* they found marked increases in dry matter production and in total protein and protein nitrogen in the plant tissue, and the increases were greatest when cattle manure and earthworms were both supplied. In the presence of *E.*

fetida the proportion of acidic and neutral amino-acids in the protein was increased, while *E. fetida* and cattle manure together resulted in a marked increase in the plants of basic amino acids. The significance of these results is difficult to assess. The typical habitat of *E. fetida* is in dung heaps, compost heaps and other concentrations of organic detritus, but it is rare in cultivated fields.

Experiments of Atlavinyte and Vanagas (1982), in which barley and rye were grown in pots and in the field, in the presence and in the absence of *Aporrectodea caliginosa* and *Lumbricus terrestris*, showed varying effects on grain quality. The calcium content of barley grain, grown in pots, was not significantly affected by the presence of earthworms, but in the field calcium content in barley grown with earthworms was c. three times that without earthworms. In both barley and rye grain, the presence of earthworms had little effect on phosphorus, potassium, sodium, nitrogen and protein content of the grain, but in field experiments when straw was added to the soil as a fertilizer, potassium content of both grain varieties was increased in the presence of earthworms.

C. Orchards

Effects of earthworms on fruit tree growth have been studied by van Rhee (1971) in orchards on reclaimed land of Ijssel Lake in Holland, which was drained in 1957 and planted with fruit trees with grass between the tree rows in 1965-1969. Earthworms (*A. caliginosa* in all plots, with *Lumbricus terrestris* or with *L. terrestris*, *L. castaneus* and *A. chlorotica*) were introduced into plots in the orchards; early effects on the soils and fruit trees, after two to four years, included a significant increase in aggregate stability in most plots, increased pore space in plots with *L. terrestris*, increased root production, up to 75% in plots with *L. terrestris*, but reduced shoot growth, relative to plots without earthworms. Eight years after earthworms were introduced van Rhee (1977a) found that litter accumulated on the surface of plots without earthworms, but was drawn into the burrows of *L. terrestris* where this species was present, that there was a 70% increase in water-stable aggregates in the presence of earthworms (by the method of Swaby 1950), that there were more fine tree roots in soils with earthworms than in soils lacking earthworms, but that fruit production increased by only about 2.5% where earthworms were present, and this increase was probably not significant. Soils contaminated with copper oxychloride and DNOC (4,6-dinitro-o-cresol), applied to fruit trees for several years were shown by van de Westeringh (1972) to lack earthworms, and though this had led to a deterioration of soil structure, decrease in porosity, and accumulation of a

270 *Relationships with Soils and Land Use*

pasture root mat on the soil surface there was no apparent effect on fruit production relative to uncontaminated orchards where earthworms were present. If there is any relationship between earthworm activity and production of perennial crops such as fruit trees, it is likely that it would relate particularly to long-term influences on soil structure and water infiltration and may only become apparent after many years.

D. Forest trees

There is little direct evidence of effects of earthworms on growth of forest trees. Zrazhevskii (1958) grew oak and green ash seedlings in pots with and without *Aporrectodea caliginosa* and found that where plant litter was added to the soil surface addition of *A. caliginosa* resulted in increases of plant weight, over a two year period, of 26% for oak and 37% for ash. He attributed the increase in growth to increased rates of decomposition and nutrient cycling due to incorporation of litter into the soil by earthworms. Marshall (1971) similarly found that black spruce seedlings grown for a year in undisturbed soil cores in pots showed increases in stem weight when *A. caliginosa* and *Lumbricus terrestris* were present, compared with treatments that lacked earthworms.

III. Production of Metabolites that Stimulate Plant Growth

Gavrilov (1962) processed extracts of the tissues of *Lumbricus terrestris*, the soil from which they were collected, casts, mucus deposits in epidermal gland cells, and coelomic fluid, and showed that they contained plant growth factors and group B vitamins which appeared to be produced by coelamoebocytes, a group of cells described by Stephenson (1930) as one of a variety of cells that float freely in the coelomic fluid.

The presence of plant growth promoting substances in *Aporrectodea caliginosa*, *Lumbricus rubellus* and *Eisenia fetida* was proposed by Nielson (1965). He killed the earthworms in water at 45°C and homogenized them in distilled water so as not to alter any materials that might be affected by high temperatures or solvents. Subsequent extraction in various solvents and chromatographic treatment of the centrifuged homogenate resulted in isolation of several indole substances which were extracted and assayed for effects on plant growth, using the Went pea test, and resulting in isolation of one compound that stimulated growth of peas. Nielson (1965) compared the increase in growth due to the earthworm extracts with that due to the known growth stimulant indolyl-3-acetic acid (IAA), plotting log concentration of the extract against growth of peas, and found that the slope of the regression exceeded that for IAA by 15° for extracts of *A. caliginosa*, 30° for *L. rubellus* and 43° for *E. fetida*. Nielson (1965) detected similar substances that stimulated plant growth in *A. longa*, *L. terrestris* and *Dendrodrilus rubidus*.

Springett and Syers (1979) examined the growth of roots of ryegrass seedlings in the presence of casts of *A. caliginosa* and *L. rubellus*. They found that herbage production increased in the presence of either species, but that in the presence of live *A. caliginosa* and its casts there was a positive correlation (r = 0.89, r^2 = 0.79) between the total length of roots (t) and the depth of penetration of the roots into the soil (d), i.e., the roots tended to grow straight downwards, as they did also in the absence of earthworms. In the presence of *L. rubellus* and its casts the roots tended to grow laterally and vertically upwards into the casts on the soil surface (see Fig. 34, p. 266), so that the correlation between t and d was 0.57 (r^2 = 0.32). Analyses showed that the total dissolved inorganic phosphorus (DIP) in the experimental soil was 6.7 µg/g, whereas in casts of *L. rubellus* it was 16.7 µg/g and in *A. caliginosa* casts it was 7.8 µg/g. Fertilizer phosphorus (KH_2PO_4 in solution) was added to the soil at two levels, increasing the DIP content of the soil to 9.0 µg/g and 37.2 µg/g respectively, and plants were then grown with and without addition to the soil surface of casts of *A. caliginosa* or *L. rubellus*. Results are shown in Table 27; they show, relative to soil without added casts, a consistent increase in root and shoot growth in the presence of *L. rubellus* casts, regardless of soil DIP levels, an increased root growth only at the highest phosphorus level and increased shoot growth at the lowest and highest phosphorus level in the presence of *A. caliginosa* casts. The soil used for the experiments was taken from a site where both earthworm species were common and the added casts came from earthworms living at the same site, so the casts could not have contained nutrients not present in the experimental soil. Springett and Syers (1979) concluded that the earthworms must alter nutrient availability, alter the plant's ability to take up nutrients, or affect the growth mechanisms of the

Table 27. Approximate root length and shoot length of rye grass plants grown for eight days in soil with and without added P and/or casts of *Aporrectodea caliginosa* or *Lumbricus rubellus*. (Approximate values taken from Springett and Syers 1979).

DIP[a] content of soil (µg/g)	6.7	9.0	37.2	6.7	9.0	37.2
	Root length (mm)			Shoot length (mm)		
No casts added	39	38	40	75	87	100
A. caliginosa casts added	41	41	60	90	91	110
L. rubellus casts added	55	53	55	112	113	115

[a] dissolved inorganic P

plants. At all DIP levels there was a marked tendency for the plant roots to be negatively geotropic, so that they grew laterally and upwards into the casts of *L. rubellus* and they did not do so with casts of *A. caliginosa*. This observtion led to the conclusion that *L. rubellus* casts probably contain an auxin-like substance, or some substance that modifies the effects of the plant's auxins, and that casts of *A. caliginosa* do not contain such substances.

Graff and Makeschin (1980) tested the effects of substances produced by *Lumbricus terrestris*, *Aporrectodea caliginosa* and *Eisenia fetida* on dry matter production of ryegrass in Germany. They kept the three species separately in soil in pots for 11 days; during this time they eluted the soil three times with distilled water and collected the eluates. The eluates were added to other pots containing soil from the same source, lacking earthworms, and planted a short time previously with ryegrass; dry matter production in these pots was c. 10% higher than in controls without added eluate, with no significant difference between eluates from the three species. In a further experiment, the earthworms were removed from the eluted pots at the end of the 11-day elution period and ryegrass was grown in the eluted soil. The grass was harvested three times and comparisons were then made between total dry matter production and total root production after the third harvest. For all three species dry matter and root production were significantly higher than in control pots that had had no earthworms. Relative to the controls, dry matter production decreased in the order *L. terrestris* > *A. caliginosa* > *E. fetida* > control, in the proportion 100 : 84 : 76 : 50 respectively, while total root production decreased in the order *A. caliginosa* > *L. terrestris* > *E. fetida* > control, in the proportion 100 : 87 : 78 : 63 respectively, and the differences in dry matter production were significant for all the earthworms. Graff and Makeschin (1980) concluded that yield-influencing substances were released into the soil by all three species, but did not speculate on the nature of the substances.

Dubash and Ganti (1964) found free amino acids that could be absorbed by plants in casts of *Metaphire posthuma* and *Pontoscolex corethrurus* in India; cystine was present in casts of *P. corethrurus* though it was not detected in the earthworm's food, and it was concluded that it might have been produced within the gut by transamination, or that it might be derived from or through the activities of the intestinal microflora.

IV. Production of Other Biologically Active Substances

Release by earthworms into the soil of vitamins and similar substances has been investigated principally by Russian and other eastern European workers. Gavrilov (1963) quoted an early investigation of Levashov, who found some evidence of secretion or excretion of B group vitamins, and of Zrazhevskii (1957) who attributed the production of some provitamin D to earthworms. Gavrilov extracted tissues of *Lumbricus terrestris*, their casts, and soil in which they had been cultured with water and tested the effects of the extracts on growth of the yeast *Saccharomyces cerevisiae*, and on the productivity of cultures of the free-living protozoan *Paramecium*, comparing the results with controls with and without added B group vitamins. Increases in production of the yeast and the protozoa in the presence of the earthworm and earthworm-affected extracts were very large, and were comparable with controls to which B group vitamins were added.

Atlavinyte and Daciulyte (1969) measured the vitamin B_{12} content of several soils in pots, planted with barley, to which they added earthworms (*Aporrectodea caliginosa, A. rosea, Lumbricus rubellus, L. terrestris*) at six population levels. The amount of vitamin B_{12} began to increase from four to 12 months after introduction of the earthworms, reaching a maximum at 12–24 months, when it was generally c. twice, and in some cases five to seven times that of controls, and then decreased, but remained higher than in the controls after three years. The four species had nearly equal effects. It was shown that total populations of microorganisms in pots with earthworms were two to three times those in controls after six and twelve months. Further similar investigations (Atlavinyte *et al.* 1971) confirmed the increases in vitamin B_{12} and showed increases up to three to five times in microbial populations, proportional to the numbers of earthworms. Eitminaviciute *et al.* (1971) related the concentration of B group vitamins to soil animal and microbial populations in four soils in the field, including seasonal changes through the year. They found particularly strong positive relationships between soil vitamin B content and microbial numbers and between microbial numbers and earthworm numbers, which were maintained as microbial and earthworm numbers fluctuated seasonally.

It seems likely that these increases in B group vitamins in soils are not directly due to the earthworms, but perhaps to the microorganisms, whose population and activity are stimulated by the presence of earthworms (p. 26).

V. Ingestion, Egestion and Viability of Seeds

Burial of seeds is known to prolong their survival, and though very large numbers of seeds are buried in soils, and earthworms have long been known to contribute to their burial, there is little quantitative information on the significance of earthworms in these processes. Grant (1983) investigated the capabilities of *Lumbricus terrestris* and *Aporrectodea longa* to ingest and bury seeds of a wide range of grassland plants, and the germination rates of seeds that had passed through the gut.

In laboratory trials Grant (1983) sowed seeds on the soil surface in pots, with or without earthworms, and recorded seed losses from the surface for up to 85 days. In a 21-day period, no seeds were removed in control pots without earthworms; 44% were removed in pots with *L. terrestris* and 12% with *A. longa*. Distinct preferences were shown for seeds of some species compared with others. Similar results were obtained in field trials. About 75%-90% of seeds ingested by *L. terrestris* and *A. longa* were recovered from gut contents or casts, the remainder apparently having been ground in the gizzard and perhaps digested. Germination trials with seeds recovered from gut contents or casts showed little reduction in viability, but there was some interspecific variability, with seeds of *Poa pratensis*, *P. trivialis* and *Dactylis glomerata* egested by *L. terrestris* less likely to germinate than those of other species. Germination of seeds egested by both earthworm species was delayed 24-48 hours relative to controls; Grant suggested that the delay may result from the poor aeration and high ammonia content within earthworm casts, as these conditions are known to induce dormacy of seeds or to delay germination. On the basis of his experiments, Grant concluded that *L. terrestris*, which forages for food at the soil surface, is probably an important agent of seed burial, which can aid seed survival in some species, while *A. longa*, which feeds at the surface and on subsurface soil, is probably more important in moving seeds close to the surface, and so promoting germination in some species. Preferential ingestion by earthworms of seeds of various plant species may contribute to frequently recorded differences between species composition of buried seeds and that of associated plant communities.

VI. Relationships with Plant Pathogens

Many microorganisms, especially those that form spores, pass unharmed through the earthworm gut and are excreted in casts (p. 23) and earthworms are thus capable of spreading plant pathogens. Hutchinson and Kamel (1956) isolated viable spores of several species of fungus from the gut of *Lumbricus terrestris* and showed that introduction of *L. terrestris* into soil that had been sterilized and then inoculated with four of the species isolated resulted in a great increase in rate of spread of the fungi compared with that where earthworms were not introduced. Thornton (1970) isolated viable spores of 15 species of phycomycetes from the gut of lumbricids, and found them also on the body surface; he concluded that the earthworms were significant in spreading fungi, and that this significance might be increased by further dispersal by birds that feed on earthworms. In India, Rao (1979) studied stem rot of papaya (*Carica papaya*) a common disease due to *Pythium aphanidermatum*, and found that a megascolecid earthworm (*Megascolex insignis*) that feeds on the rotting roots of infected plants had viable spores of *P. aphanidermatum* in its gut, and probably contributes to spread of the disease. The general habit in earthworms of feeding on partially decayed plant tissue (p. 23) may commonly result in spread of disease organisms through the soil from plants that have been attacked by pathogens. Stephenson (1930) mentions some examples; Baweja (1939) and Khambata and Bhat (1957) implicated earthworms in the spread of *Pythium* and *Fusarium*; Hoffman and Purdy (1964) found viable teliospores of dwarf bunt (*Tilletia controversa*) in the gut of earthworms.

Passage of cysts of the potato root nematode (*Heterodera rostochiensis*) with soil through the gut of *Aporrectodea longa* was shown by Ellenby (1945) to greatly increase the hatching rate of the larvae in the cysts; an average of 38 larvae per cyst hatched from 52 cysts that were taken from casts of *A. longa* while 16 per cyst hatched from cysts isolated from the same soil that had not passed through the earthworms. The mechanism that provoked the higher hatch rate was not determined, but Ellenby suggested that it might be due to partial digestion of the outer wall of the cyst by an enzyme in the earthworm's gut. In contrast to these data, Yeates (1981) recorded a reduction of 37%–66% in total nematode populations (including pathogens) when lumbricid earthworms were introduced to three New Zealand pasture soils. The number of major nematode taxa was affected differentially, with Tylenchida generally decreasing and Dorylaimida increasing; bacterial feeding genera declined in two of the three soils, while root-feeding *Paratylenchus* generally increased, reflecting greater root production and thus increased food supply for this genus. Yeates (1976, 1978) had previously recorded a three-fold increase in earthworm abundance and biomass, accompanied by a 1.2 times increase in nematode populations, in New Zealand pastures irrigated with effluent from a dairy farm milking shed. He considered it likely that the earthworms were feeding on nematodes. Piearce and Phillips (1980) similarly concluded that nematodes made a significant contribution to the nutrition of *Lumbricus terrestris*, while Dash et al. (1980), in India, reported that *Lampito mauritii* (Megascolecidae) probably derived some of its nutrition from ingested nematodes. Yeates (1981) suggests that the overall reduction of nematode populations by earthworms may have a significant influence on damage to plants by parasitic nematodes that spend part of their life cycles in the soil. This may be so, but it must be realized that plant parasitic nematodes generally have very high levels of fecundity and correspondingly high mortality, with only a small proportion of hatchlings required to survive and

VII. Earthworms as Plant Pests

parasitize their host plants to ensure high population levels at the next generation.

Before the publication of Darwin's account of the activities of earthworms (Darwin 1837, 1881) it was commonly believed that earthworms feed on living plants, and they were widely regarded as plant pests; this was so despite the contrary evidence presented in Gilbert White's "The Natural History of Selborne" (White 1789). Graff (1983) summarized published work that attempted to disprove Darwin's assessment of the beneficial effects of earthworms, especially claims that earthworms are harmful to plants in flower pots.

There may be some circumstances in which presence of earthworms adversely affects horticultural plants, as was illustrated by Heungens (1969) in Belgium. Ornamental azaleas are grown in a conifer litter layer 10-20 cm deep, and Heungens showed that the presence of lumbricid earthworms greatly accelerates physical breakdown of this litter layer, with consequent reduction of azalea growth. Azaleas grow best in a well aerated medium; in experimental containers with earthworms, air-filled pores in the medium were reduced to *c.* 50% of the initial level after three months, while in control plots without earthworms air-filled pore space was reduced to *c.* 73%-88%. There were corresponding differences in water content, positively correlated with numbers of earthworms introduced to the containers. The combination of reduced aeration and increased moisture adversely affected growth of the plants and necessitated replanting in new conifer litter at more frequent intervals than when earthworms were absent.

Stephenson (1930) recorded several reports from India of earthworms as plant pests, including one case where *Metapheretima elongata* was said to damage the betel-vine by excessive burrowing in the soil around its roots, several where *Malabaria paludicola*, *Aphanascus oryzivorus* and *Criodrilus* sp. were said to attack the roots of paddy rice, and another where *Perionyx* sp. was found burrowing in the stems of cardamom plants. Further reports of damage to rice have come from the Philippines (Otanes and Sison 1947), Japan (Inoue and Kondo 1962), and China (Chen and Liu 1963), of destruction of tobacco seedlings in Bulgaria (Trifonov 1957) and India (Patel and Patel 1959), damage to vegetable crops in India (Puttarudrian and Sastry 1961), while Edwards and Lofty (1977) recorded destruction of lettuces by earthworms that pulled fragments of the leaves into their burrows, and accumulations of casts in cereal stooks to the point where it was difficult to obtain clean grain. It seems likely that earthworms do occasionally damage healthy plants, but it is probable that in most cases the plants are already attacked, and partly decayed by other organisms, and the earthworms attack the partly decayed tissue.

Seeds have been shown to be ingested by *Lumbricus terrestris* and some other species (p. 273). McRill and Sagar (1973) exposed seeds, in lots of 10, of 14 common British grassland plants to *L. terrestris* in cultures and found that up to 60% of individual seed species were ingested; recovery in casts ranged from zero (*Agrostis tenuis*) to 80% (*Dactylis glomerata*), while germination of seeds of four species from the casts ranged from 15% to 90%. McRill (1974) found seeds of many species of weeds in *L. terrestris* casts in grassland and reported that many weed seedlings were located where casts were found or had previously been; the earthworms were implicated in the spread of weeds, and in providing an apparently favourable microenvironment in their casts for establishment of the seedlings.

14
Earthworms and Land Use Practices

I. Land Clearance and Revegetation

Until about the middle of the nineteenth century agricultural machinery was not capable of cultivating natural grasslands, and it was usual to clear forested land to develop new areas for agricultural and pastoral farming. Forest clearance for land development still continues throughout much of the world; as a background to the changes in the earthworm fauna that result from forest clearance, it is of interest first to consider the effects of the development of grasslands on Lumbricidae in Europe, where climax forests formerly covered most of the present farm lands and where the endemic earthworms are Lumbricidae. European lumbricids are now found in most agricultural and pastoral lands in temperate regions of the world.

Bouché (1971, 1972) compared pastoral lands with undisturbed forested areas in France, and described changes in their lumbricid earthworm communities in relation to his classification of morpho-ecological categories of earthworms (p. 102). Deforestation particularly affected the épigées, whose habitats are in the litter layers or above-ground, completely eliminating the corticoles, whose under-bark habitat disappears with the trees, and drastically reducing the straminicoles (litter dwellers), although some may survive in patches where litter accumulates (e.g. around hedges). Lack of the cover formerly available on the forest floor and the resultant more variable microclimatic conditions must also contribute to the decline of épigées, but drastic changes in the nature, continuity of supply and quantity of available food are probably most important. The anéciques, which, in forest soils, live in burrows but come to the surface to feed, often remain after deforestation. Bouché (1971) pointed out that, in grazed pastures, the dung of herbivorous mammals provides an above-ground food supply and microenvironment similar to that of forest litter; it is ephemeral and therefore would not be expected to support épigées, but for the anéciques, such as *Lumbricus terrestris*, that live in burrows and feed at the surface, it provides an acceptable substitute food supply. The phenology of some litter-inhabiting species makes it possible for them to survive deforestation; e.g., Bouché (1977b) described how *L. castaneus* can persist in ephemeral habitats, since it has two generations per year, and cocoons can pass through adverse periods, when all post-embryonic stages of the species are eliminated, to hatch when conditions are favourable and exploit dung or grass litter as a food source. The endogées, which live deep in the soil, feeding on decaying roots, plant debris that falls into cracks, or ingesting soil and digesting its organic matter, are often little affected by deforestation.

Nordström and Rundgren (1973) compared species associations of lumbricids in forests with those in agricultural lands in northern and northwestern Europe. In forest habitats they found the earthworm biomass to be < 5 g/m^2 in "heath" type beech on podzols, up to about 50 g/m^2 in "meadow" type beech on brown forest soils and rendzinas in southern Scandinavia, Germany and the Netherlands, with *Lumbricus rubellus*, *Dendrobaena octaedra* and *Octolasion lacteum* dominant in the podzols and *L. rubellus* and *Aporrectodea caliginosa* dominant in the brown forest soils and rendzinas; in the deciduous forests of southern Scandinavia, Germany and Great Britain the biomass was often very high (> 75 g/m^2), with *A. caliginosa*, *A. rosea*, *Lumbricus* spp. abundant and *O. lacteum* and *Dendrobaena* spp. also common; coniferous woodlands in Germany and Scandinavia had generally low biomass (usually < 10 g/m^2), with *D. octaedra* usually dominant. Agricultural lands had variable earthworm associations, varying from four to seven to eight species, with biomass often c. 50 g/m^2, but up to 120 g/m^2, dominated by *Allolobophora* spp., commonly associated with *L. terrestris* and *L. rubellus*, and with *O. lacteum* in Germany or *O. cyaneum* in England. These species are predominantly anéciques in Bouché's terms, while *Dendrobaena* spp., which are épigées, were rarely found.

Mean earthworm populations in mixed forests in Poland were estimated by Nowak (1975) as 133/m^2, compared with 355/m^2 in pastures, but mean biomass in forests (64.3 g/m^2) was not much less than in pastures (72.7 g/m^2) because small species are most abundant in pastures, while larger species (especially *L. terrestris*) are more common in forests. The mean wet weight of individual earthworms was 210 mg in pastures, 480 mg in mixed forests, and 670 mg in oak forests. In pasture and meadow ecosystems Nowak found that earthworm numbers and biomass increased with increasing productivity (dry matter production). Similarly, in Iceland, Bengtson *et al.* (1975) found highest densities and biomass of earthworms in hay fields and abandoned farm lands, with *A. caliginosa, D. rubida* and *L. rubellus* the most common species, while in heath lands and birch wood densities and biomass were much lower and the dominant species were *Dendrodrilus rubidus* and *Dendrobaena octaedra*. Unlike the Polish woodlands, *L. terrestris* was not found in Icelandic birch forests, but was associated with man-modified habitats.

Deforestation in Europe effects changes in species associations and populations within a narrow spectrum of lumbricid species. Far more drastic changes have resulted from deforestation and contemporaneous introduction of lumbricids, where the original earthworms were not lumbricids, e.g., in New Zealand. Lee (1959, 1961) studied a series of New Zealand sites where low shrubs (*Leptospermum* dominant) and fern growing on andosols, derived from deep rhyolitic pumice showers about 2000 years old, were in various stages of clearance and conversion to pastures. Under undisturbed shrubs a sparse population of the endemic *Rhododrilus similis* (Acanthodrilidae) was found. Removal of the natural vegetation resulted in rapid extinction of *R. similis*, and in pastures (rye grass and clovers) up to about three years old no earthworms could be found. Thereafter the first species to appear was *Octolasion cyaneum*, and populations of about $25/m^2$ were commonly found; four to five years after pasture was established *Lumbricus rubellus* was commonly found, with *O. cyaneum*. As the fertility of the pasture was built up the humus-stained topsoil layer increased in depth and was invaded by *Aporrectodea caliginosa* and *A. longa*, so that in pastures 25 years old one or both of these species dominated the populations and numbers reached $250-500/m^2$.

In an area of shrub land (*Leptospermum* spp./*Ulex europaeus* dominant, but formerly *Nothofagus* forest) on podzolic soils, that was cleared and sown to rye grass/white clover pasture, Miller *et al*. (1955) found that under the shrub vegetation the earthworms present were the endemic megascolecoids *Eodrilus pallidus* (litter-inhabiting), *Maoridrilus ruber* (topsoil) and *Octochaetus multiporus* (subsoil). Within a year of the change to pasture, only *O. multiporus* could be found, and after about four to five years *A. caliginosa* and *L. rubellus* invaded the empty topsoil niche and over a period of a few years increased to large numbers, with *O. multiporus* still present in the subsoil. This pattern of succession of lumbricid species is common to areas cleared from native vegetation and planted in pasture throughout New Zealand, with, apparently, a clear break between the disappearance of litter-inhabiting and topsoil-inhabiting endemic species before the appearance of the lumbricids, which usually invade from surrounding areas developed previously. In the least fertile soils with pastures of low productivity, the succession stops with a low population density, dominated by *O. cyaneum*. More productive pastures have populations that include progressively higher proportions of *L. rubellus* and species of *Aporrectodea*. Watkin and Wheeler (1966) correlated earthworm numbers (*A. caliginosa* and *L. rubellus* dominant) with dry matter production of highly productive New Zealand pastures; no relationship was evident until three years after pasture establishment, when a significant correlation ($P < 0.05$) was shown, and after five years the level of significance had increased ($P < 0.01$). They found that under pure grass swards, grazed by sheep, the population was dominated by *A. caliginosa*, while under mixed grass/clover swards *L. rubellus* was dominant, with total populations varying through the year from 150/m² to 460/m² under the grass and from 355/m² to 800/m² under the grass/clover swards.

Very large areas in New Zealand have been planted with exotic forest trees (mainly pines, and most commonly *Pinus radiata*), sometimes after clearance of endemic scrub, sometimes by over-planting of depleted endemic grasslands. The largest plantations are on andosols from recent volcanic ash, in the North Island, where the original earthworm population was sparse, including *Diporochaeta obtusa* (litter species) and *Rhododrilus similis*, with small numbers of other species of *Rhododrilus* and *Deinodrilus agilis*; Lee (1958) recorded population densities of 99/m² and 25–74/m² at two sites under endemic grasslands on similar soils. In mature pine forests all of these species can still be found, but the lumbricids *A. caliginosa* and *L. rubellus* are also present in large numbers, with some *O. cyaneum*, living in the humified topsoil with the endemic species.

Wood (1974) compared the earthworms of unmodified forest and alpine grassland habitats with those of disturbed and modified habitats in the Kosciusko National Park (New South Wales). In unmodified habitats he found only megascolecids, including species that live in litter and topsoil, in the subsoil, or in permanent burrows penetrating from the topsoil to the subsoil (p. 166). Where *Eucalyptus* forest had been felled, but not burned, and the ground cover consisted of endemic plant species, he found endemic topsoil and subsoil megascolecids, but no lumbricids. Where forest had been felled and burned and the ground cover consisted of exotic grasses and weeds, in lawns and gardens, and in roadside clearings, he found the lumbricids *L. rubellus*, *O. cyaneum*, *A. rosea* and *A. longa*, and a few remaining endemic subsoil earthworms (*Notoscolex montikosckiuskoi*), a situation very similar to that commonly found in New Zealand. Similarly, Ljungström and Reinecke (1969) found that endemic microchaetine species (Glossoscolecidae) in undisturbed South African soils were replaced by lumbricids when the soils were cultivated, except that some large subsoil-inhabiting microchaetines often survived land clearance and revegetation.

Associations of certain lumbricid species with humans are discussed in Chapter 8. It is of interest to note here that a similar spectrum of lumbricid species has been favoured by forest clearance and land development in temperate regions of the southern hemisphere and in northern Europe, which was the origin of the most widespread species.

II. Cultivation

"The plough is one of the most ancient and most valuable of man's inventions; but long before he existed the land was regularly ploughed, and still continues to be thus ploughed by earthworms" (Darwin 1881).

Conventional mechanical tillage practices such as ploughing, rotary-hoeing and harrowing have been shown by many authors to lead to a reduction in earthworm populations and loss of the beneficial effects on soil properties implied in the above quotation from Darwin. In contrast, some experiments have shown that earthworm numbers increase with cultivation.

Low (1972) reported on observations and measurements made in England of effects of cultivation over 25 years; numbers of earthworms in old arable land were only 11%–16% of those in old grassland on the same soil series, and after three years' cultivation of old grassland, numbers were reduced by 50%. In 100-year old pastures he found populations of 50–180/m^2, while in 90-year old arable land at an adjacent site populations were 1–20/m^2 (Low 1976). Similarly, it has been found that lumbricid populations are generally lower in arable fields or cultivated areas than in permanent grasslands in Hungary (Zicsi 1958a, 1969), Russia (Ponomareva 1950), Latvia (Atlavinyte 1974b), Kazakhstan (Dzangaliev and Belousova 1969), Great Britain (Guild 1948, Evans and Guild 1948), Germany (Finck 1952, Graff 1953b), Australia (Barley 1959a, 1961), and the United States (Hopp 1946, Hopp and Hopkins 1946, Slater and Hopp 1947).

Mechanical damage, due to tillage implements, has often been assumed to be the major cause of mortality and population decline, but the loss of surface litter and general decline in soil organic matter content that results from long-continued cultivation and cropping, and leads to a reduction in food supply for earthworms may be more important.

Edwards and Lofty (1977) compared the effects of increasing intensity of cultivation of grassed plots on earthworms and showed that the more the soil was cultivated over a two-year period, the larger was the earthworm population. It is likely that two effects were confused here. If an established grassland is ploughed a large quantity of plant material, previously above ground and unavailable to the "decomposer industry", is mixed into the soil to become available as a substrate for decomposer organisms. It would be expected that earthworms whose primary source of food is partially decomposed plant tissue (pp. 266-269), would flourish as a result, and would continue to flourish, along with all other decomposer organisms, until the flush of food provided by the ploughing-in of the plants was exhausted and a new equilibrium was established. The capacity for population growth, given luxury rates of food supply, of small invertebrates and microorganisms, and their corresponding ability to decline in numbers without totally dying out, going into resting stages when their food supply declines, is basic to their survival. Cultivation over a few years probably affects the environment for earthworms similarly to mulching (see following paragraphs), both by incorporation of plant material and by breaking the surface soil into coherent aggregates, but long-continued cultivation usually results in loss of aggregate structure and reduction of soil organic matter content, and these must result in lower populations of earthworms.

In recent years, largely because of increasing fuel costs for agricultural machinery, there has been an upsurge of interest in and development of cultivation methods that require less frequent and less energetic tillage; these methods are usually described as "minimum-tillage", "direct drilling", or "zero-tillage", and "stubble-mulching", in which crop residues are left on the surface and the soil below is tilled with special implements, called "subtillers". Effects of these tillage practices on earthworm populations and consequent effects on soil structure, aeration and water infiltration have been extensively studied.

Teotia *et al.* (1950) and McCalla (1953) found that earthworm populations were much higher in the top 20 cm of USA croplands with stubble mulching than when the soils were ploughed, and related the increased earthworm populations to improvement in the stability of structural aggregates. The relative effects of deep or shallow tillage on earthworm populations were first carefully investigated by Zicsi (1967, 1969) in Hungary. His treatments comprised shallow ploughing, shallow cultivation with discs, shallow ploughing and cultivation combined, deep cultivation by subsoil loosening (subtilling) and deep tilling by plough. His investigations showed that one year after shallow cultivation with discs the mean earthworm population at four sites was 83.8% of that before cultivation, compared with 49.2% at corresponding sites that were ploughed, and corresponding figures after two years were 130.8% with shallow cultivation and 68.2% with ploughing. He further showed that, during summer drought, with increasing intensity and depth of cultivation, an increasing proportion of the earthworm population was found to be inactive and in a resting stage (Table 28), so that drastic tillage methods not only tend to reduce earthworm populations, but also increase their susceptibility to the effects of seasonal differences in soil moisture, and so reduce the period during the year when they are able to influence soil conditions for plant growth.

Table 28. Earthworm populations and percentage in resting stages during summer drought in plots subjected to various tillage practices. (From Zicsi, 1967).

Cultivation methods	Earthworm population no./m²	Proportion of earthworm population in resting stages (%)
Shallow ploughing	85.5	55.5
Shallow discing	67.0	50.1
Combination of two above	55.5	56.7
Subtilling	45.0	43.4
Deep ploughing	61.0	94.3

Edwards and Lofty (1977, 1978, 1980) have compared the effects on earthworm populations of direct drilling with those of ploughing and on growth of barley through six successive years of crops. They showed that populations and biomass were higher in direct drilled than in ploughed plots, but the effects were different for various earthworm species. *Lumbricus terrestris* populations were 1.5–6.0 times higher and their biomass was 1.7–6.7 times higher in direct drilled than in ploughed plots, while populations of *L. castaneus*, *A. caliginosa*, *A. longa*, *A. rosea*, *A. chlorotica* and *Octolasion cyaneum* were 1.1–2.6 times higher and their biomass 1.2–3.1 times higher in direct drilled than in ploughed plots. Experiments showed that *A. longa*, or *A. longa* and *L. terrestris* together (pp. 266–269), had the greatest effect on barley growth. If soils were sterilized, subsequent addition of earthworms promoted root growth of barley in direct drilled plots, but growth of roots in the plots with earthworms did not differ significantly from similarly sterilized plots that were ploughed, but lacked earthworms. In longer term experiments earthworm populations were estimated for up to eight years in plots or fields that were ploughed, chisel-ploughed, or direct drilled and sown to cereals each year (Edwards and Lofty 1982b). Deep-burrowing species (*L. terrestris* and *A. longa*) increased in direct drilled plots up to 17.5 times the numbers in ploughed plots, while shallow-burrowing species (*A. caliginosa* and *A. chlorotica*) were 3.4 times as numerous. In chisel-ploughed plots populations were intermediate between those of direct-drilled and those of ploughed plots. Barnes and Ellis (1979) similarly showed that earthworm populations in English wheat and barley croplands increase in direct drilled relative to ploughed plots, and that the effect increases with successive years' treatment, but they were not able to show any corresponding increase in crop yield. They did show a significant increase in large pores (> 1.5 mm diameter) at 20 cm and 30 cm depth, apparently due to earthworms, in the soil in direct drilled compared with ploughed soil; this effect may only be significant in terms of increased plant growth where the pattern of rainfall is sporadic and earthworm burrows function as channels for rapid infiltration during heavy showers (pp. 188–195).

The evidence for or against the beneficial effects of increased earthworm populations under conditions of reduced mechanical tillage is equivocal. There may well be circumstances, especially highly seasonal rainfall, where large earthworm populations are beneficial, but more work is necessary to resolve the question.

In the less industrialized countries of the world, where shifting cultivation with hand tools is the normal land use pattern, earthworms may be more important to the maintenance of soil fertility.

Critchley *et al.* (1979) monitored cast production in Nigeria, comparing plots that were cleared, cultivated and planted with cowpeas (*Vigna unguiculata*) with "bush" plots that had lain fallow for five to eight years and had reverted to forest. The dominant earthworms were the eudrilids *Hyperiodrilus africanus* and *Eudrilus eugeniae*. In the bush plots *H. africanus* produced about nine times as many casts as *E. eugeniae*. In newly cultivated soils *H. africanus* casts outnumbered those of *E. eugeniae* and were concentrated particularly under the crop canopy. All casting ceased during the first dry season after planting, and in the next wet season *E. eugeniae* casts were more frequent in cultivated plots than those of *H. africanus*; by the end of the wet season *E. eugeniae* casts were seven times more frequent than those of *H. africanus*. Critchley *et al.* (1979) related these changes to differences in soil temperatures, which fluctuated between 26°C and 32°C in bush plots and between 23°C and 41°C in cultivated treatments at 10 cm depth, and soil moisture contents, which were consistently higher in bush plots than in cultivated plots. Wilkinson (1975) investigated earthworm populations and related them to changes in rates of water infiltration in the savanna lands of northern Nigeria, under land use rotations that involved two, three or six years of fallow, followed by three years of crops. Measurements of water infiltration rates and structural aggregates showed that improvement of soil physical conditions was strongly correlated with, and primarily due to intensity of earthworm activity during the period of fallow. By contrast, Lal (1974) compared earthworm populations and maize production on plots subjected for two years to zero-tillage or conventional tillage (ploughing) in Nigeria. He found that, at the end of the two years, although the zero-tillage plots had 2400 earthworm casts per square metre compared with 100/m^2 in ploughed plots, infiltration rates were respectively 36 cm/h and 21 cm/h, and the soils of the zero-tillage plots had higher organic C, higher cation exchange capacity, exchangeable Ca and Mg, nitrate-N and available P than those of ploughed plots; there was no significant difference in maize production between the tillage treatments.

It might be concluded that conventional mechanical tillage can replace any beneficial effect of earthworms on plant growth under intensive cropping, although this may depend on the capability of particular soils to maintain their porosity and structure when subjected to frequent tillage over a long period. On the other hand, it might be concluded that the increased earthworm populations where minimum tillage methods are applied may reduce the number of tillage operations necessary, with a consequent saving in energy input.

III. Mulching

The ancient agricultural practice of mulching, where animal dung and/or plant residues are spread on the soil surface, has the dual purpose of insulating the soil from extremes of temperature and moisture variation and of providing plant nutrients and humic materials for incorporation into the soil from the decomposition of the added organic matter.

Clean cultivation of crop lands after late summer harvesting, as commonly practised in the USA at the time, was shown by Hopp and Linder (1947), Slater and Hopp (1947), Hopp and Slater (1949b) in Maryland and Ohio, to result in freezing of the top 10 cm or more of the soil in the winter, leading to death of lumbricid earthworms, and consequent deterioration in soil aggregate stability, large pore space, and water infiltration. Mulching with straw prevented freezing and resulted in survival of the earthworms and retention of favourable soil structure. Over a three-year period Graff (1961) found that the mean population of *Aporrectodea caliginosa* was $138/m^2$ in straw-mulched cultivated plots, compared with $92/m^2$ in control plots that received no mulch. He experimented with various kinds of mulches (Graff 1969, 1970) and found that straw and compost, which provide food for earthworms as well as surface insulation, are more effective than mulches of peat or plastic film in that earthworm numbers increase, large pore space is greater, more surface casts are produced, and these provide increased quantities of available plant nutrients in soils mulched with straw or compost compared with those with peat or plastic film. Similarly, in Dutch orchards, Hoeksema and Jongerius (1959) showed that lumbricid earthworm numbers were greatly increased, with corresponding increases in soil macroporosity, moisture holding capacity, infiltration rates and stability of aggregates, when ground cover of grass was mowed at frequent intervals and allowed to lie as a mulch on the ground surface. Bosse (1967) in Germany, showed that the structure, aeration and infiltration of water in vineyard soils were greatly improved in the presence of much enlarged lumbricid populations where mulches of straw, green mulch, and compost were applied, compared with vineyards that were tilled throughout the year, in the conventional manner.

Edwards and Lofty (1979) compared the effects of four methods of handling the straw residues from cereal crops on earthworm numbers and biomass. The four treatments, for three successive years before sampling, were: (a) straw chopped and spread evenly on soil surface, (b) straw baled and removed, stubble not burnt, (c) straw heaped along rows and burnt, and (d) straw spread evenly and burnt. There was little difference in numbers between the two unburnt treatments, nor between the two burnt treatments, but numbers were 15%-20% lower in the latter. The large, deep-burrowing *Lumbricus terrestris* was common in unburnt treatments and almost completely absent in burnt treatments; *Aporrectodea longa*, also deep-burrowing, was unaffected; *A. caliginosa*, a shallow-burrowing species, decreased markedly under all treatments where straw was removed, while *Allolobophora chlorotica*, also shallow-burrowing, increased, more or less compensating for the decrease in *A. caliginosa*. The overall effect was a considerable change in proportions of species, a small effect on total numbers, and virtually no effect on biomass. In the long-term, continued removal of straw must lead to a decrease in soil organic matter, with consequent decrease in earthworm populations; the changes in proportions of the earthworm species present are likely to result in deterioration of soil structure, macroporosity and water infiltration.

In irrigated peach and pear orchards in the Goulburn Valley (Victoria, Australia), Tisdall (1978) collected earthworms in three separate years, and found mean populations of 150/m^2 at 158 sites. The dominant species was *Microscolex dubius* (Acanthodrilidae), with a minor proportion of lumbricids, including *Aporrectodea caliginosa*, *A. longa*, *A. rosea* and *Lumbricus rubellus*. Populations were high only where plant litter on the surface was plentiful (11-60 g/m^2) and where summer irrigation was frequent and sufficient to maintain a soil suction < 100 kPa for > two-thirds of the time. In one block of peach trees, straw (68 t/ha) was mixed into the soil, followed by annual surface applications of straw and sheep dung (5.5 t/ha of each), without cultivation, and irrigation was frequent and aimed to keep soil suction at < 30 kPa throughout the growing season. The earthworm population (Lumbricidae) in this soil three years after planting, was *c.* 2000/m^2, compared with a mean of 151/m^2 for all 158 sites sampled, air-filled porosity was 19% at 4 kPa suction, compared with a mean of 5% for all 158 sites, and the time required for infiltration of 50 mm water was 1 min, compared with a mean of 83 min for all 158 sites. Tisdall concluded that, with carefully planned addition of mulching materials and control of soil moisture, earthworms can replace the conventional cultivation between tree rows and maintain a stable, well-structured soil.

There may be differences in species composition of earthworm communities, related to types of mulching materials used, as was illustrated by Kühle (1983) for lumbricids in orchards in Germany. Mulching with wood chips resulted in maximum abundance (540/m^2, wet weight 208 g/m^2), and particularly increased *A. caliginosa* numbers (154/m^2, 82.9 g/m^2), when compared with green grass-mulching (272–362/m^2, 94–114 g/m^2 including *A. caliginosa* at 24–51/m^2, 23.4–42.5 g/m^2) and bare soil surface treatment (up to 64/m^2, 9.6 g/m^2, with no *A. caliginosa*). *Lumbricus terrestris* populations were particularly favoured by green-mulching applied to a grassed ground layer, while *Helodrilus antipae* was also favoured by green-mulching with grass, and was absent from plots with a mulch of wood chips.

In the labour-intensive non-mechanized cropping systems of less developed countries, mulches and animal manure are commonly applied by hand. Mba (1978), in Nigeria, has studied population changes of the earthworms, and effects on soil properties of the common western African topsoil-inhabiting earthworm, *Eudrilis eugeniae* (Eudrilidae). A series of treatments of an acid sandy subsoil (pH 5.2, 80% sand) including mixing in of 2% cow-dung, goat-dung, or poultry-dung, with or without surface mulches of grass, cassava leaves, plantain pseudotrunk, and cocoyam leaves, was set up and maintained in wooden crates for c. 20 weeks, with four clitellate specimens of *E. eugeniae* added to each crate at the beginning of the experiment. *Eudrilus eugeniae* populations increased in all treatments, with minimum mean populations after 20 weeks of 27.3 per culture crate in the presence of the plantain mulch, and maximum of 280.7 per culture crate in the presence of the cassava mulch. The three kinds of animal dung were about equal in their effects on populations. Positive correlations were shown between final worm populations and reduction in soil bulk density ($P < 0.1$), and final carbon and nitrogen contents of the soil also related to the bulk density ($P < 0.02$, $P < 0.01$ respectively). The final earthworm population and biomass per culture crate were correlated positively with the nitrogen content of the mulch ($P < 0.02$, $P < 0.05$). Mba concluded that *E. eugeniae* was a significant agent for the incorporation of animal manures and plant mulches, and for the improvement of soil fertility in Nigerian conditions, but that its significance depended on the provision of food (mulch of suitable quality) especially in terms of its nitrogen content.

IV. Irrigation

Gerard (1960) compared lumbricid populations of irrigated and non-irrigated plots on sandy soils in southern England. The deep-burrowing species *Lumbricus terrestris* and *Aporrectodea longa* were not affected, with populations of 26/m² and 0.3/m² respectively in the irrigated compared with 25/m² and 0.1/m² in the non-irrigated plots, but populations of the shallow-burrowing species *Allolobophora chlorotica, Aporrectodea caliginosa* and *A. rosea* were 9/m², 4/m², and 4/m² respectively in irrigated plots, compared with 0.2/m², 0/m² and 0/m² in non-irrigated plots.

Lee (1959) recorded lumbricid populations of *c.* 125/m² in non-irrigated plots on seasonally dry stony soils at Winchmore, New Zealand, compared with *c.* 375/m² on plots that had been irrigated for eight years to maintain total soil moisture contents at a minimum of 11% and of 20%. *Aporrectodea caliginosa, L. rubellus* and *Octolasion cyaneum* were present in all the plots, with *A. caliginosa* the dominant species on the irrigated plots. In Australian orchard soils, Tisdall (1978) found that a combination of irrigation and mulching was necessary to maintain high populations of lumbricids (p. 109).

Earthworms have been introduced into irrigated desert soils in the USSR, where there were none before irrigation. Ghilarov (1965) reported on the early work of Dimo, who reported that populations reached *c.* 100/m², and were responsible for an increase in porosity equal to *c.* 1% of the volume of the 0–100 cm soil layer, 10–18 years after earthworms were introduced. Ghilarov (1965) also noted that, in Tadjikistan, Valiachmedov and Perel' showed that the only earthworm present in non-irrigated dark grey desert soils (sierozems) was *Perelia kaznekovi*, while in irrigated soils there were up to four more lumbricid species, with *A. caliginosa* usually dominant. In the light grey sierozems of the Golodnaja Steppe and of Tadjikistan there were no earthworms in non-irrigated soils. Ghilarov and Mamajev (1967) noted that earthworms spread along the banks of rivers and irrigation ditches into irrigated areas; earthworms were introduced in soils of oases that were irrigated from wells, and the most successful species were *A. caliginosa* and *A. rosea*.

The effects of earthworms on water infiltration in irrigated soils are discussed on pp. 188–195.

V. Fertilizers

Fertilizers are conveniently divided into inorganic and organic groups, the former mined, or synthesized in chemical factories, and the latter consisting of animal dung and decayed plant debris, alone or in various combinations, and closely related to some mulching materials.

The effects of various inorganic fertilizers on earthworms apparently vary from site to site, and have led to opposing claims that they are harmful or that they are beneficial to earthworms. Khalaf El-Duweini and Ghabbour

(1964), in Egypt, checked the survival of groups of five earthworms after 24 hours' immersion in dilute buffer solutions, ranging in pH from 2.2 to 8.0, and rising in steps of 0.2, and in sodium borate solution of pH 9.16. The lower limit of pH for survival was 5.4 for *Aporrectodea caliginosa*, 6.4 for *Metaphire californica*, and 5.2 for *Alma stuhlmanni*; all individuals survived at higher pH, up to 9.16. Egyptian soils are mostly slightly alkaline to about neutral in reaction, as are those of much of Australia, and other dry regions, but many of the soils of moist temperate regions are in the pH range from neutral to more acid, and may often have pH 4–5.

The initial soil pH and the pH of added fertilizers may be most important in determining the outcome for earthworms of fertilizer treatments. Khalaf El-Duweini and Ghabbour (1964) tested the effects of various electrolytes on groups of five earthworms of each of the same three species (p. 61 and Table 4). The salts used in the experiments included $(NH_4)_2SO_4$, which was not in particular found more damaging than other salts. In New Zealand, where soil pH is commonly < 6, $(NH_4)_2SO_4$ is preferred as a nitrogen fertilizer on bowling greens, as it further lowers the pH and rids the turf of the common lumbricid earthworms. Richardson (1938) reported that it eliminated earthworms from plots at Rothamsted, but Edwards and Lofty (1982a), who also worked at Rothamsted found that additions of farmyard manure with $(NH_4)_2SO_4$ to arable land resulted in larger earthworm populations than did farmyard manure with $NaNO_3$. Edwards and Lofty (1982a) sampled earthworm populations from long-term plots, where fertilizer treatments have been tested regularly since 1836 or 1843 on continuous pasture or annual wheat crops, and on short term experimental plots, where fertilizer treatments had been applied for only two years. Fertilizer treatments on the long-term plots included farmyard manure with added $(NH_4)_2SO_4$ or $NaNO_3$, and combinations of nitrogen, phosphorus, potassium, sodium and magnesium in inorganic forms. On the short-term plots sewage cake or sewage sludge were used at rates equivalent to 200, 402 and 536 kilograms of nitrogen per hectare per year. All forms of applied nitrogen resulted in increased earthworm populations relative to untreated controls, with relative increases greater in arable land that in grassland. Organic fertilizers increased earthworm populations more than inorganic fertilizers in arable lands for equivalent increments of N, and this was attributed to the additional food (organic matter) for earthworms provided by the farmyard manure and sewage products. *Lumbricus terrestris*, which feeds on surface organic matter, particularly increased in numbers compared with *Aporrectodea longa*, *A. caliginosa* and *Allolobophora chlorotica*.

Huhta *et al.* (1967) found that addition of a fertilizer mixture, at 800 kg/ha, including 11.3% nitrogen, 3.0% phosphorus, 2.5% potassium, 5% calcium, 4% sulphur, and 3% chlorine, to acidic soils under pine forest in

Finland, led to almost complete elimination of earthworms, and they did not reappear until two and a half years later. On the other hand, Huhta (1979) found that addition of agricultural lime at 1 kg/m^2 (10 t/ha) to plots in spruce forests on acidic soils resulted, six years later, in a population density of 302/m^2, with a biomass (dry weight) of 8.7 g/m^2, compared with 105/m^2 and 0.7 g/m^2 in control plots, with accompanying increase in species diversity, especially in the dominance of *A. caliginosa* in the limed plots and its absence from controls. Hanschko (1958) similarly found that in forests of central Germany, plots treated with lime at 2 t/ha showed an increase in mean earthworm biomass three years after treatment, from 7 g/m^2 to 20 g/m^2 under beech and from 0 g/m^2 to 3 g/m^2 under spruce. Edwards and Lofty (1977) give several examples of population increases due to lime, and conclude that they are probably confined to soils with pH < 4.5-5.0, because common lumbricids are not much affected by pH variations above this level.

Effects of superphosphate on earthworms are also somewhat confused, perhaps again because this fertilizer contains much free sulphate and is strongly acidic, so that direct contact with earthworms or addition to acidic soils can cause death of earthworms. However, it is well established that increased fertility due to top-dressing with superphosphate, often with lime, results in large increases in pasture production, hence in availability of litter and dead roots as food for earthworms, and hence results in very large earthworm populations, up to *c.* 900/m^2 lumbricids, with a biomass (fresh weight) of *c.* 270 g/m^2 in New Zealand pastures (Sears and Evans 1953). The casts of earthworms, especially of some lumbricids provide an enriched source of readily available P for plants in fertile pastures (p. 222).

Zajonc (1970) found that application to pastures in Slovakia of a fertilizer mixture at a rate that provided N 100 kg/ha, P 60 kg/ha and K 80 kg/ha produced optimum conditions for lumbricid earthworms, which reached a maximum of 692/m^2. Rates of N > 200 kg/ha resulted in reduction of species diversity in these and in subsequent experiments (Zajonc 1975). Similarly Gerard and Hay (1979) found highest populations of lumbricids (8 spp.) in soils under barley where fertilizer nitrogen was 50 kg/ha or 100 kg/ha, but significantly lower numbers at N 150 kg/ha.

Animal dung, often mixed with straw from stables and known as farmyard manure is the most common form of organic fertilizer. Satchell (1967) reported that a field at Rothamsted that received 35 t/ha of farmyard manure and 0.75 t/ha of guano every fourth year had a lumbricid earthworm population three times that of an unmanured plot. He associated the higher populations with the relatively high protein content of the added organic matter, perhaps associated with a relatively high sugar content, which is known to be attractive to earthworms. Many investigations have shown similar increases in earthworm populations, with associated enhancement of soil fertility, when farmyard manure is mixed into agricultural soils.

Mixtures of organic and inorganic fertilizers, especially when associated with sowing of lucerne, were shown by Valiachmedov (1969) to favour lumbricid earthworms, and he concluded that it should be possible to regulate earthworm populations at optimum levels of influence on soil fertility by judicious manipulation of fertilizer inputs and sowing of legumes.

There are indefinite boundaries between the application of organic mulches on the soil surface and their incorporation by earthworms, the use of farmyard manure and its incorporation into the soil as a fertilizer, and the practice of disposing of organic wastes by spreading them on the soil as slurries or solids, in the hope that they will be incorporated into the soil by earthworms and other soil organisms, to be decomposed and consequently rendered innocuous.

VI. Biocides and Earthworms

A. Toxicity of Biocides to Earthworms

The development of new biocides to control microbial, fungal, animal and plant pests in agricultural and pastoral industries has been a feature of modern agricultural chemistry. Their use has made it possible to greatly increase food production, and it would not be possible to maintain or increase present levels of nutrition for the expanding human population without these potent chemicals. However, it is important to understand that their effects are not necessarily confined to species that do harm to the crops or domestic animals whose protection is intended. Some of the most effective "pesticides" are in fact broad-spectrum biocides, whose effects include the accidental destruction of beneficial species. From the preceding account of the importance of earthworms in promoting soil fertility and plant growth it is obvious that care should be exercized to select compounds that do not excessively damage earthworm populations.

The range of biocides now available is very wide, and new compounds and application methods are constantly appearing. Comparatively few have been tested on earthworms; in Table 29 I have summarized the known effects of 84 compounds, of a bacterium, *Bacillus thuringiensis* and of four materials traditionally used to eliminate earthworms (vermicides). Insecticides, acaricides and nematicides are grouped together because many are used

Table 29. Summary of known toxic effects on earthworms of some commonly used biocides.[a]

Biocide	Toxicity	References[b]	Notes
A. INSECTICIDES, ACARICIDES AND NEMATICIDES			
1. Organochlorine Compounds			
Aldrin	Very little effect at normal dose rates; toxic at very high dose rates	1,4,5,6,9	83% survival, but drastic reduction in reproduction of *Eisenia* sp. after 8 weeks' exposure at 8.4 kg/ha a.i. (Ref.5)
BHC	Very little effect at normal dose rates; more toxic in liquid than in powder application; toxic at very high dose rates	1,2,3,4,5,9	96% survival of *Eisenia* sp. after 8 weeks' exposure at 5.6 kg/ha a.i. (Ref.5)
Chlordane	Very toxic, used widely for earthworm control	1,4,5,6,7,8,9	LD_{50} = 5.6 kg/ha a.i. for *Eisenia* sp., as dust (Ref.5) LD_{50} > 35 kg/ha a.i. for *L. terrestris* (Ref.7) All earthworms eliminated in 2 months by 48% emulsion at 44.8 kg/ha a.i. (Ref.6); in 18 months by 11.2 kg/ha a.i. (Ref.8)
DDT	Very little effect at normal dose rates; toxic at very high dose rates	1,4,5,9,12	*Eudrilus* spp. (Nigeria) avoid surface soil treated with DDT; reduced surface casting associated with loss of fertility (Refs 10,11). 90% survival, no effect on reproduction of *Eisenia* sp. after 8 weeks' exposure at 16.8 kg/ha a.i. (Ref.5)

Table 29—continued

Biocide	Toxicity	References[b]	Notes
Dieldrin	Little effect at normal dose rates; long-term use may reduce populations; high dose rates used for earthworm control (16.8 kg/ha a.i.)	1,4,5,8,13,14	100% survival of adults, but large reduction in reproductive rate of *Eisenia* sp. after 8 weeks' exposure at 5.6 kg/ha a.i. (Ref.5) LD_{50} = 3.1 kg/ha a.i. for *Eisenia* sp. (Ref.5) *L. terrestris* exterminated in 18 months by 11.2 kg/ha a.i. (Ref.8)
Endrin	Variable. Some reports of little effect, some of high toxicity at normal dose rates	4,9	100% survival of adults, but reduction in reproductive rate about equal to dieldrin for *Eisenia* sp. after 8 weeks' exposure at 5.6 kg/ha a.i. (Ref.5)
Heptachlor	Very toxic, similar to chlordane; used for earthworm control	4,9	93% survival of adult *Eisenia* sp. but reproductive rate reduced to zero after 8 weeks' exposure at 8.4 kg/ha a.i. (Ref.5)
Isobenzan	Variable, from no effect to highly toxic	4,15,16	No effect at 2.24 kg/ha a.i. in Englnd (Ref.15), but complete extermination for 3 years on one soil in New Zealand (Ref.16)
Toxaphene (= campheclor)	Toxic; used for earthworm control	5,9	76% survival of adult *Eisenia* sp., but reproductive rate reduced to zero after 8 weeks' exposure at 33.6 kg/ha a.i. (Ref.5)

14. Earthworms and Land Use Practices   295

2. *Organophosphorus Compounds*			
Azinphosmethyl	Moderately toxic	4,9	100% survival of adults, large reduction in reproductive rate of *Eisenia* sp. after 8 weeks' exposure at 5.6 kg/ha a.i. (Ref.5)
Bay 92114	Probably not toxic at normal dose rates	7	$LD_{50} > 35$ kg/ha a.i. for *L. terrestris* (Ref.7)
CGA 12223	Probably not toxic at normal dose rates	7	$LD_{50} > 35$ kg/ha a.i. for *L. terrestris* (Ref.7)
Isafos	Slightly toxic, probably not significant at normal dose rates	4,17,18	Field trials show short-term reductions up to 50% in populations at 4.5 and 9 kg/ha a.i. (Ref.17)
Chlorpyrifos	Probably not toxic at normal dose rates	4,7	$LD_{50} > 35$ kg/ha a.i. for *L. terrestris* (Ref.7)
Disulfoton	Slightly toxic, probably not significant at normal dose rates	4,7	$LD_{50} > 35$ kg/ha a.i. for *L. terrestris* (Ref.7)
Fonofos	Moderately toxic; should be used with caution	4,7	$LD_{50} = 28$ kg/ha a.i. for *L. terrestris* (Ref.7)
Ethoprop (= ethoprophos)	Very toxic at normal dose rates	7	$LD_{50} = 5.08$ kg/ha a.i. for *L. terrestris* (Ref.7)
Fenitrothion	Slightly toxic, probably not significant	4,12	Surface application (granulate) at 2.24 kg a.i./ha had no significant effect on *A. caliginosa* population (Ref.12)

Table 29—continued

Biocide	Toxicity	References[b]	Notes
Fensulfothion	Heavy initial kills have been recorded but effects vary and apparently not toxic to some species	4,7,9,12,18,21,22	79% reduction in population, 92% reduction in biomass due to 3.4 kg/ha a.i. (Ref. 22) Variable effects on different species (Ref. 21) Effects as above for fenitrothion (Ref. 12) $LD_{50} > 35$ kg/ha a.i. for *L. terrestris* (Ref. 7)
Isofenphos	Probably not toxic at normal dose rates	18	
Karathion	Moderately toxic	19	Earthworm mortality 48% for damp soil, 20% for dry soil, 3 days after spraying in orchards (Ref. 19)
Malathion (= maldison)	Non-toxic at normal dose rates	4,5,20	90% survival of adult *Eisenia* sp., no effect on reproductive rate after 8 weeks' exposure at 8.4 kg/ha a.i. (Ref. 5)
Menazon	Non-toxic at normal dose rates	4	
Fenamiphos	Slightly toxic, probably not significant at normal dose rates	7	$LD_{50} > 35$ kg/ha a.i. for *L. terrestris* (Ref. 7)
Parathion	Moderately toxic at normal dose rates, very toxic at higher rates, used for earthworm control	1,4,6,19,20,23,24,25	Earthworm mortality 33% for damp soil, 2% for dry soil, 3 days after spraying in orchards (Ref. 19)

14. Earthworms and Land Use Practices

Phorate	Very toxic at normal dose rates	4,7,26,45	Turf free of earthworms (*Amynthas hupeiensis*) 4 years after heavy treatment (Ref.6) Orchard plots sprayed 1–5 times/y had higher populations than unsprayed plots; plots sprayed 6–10 times/y had very low populations (Ref.25) 81% reduction in population with 4.5 kg/ha a.i. (Ref.26) $LD_{50} > 33.17$ kg/ha a.i. for *L. terrestris* (Ref.7) 100% mortality in pasture soils after 12 years continued heavy dosage (3.3 kg a.i. ha^{-1} month^{-1}) (Ref.45)
Phosalone	Probably not toxic at normal dose rates	7	$LD_{50} > 35$ kg/ha a.i. for *L. terrestris* (Ref.7)
Terbufos	Probably not toxic at normal dose rates	4,7	$LD_{50} > 35$ kg/ha a.i. for *L. terrestris* (Ref.7)
Trichlorfon	Probably not toxic at normal dose rates	4	
Thionazin	Moderately toxic at normal dose rates	4	
3. *Carbamates*			
Aldicarb	Very toxic	4,7,23	$LD_{50} = 6.40$ kg/ha a.i. for *L. terrestris* (Ref.7)

Table 29—continued

Biocide	Toxicity	References[b]	Notes
Bufencarb	Very toxic	7	LD_{50} = 10.16 kg/ha a.i. for *L. terrestris* (Ref.7)
Carbaryl	Very toxic, used for earthworm control	4,9,13,14,23,27, 30,31	*E. fetida* not affected but *A. caliginosa, L. rubellus* very sensitive, *A. chlorotica* moderately sensitive (Ref.23) 60% reduction in pasture populations at 2.25 or 4.5 kg/ha a.i.; 66% reduction in forest clearings at 5.6 kg/ha a.i., effective at least 14 months (Ref.9) Effective control, but little persistence (Refs 14,27) High mortality with direct contact at concentrations of 0.1% (Ref.30)
Carbofuran	Very toxic at normal dose rates, but effects may be temporary	4,7,9,12,18,23, 28,29	LD_{50} = 9.06 kg/ha a.i. for *L. terrestris* (Ref.7) 60% reduction of *A. caliginosa* population (76% for mature, 50% for immature) after 5 weeks with 2.24 kg/ha a.i. (Ref.12) > 80% reduction of *A. caliginosa/L. rubellus* populations, but recovery after 20 months (Ref.29) 95% reduction in pasture populations at 4.5 kg/ha a.i. (Ref.9)
Dithiocarbamate (= PHMD)	Moderately toxic to very toxic, varying between species	23	

Landrin	Probably not toxic at normal dose rates	7	$LD_{50} > 35$ kg/ha a.i. for *L. terrestris* (Ref.7)
Methomyl	Very toxic (Ref.7); essentially non-toxic (Ref.23)	7,23	$LD_{50} = 11.65$ kg/ha a.i. for *L. terrestris* (Ref.7)
Oxamyl	Very toxic	7	$LD_{50} = 14.63$ kg/ha a.i. for *L. terrestris* (Ref.7)
Promecarb	Probably not toxic at normal dose rates	7	$LD_{50} > 35$ kg/ha a.i. for *L. terrestris* (Ref.7)
Propoxur	Toxic at normal dose rates	7	$LD_{50} = 20.19$ kg/ha a.i. for *L. terrestris* (Ref.7)
Tirpate	Very toxic	7	$LD_{50} = 8.47$ kg/ha a.i. for *L. terrestris* (Ref.7)
4. Bacterial Agents			
Bacillus thuringiensis	Probably non-toxic at normal dose rates (Ref.44); slight increase in mortality possible in some species (Ref.46)	44,46	Two proprietary preparations applied at 100 times normal rate had no deleterious effects (Ref.44) Slightly increased mortality of *A. caliginosa* in pots at 1x, 2x normal dose, not at 4x (Ref.46)

300 *Relationships with Soils and Land Use*

Table 29—continued

Biocide	Toxicity	References[b]	Notes
B. *FUNGICIDES*			
Benomyl	Very toxic at normal dose rates; species differ in susceptibility	25,32,33,34,35, 36,37,38	Orchard spraying 1–4 times/y had little effect (Ref.25) LD_{50} = c. 10μg/worm for *L. terrestris* (Ref.35). Feeding habits affected — earthworms will not eat heavily contaminated litter (Refs.32,35). Populations much reduced in orchard soils and take 1–3 yrs for recovery after spraying (Refs 33,34,35,36); in pastures casting rate reduced, pasture production reduced (Refs 37,38) In pastures, *L. terrestris* eliminated but *L. festivus* population greatly increased 1 yr after application of 5 kg a.i./ha (Ref.50)
Binapacryl	Probably little significant toxicity at normal dose rates	34	Rate of litter incorporation decreases — may affect feeding habits as for benomyl (Ref.34)
Carbendazim	Very toxic at normal dose rates	34,35,37	LD_{50} = c. 10 μg/worm for *L. terrestris* (Ref.35) Benomyl and thiophanate-methyl (q.v.) convert to carbendazim after application. Effects similar to benomyl (Refs 35,37)
Copper oxychloride	Very toxic with long-continued use	4,39,40	Complete extermination of earthworms in orchard soils after many years use (Ref.40)

Ditalimfos	Probably little significant toxicity at normal dose rates	34	Rate of litter incorporation decreases — may affect feeding habits as for benomyl (Ref.34)
Thiabendazole	Very toxic at normal dose rates	35	Toxicity similar to benomyl; $LD_{50} = c.$ 10 µg/worm for *L. terrestris* (Ref.35)
Thiophanate-methyl	Very toxic at normal dose rates	33,35	Effects similar to benomyl; $LD_{50} = c.$ µg/worm for *L. terrestris*; drastic population reduction of nine lumbricid species in sprayed orchard (Ref.35)
C. *FUMIGANTS* Chloropicrin Dichloropropane- dichloropropene Metham sodium Methyl bromide	Very toxic at normal dose rates	4	Lethal even for earthworms that live in deep burrows (Ref.4)

Table 29—continued

Biocide	Toxicity	References[b]	Notes
D. HERBICIDES			
Amitrole			
Asulam			
Atrazine			
Aziprotyne			
Bromacil			
Chlorpropham			
Dalapon			
Diphenamid			
Diuron	Probably no significant toxicity at normal dose rates used in agriculture, but some are toxic at high rates sometimes used for complete suppression of plant growth	4,18,41,42,43, 46,47	Immersion in solutions at high concentrations relative to field usage causes high mortality. *E. fetida* particularly susceptible. In field use of herbicides some evidence of short-term increase in populations due to increased food available from dead plants, then of long-term decline with continued herbicide use, due to reduction of organic matter-input to soil. Some toxicity apparent, especially for amitrole, atrazine, bromacil, chlorpropham, dalapon, diphenamid, diuron, when applied at very high rates (Ref. 47)
Endothal			
Glyphosate			
Hexazinone			
Linuron			
MCPB			
Methabenzthiazuron			
Metribuzin			
Monuron			
Nitrofen			
Oxadiazon			
Paraquat			
Prometryn			
Propachlor			
Propazine			
Propham			
Simazine			
TCA			
Terbacil			
Trifluralin			

2,3,6-TBA			
2,4-D			
2-4 DES			
E. *VERMICIDES*			
Ammonium sulphate	Very toxic on acidic soils, not effective on neutral or alkaline soils	49	Applied at rate of c. 15 g/m² eliminates earthworms from soils with acid pH
Lead arsenate	Very toxic and high environmental hazard	49	Lethal at dose rates of c. 25 g/m², applied to soil surface
Mercuric chloride	Very toxic and very high environmental hazard	49	Lethal at dose rates of c. 1 g/m², applied to soil surface
Mowrah meal	Toxic, with low environmental hazard	49	Lethal at dose rates of c. 75 g/m², but toxicity declines with storage

[a] Names used for biocides are as listed in: Commonwealth (Australia) Department of Health (1983). "Pesticides. Synonyms and Chemical Names". 6th Edition. Australian Govt Publishing Service, Canberra.

[b] 1. Bauer 1964; 2. Ghabbour 1966; 3. Zachariae and Ebert 1970; 4. Edwards and Lofty 1977; 5. Hopkins and Kirk 1957; 6. Schread 1952; 7. Ruppel and Laughlin 1976; 8. Doane 1962; 9. Brown 1978; 10. Perfect *et al.* 1977; 11. Cook *et al.* 1980; 12. Martin 1976; 13. Randell *et al.* 1972; 14. Legg 1969; 15. Edwards 1965; 16. Kelsey and Arlidge 1968; 17. Edwards *et al.* 1968; 18. Finlayson *et al.* 1975; 19. van der Drift 1963; 20. Hyche 1956; 21. Griffiths *et al.* 1967; 22. Thompson 1971; 23. Stenersen 1979a; 24. Stenersen 1979c; 25. Tisdall 1978; 26. Edwards and Thompson 1973; 27. Heungens 1966; 28. Stenersen *et al.* 1973; 29. Martin 1978; 30. an der Lan and Aspöck 1962; 31. Atlavinyte 1975; 32. Stringer and Wright 1973; 33. Stringer and Lyons 1974; 34. Cook and Swait 1975; 35. Wright 1977; 36. Black and Neely 1975; 37. Keogh and Whitehead 1975; 38. Keogh 1979; 39. van Rhee 1967; 40. van de Westeringh 1972; 41. Martin and Wiggans 1959; 42. Ghabbour and Imam 1967; 43. Summers 1980; 44. Benz and Altwegg 1975; 45 Clements 1982; 46. Atlavinyte *et al.* 1982; 47. Caseley and Eno 1966; 48. Martin 1982b; 49. Walton 1928; 50. Edwards and Brown 1980.

interchangeably for various invertebrate pests; most nematicides especially are particularly potent against a wide range of animal and often microbial species. Fungicides and fumigants are listed separately, but they are often broad-spectrum biocides and may have drastic effects on invertebrate populations.

References listed in Table 29 include comprehensive reviews of Bauer (1964) and Edwards and Lofty (1977). In general, source material in their views is not included in the reference list with the table.

In some circumstances, e.g., on sports fields and ornamental lawns, and occasionally on heavily grazed and animal-trampled pastures, the surface casts of earthworms create a nuisance, or result in puddling of the soil surface (p. 179), and it is desirable to reduce or eliminate earthworm populations. Some of the biocides listed are useful for this purpose, as well as the more traditionally used vermicides.

Bouché (1984) has drawn attention to the problem of interpreting data on the toxicity to earthworms of biocides, due to the variability of circumstances in which tests have been conducted. He recommended the use of a standard synthetic medium (artisol) for tests, consisting of amorphous hydrated silica and water in a substrate of glass beads, and a standard earthworm species (*Eisenia fetida*). In view of the current interest and potential importance of earthworms in reduced-tillage agriculture, as agents of waste disposal and as a source of protein for animal food, it is essential that standards for toxicity tests should be established, and it seems likely that those proposed by Bouché (1984) may be accepted in western Europe. It should be appreciated, however, that in practice, many species of earthworms will continue to be exposed to biocides in more or less natural and highly variable situations, where their reactions will not necessarily correspond to those determined for a particular species in standard conditions. As is the case for reactions between earthworms and heavy metals (p. 240) it can not be presumed that the wide variety of species of earthworms, with their various capabilities for physiological, biochemical and behavioural adaptations, can be regarded as a kind of standard laboratory test material.

The information in Table 29 may be summarized as follows.

1. Insecticides, Acaricides and Nematicides

a. *Organochlorine compounds.* Chlordane, heptachlor and toxaphene are very toxic and endrin is fairly toxic to earthworms at normal dose rates, while very high dose rates of the other six compounds listed are toxic, although at normal dose rates they probably do little harm. These compounds are stable and persist for very long periods in soils (Jackson 1983); earthworms accumulate them in their tissues and they may present a hazard to predators of earthworms (p. 313).

b. *Organophosphorus compounds.* Ethoprop and phorate are particularly toxic to earthworms and fensulphothion is very toxic to some species; of the other 23 compounds listed some are moderately toxic and should be used with caution where earthworm survival is important, while some are slightly toxic, but many apparently do no significant damage at normal dose rates. Organophosphorus compounds are generally less persistent in soils than organochlorine compounds, and they are not known to be accumulated by earthworms.

c. *Carbamates.* Most carbamates are very toxic to earthworms, and though many are not persistent in soils they can greatly reduce populations when used at normal dose rates.

d. Bacillus thuringiensis. This insect pathogen successfully used for control of many pest species, was shown by Benz and Altwegg (1975) to have little or no effect on earthworms when applied by the usual method (as a foliar spray) at the accepted dose rates for insect control. It has been shown, however, by Smirnoff and Heimpel (1961) that *B. thuringiensis* is capable of penetrating the gut wall into the coelom of earthworms and causing fatal septicaemia, but this could not be demonstrated under field conditions.

2. Fungicides

Copper oxychloride, which was formerly widely used as a fungicide in orchards, but is now not as commonly used, accumulates in the soil and is very toxic with long-continued use; of the remainder listed, benomyl, carbendazim and thiophanate-methyl are closely related carbamates, and like most of the carbamates used as insectides and nematicides, are very toxic to earthworms; thiabendazole is not a carbamate but is very toxic, and only benlate, binapicryl and ditalimfos among those listed are apparently relatively harmless.

3. Fumigants

These materials are used for partial or complete soil sterilization and are very toxic to earthworms.

4. Herbicides

None of the 33 herbicides for which information is available is known to be significantly toxic to earthworms when used at dose rates common in agriculture, nor to be accumulated in their tissues. With the increasing use of herbicides in crop production associated with minimum-tillage or zero-tillage practices it is important that other herbicides should be examined for their effects on earthworms, as there is much evidence that the maintenance of earthworm populations is desirable when these tillage practices are used (pp. 282-286). Increases in earthworm populations have been recorded following herbicide applications, probably due to increased availability of food from the decaying tissue of the affected plants (see Edwards and Lofty 1977); with continued use, there must be a reduction in total production of plant tissue, that will eventually result in a reduction of food available to earthworms, and consequently to reduction of earthworm populations. In extreme cases, where herbicides are used to maintain bare ground over extended periods, earthworms must eventually be eliminated because of lack of food.

5. Vermicides

Lead arsenate and mercuric chloride are potent poisons of man and other animals; they were formerly widely used as pest control chemicals but their place has now been taken by less hazardous organochlorine or organophosphorus materials. Ammonium sulphate is widely used as a nitrogenous fertilizer and presents no special hazards. Its effect on earthworms results from lowering of soil pH to a level below the optimum range for common species, so it is most effective in eliminating earthworms on soils that are already acidic and it is not effective in alkaline soils. Ammonium sulphate has the added benefit for turf of stimulating growth of fine grasses and depressing clover growth, but if used for many years on acidic soils can result eventually in reduction of pH to a level at which damage is done to the grass. Mowrah meal is the ground oil-free seeds of several species of Indonesian trees of the genus *Madhuca*; it is not toxic to man, but it has the disadvantage that it slowly loses its potency during storage.

The impact of biocides that are toxic to earthworms can be reduced by careful application to soils. Ruppel and Laughlin (1976) applied a wide range of carbamate and organophosphorus biocides to croplands, using four methods of application, as follows:
(a) "in furrow" application, where granules of the biocide were placed in

a 2.5 cm wide band about 5 cm below and 5 cm to one side of the seed at planting;
(b) "band-under" application, where granules were placed in a 5 cm wide band about 10 cm below the seed at planting;
(c) "roto-till" application, where granules were placed on the surface in bands c. 18 cm wide, or were broadcast on the surface, and then roto-tilled into the top 10–15 cm of the soil;
(d) "band-over" application, where granules or biocide sprays were applied in bands c. 18 cm wide and then mixed with the surface soil by the planting drill.

Even with the very toxic carbamates aldicarb, carbofuran and tirpate, and with ethoprop (organophosphorus), earthworm mortality was much lower with the localized effect of "in furrow" application than with the other methods.

B. Mechanisms of Action of Biocides on Earthworms

Wright (1977) immersed individuals of *L. terrestris* for 1 minute in solutions with various concentrations of the carbamate fungicides benomyl, carbendazim, thiophanate-methyl and thiabendazole; those exposed to 0.5% and 0.05% (i.e. 5000 µg/mL and 500 µg/mL) concentrations died within 13 days and 27 days respectively with all four fungicides, while those exposed to lower concentrations survived, but stopped feeding for up to about 14 days. Ingestion of about 25 µg of thiabendazole by individual worms resulted in cessation of feeding and death in about 14 days. Carbamates used as insecticides are generally considered to cause death through cholinesterase inhibition, and thiabendazole, which has no carbamate group, paralleled the behaviour of the carbamates in Wright's experiments. Carbendazim is known to cause mitotic failure, and hence failure of nucleic acid synthesis in fungi and in rats, and Wright (1977) considered it possible that the same mechanism may operate in *L. terrestris*; if death were due to mitotic failure this could account for the delay between exposure and death.

There are considerable interspecific differences in the toxicity of individual biocides to earthworms. Stenersen (1979a) dipped specimens of the common lumbricids *E. fetida*, *A. caliginosa*, *A. chlorotica* and *L. rubellus* for 30 minutes in biocide solutions of various concentrations, transferred them to uncontaminated soil, and determined the highest concentrations that caused < 50% mortality and the lowest concentrations that cause > 50% mortality. The LC_{50} must lie between these concentration limits, which are listed in Table 30.

Table 30. Toxicity to four lumbricid species of biocide solutions. The highest concentrations causing < 50% mortality and the lowest causing > 50% mortality are shown. LC_{50} lies between these concentrations. (From Stenersen 1979a).

Species	Aldicarb	Oxamyl	Carbaryl	Biocide (µg/ml) Carbofuran	Parathion	Trichloronate	Paraoxon	Lindane
Eisenia fetida	25–50	>200	200–800	400–800	400–800	800	100–200	400–800
Aporrectodea caliginosa	3.1–6.3	>200	3.1–6.3	0.39–0.78	—	—	—	400–800
Allolobophora chlorotica	—	>200	~200	5–10	—	—	—	400–800
Lumbricus rubellus	12.5–25	>200	25–50	5–10	—	—	—	400–800

Eisenia fetida was the most tolerant of the species tested, but was most sensitive to aldicarb, and this material was highly toxic also to *A. caliginosa* and *L. rubellus*. *Aporrectodea caliginosa* was very sensitive to carbofuran and to carbaryl, while *A. chlorotica* and *L. rubellus* were most sensitive to carbofuran. Oxamyl, paraoxon, parathion, trichloronate and lindane displayed comparatively low toxicity to all species. Aldicarb, oxamyl, carbaryl, carbofuran, parathion, trichloronate and paraoxon are recognized as cholinesterase inhibitors. Stenersen (1979a) considered that cholinesterase inhibition might not be lethal to the species tested, since long lasting disturbance of the nervous system, including excision of the brain, is tolerated in laboratory conditions, and respiration does not depend on muscular coordination of special respiratory organs, as in insects and vertebrates. However, respiration does depend on the maintenance of blood flow through a sub-cuticular capillary network, and since blood circulation in earthworms is due to rhythmic peristaltic contractions of muscles in the walls of the major blood vessels, death could be from anoxia if these muscles lost their coordination. When *E. fetida* were kept for five days in soil containing aldicarb, oxamyl or parathion at dose rates of 1–32 μg/kg soil, Stenersen (1979a) found that they lost water, especially with aldicarb, which caused some water loss even at 1 μg/kg, rising to a maximum of c. 70% loss of body weight at 16 μg/kg, and also induced tonic tremors; these are the characteristic symptoms of cholinesterase inhibition. Carbofuran and carbaryl caused long lasting immobility and rigidity in all species tested, and these effects might be expected to be fatal in field conditions. Tests in the laboratory showed that *E. fetida* suffered only moderate mortality, even with high doses of carbofuran and carbaryl.

Further investigations (Stenersen 1979c) showed a lack of correlation between initial cholinesterase inhibition and toxicity of the biocides to *E. fetida*, leading to the discovery of two cholinesterases, E1 and E2, in this species, each with differing sensitivity and duration of inhibition when treated with various cholinesterase-inhibiting biocides. *Eisenia fetida* can survive almost complete inhibition of all cholinesterase activity, if it is transient, and reactivation of either the E1 or E2 form allows the worm to maintain itself until full recovery of function of both forms is restored. Stenersen (1979b) injected 1-ethyl ^{14}C-labelled parathion into *E. fetida*, and was able to show that the parathion was slowly converted to paraoxon, which was rapidly hydrolysed to diethyl hydrogen phosphate, and thus detoxified. The earthworms were kept in sand, and five days after injection of the parathion 70.3% of the ^{14}C-labelled material extracted from the sand and 80.4% of that extracted from the worms was in unmetabolized parathion. Nakatsugawa and Nelson (1972) similarly found diethyl hydrogen phosphate to be a parathion metabolite in *Lumbricus terrestris*.

C. Accumulation of Biocides

Organochlorine insecticides are persistent and because they are relatively harmless to earthworms and are lipophilic they can be concentrated in fat in the tissues of earthworms relative to their concentrations in soil. Other types of biocides are apparently not accumulated by earthworms (Edwards et al. 1968).

Beyer and Gish (1980) sprayed dieldrin, heptachlor and DDT, each at rates (kg/ha) of 0.6, 2.2 and 9.0, onto plots in a hay field on an alluvial loam in Maryland, and for 11 years thereafter analysed soil and earthworms (*Aporrectodea* spp. and *Lumbricus terrestris*) for residues of the pesticides. The average ratios of concentrations in earthworms (dry weight) compared with those in the soil (upper 3–5 cm) were: dieldrin, 8; heptachlor epoxide, 10; total DDT, 5. The average time for 50% reduction of residue concentrations in the soil were: dieldrin, 5.1 y; heptachlor epoxide, 3.2 y; total DDT, 3.2 y; the corresponding times for earthworms were: dieldrin, 2.6 y; heptachlor epoxide, 3.0 y; total DDT, 3.2 y.

Cook et al. (1980) found much higher concentration ratios (30–45 on a dry weight basis) of DDT and its breakdown products DDE and TDE in tissues of *Eudrilus eugeniae* and *Hyperiodrilus* sp. relative to soil concentrations in DDT-treated plots in Nigeria.

It is apparent from several investigations that organochlorine insecticides accumulate and decay at differing rates in different soils, and that concentrations vary between earthworm species. Wheatley and Hardman (1968) found that residues of γBHC, aldrin and dieldrin in an arable soil were consistently higher in the smaller, shallow-living species *Aporrectodea caliginosa*, *A rosea* and *Allolobophora chlorotica* than in the larger deep-living species *Aporrectodea longa*, *Lumbricus terrestris* and *Octolasion cyaneum*. Concentration ratios between earthworms and the soil were not linear, but decreased with increasing soil concentrations. Edwards and Lofty (1977) have reviewed the literature on the subject up to about 1972.

Accumulation of pesticides by earthworms also varies with soil properties. For example, Davis (1971) measured dieldrin concentration in *A. caliginosa* that had lived for four weeks in dieldrin contaminated soils (five soil types); multiple regression analysis of the data showed that 95.3% of the variance of dieldrin concentration in the earthworms was explained by variations in soil pH and in the reciprocal of soil organic matter content (%).

D. Effects of Biocides on Earthworm Behaviour

Although their abundance may not be much affected by many of the commonly used biocides, there is evidence that earthworms avoid surface litter and soil layers contaminated by biocides and that litter incorporation and surface casting rates are reduced when biocides are applied.

In New Zealand benomyl and carbendazim are sprayed on pastures to control the litter-inhabiting fungus *Pithomyces chartarum*, whose spores contain a toxin that causes serious damage to the livers of grazing mammals. Seven weeks after plots in rye grass/white clover pastures were sprayed with these fungicides at 560 g a.i./ha, Keogh and Whitehead (1975) found that the dry weight of earthworm casts on the ground surface was 66.8 g/m^2 in benomyl sprayed plots and 96.8 g/m^2 in carbendazim sprayed plots, compared with 345.2 g/m^2 in unsprayed control plots (least significant difference is 109.6 g/m^2 for $P < 0.01$); the dry weight of litter at the ground surface was 5.15 kg/m^2 for benomyl plots and 4.72 kg/m^2 for carbendazim plots, compared with 3.32 kg/m^2 for unsprayed controls (least significant difference is 0.72 kg/m^2 for $P < 0.01$). Estimates of abundance of earthworms 11 weeks later showed no significant differences between sprayed and control plots. Benomyl and carbendazim are potentially very toxic to earthworms (Table 29), but the earthworms were apparently repelled by the fungicides and avoided feeding on the contaminated litter or casting at the surface for sufficient time for fungicide concentrations to fall to non-toxic levels. In England, Cook and Swait (1975) similarly found that apple leaf litter disappeared from the ground surface more slowly where orchard trees were sprayed with benomyl, carbendazim and thiophenate-methyl than where alternative sprays, e.g., ditalimfos, binapicryl, were used. Contrary to the lack of reduction of earthworm populations found by Keogh and Whitehead (1975), Cook and Swait found a significant decrease where benomyl sprays were used, and the decrease became more marked with increasing frequency of spray treatments. Population density increased when benomyl sprays were discontinued, but recovery was still not complete after three years without benomyl. Wright (1977) also found that litter accumulated, surface casting decreased, and earthworm abundance and biomass declined markedly in orchards where benomyl and related fungicides were used. Of eight lumbricid species present, Wright found that all declined in numbers and biomass, but *L. terrestris* was particularly affected, probably because it lives in a permanent deep burrow and feeds on dead leaves dragged from the surface into the burrow.

Feeding trials, using leaf material with known amounts of spray deposits of benomyl, carbendazim or thiophenate-methyl showed that feeding rates were significantly reduced when the deposits reached 0.8 μg/cm^2 of leaf surface, and that feeding ceased when deposits were 1.75 μg/cm^2, an amount of the order expected after one commercial spraying. The amounts of fungicides used in orchards are much greater than those (one dose of 560 g/ha) used on pastures by Keogh and Whitehead (1975). In the orchards investigated by Wright (1977), spraying programmes involved five or seven sprays per year, each of 280 g/ha benomyl or 780 g/ha thiophenate methyl, i.e., total benomyl applied was 2.5-3.5 times that applied by Keogh and Whitehead, and concentrations in the litter must have remained high for much longer than in Keogh and Whitehead's experiments, with much more drastic effects on earthworm populations.

Treatment with DDT at high rates of application for leaf-feeding insect pests of cowpea (*Vigna unguiculata*) in Nigeria (Perfect *et al.* 1977, Cook *et al.* 1980) resulted in a reduction of surface casting by *Hyperiodrilus* spp. and *Eudrilus eugeniae*. Initial clearing and cultivation of the plots reduced the casting rates (casts per square metre per day) to < 2 but by the end of two cropping seasons the rate reached 83 in untreated plots, while it never exceeded 4 in DDT-treated plots. Surface casting by *E. eugeniae* virtually ceased in DDT-treated plots. Population estimates showed no significant differences in earthworm numbers or biomass between treated and untreated plots, and the difference in surface cast accumulation must have resulted from a change in the proportion of surface-casting to subsurface-casting. In laboratory trials with soil uniformly contaminated with up to 5 μg/g DDT, surface casting by *Eudrilus* spp. was stimulated, while for *Hyperiodrilus* sp. surface casting was stimulated with DDT levels up to 1 μg/g soil. In the field plots mean DDT levels in the top 50 cm of soil were *c.* 5 μg/g, but concentrations would have been much higher than this near the surface. At higher DDT levels, in laboratory trials, surface casting activity was reduced, and no surface casts were produced by *Hyperiodrilus* sp. at DDT concentrations > 125 μg/g nor by *Eudrilus* spp. at concentrations > 625 μg/g. Reduction in surface casting rates and a change to subsurface casting, especially in a region with very heavy seasonal rainfall like that studied in Nigeria, could have a profound effect on soil structure and fertility. Cook *et al.* (1980) considered that this could have contributed to an overall decline in fertility observed in the treated plots by Perfect *et al.* (1977).

E. Dispersal of Biocides in Soil by Earthworms

Stockdill (1966) demonstrated that earthworms can enhance the effects of insecticides on soil-inhabiting insect pests by mixing the insecticides into the soil. In pastures in New Zealand, where it was the practice at the time to use DDT, applied as a powder to the soil surface with superphosphate fertilizer, to control the root-feeding larvae of *Costelytra zealandica* (Scarabaeidae:Melolonthinae), Stockdill measured DDT concentrations and *C. zealandica* larva densities in surface soil layers of a pasture treated two years previously with DDT. His results are summarized in Table 31, and show

Table 31. DDT concentrations and populations of *Costelytra zealandica* (no./m² surface area) in the soil of a pasture in Otago, New Zealand, two years after DDT application. (From Stockdill 1966).

	Earthworms absent	Earthworms present (645/m²)
C. zealandica larvae (no./m²)	430	0
DDT concentrations (µg/g)		
Soil depth (cm) 0–1.25	20.0	11.0
1.25–2.5	1.8	6.4
2.5 –3.75	0.1	0.2

a very marked effect associated with the presence of earthworms (predominantly *Aporrectodea caliginosa*).

F. Transfer of Biocides along Food Chains

Earthworms are an important component of the diet of many birds, reptiles, amphibians and small mammals, as well as predatory invertebrates (Chapter 7). Accumulation of biocides in earthworm tissues results in transfer to their predators, and may result in dangerous levels of concentrations in predators, especially of organochlorine insecticides. The little quantitative information available on this subject is summarized below.

Brown (1978) studied the flow of DDT and its metabolites through a food chain on the campus of the University of Michigan, USA, where DDT was applied at very high rates of 0.45-2.25 kg per tree to elm trees to control the bark beetle vectors of Dutch elm disease. At this level of application, soil contamination was 5-10 µg/g DDT, and earthworms in the soil concentrated this to levels of 50-200 µg/g in their tissues; 100 earthworms contained c. 3 mg of insecticide, c. 1 mg in the form of the metabolite DDE. Earthworms are a staple item in the diet of robins (*Turdus migratorius*); 3 mg of DDT was considered to be lethal to a robin, and following the treatment of the elm trees with DDT the robin population of the campus fell from c. 370 to c. 15, none of which successfully raised young. In Wisconsin towns where DDT was similarly applied to elm trees the average annual mortality of robins was found to be 87%, compared with a normal mortality of c. 50%. The surviving population in affected areas fell to c. 20% of that previous to the DDT-treatment; as territories became vacant more robins moved from surrounding areas and were exposed to poisoning.

Hill *et al.* (1975) determined the toxicity of DDT, dieldrin and heptachlor for captive bobwhite quail (*Colinus virginianus*), Japanese quail (*Coturnix*

coturnix), ring-necked pheasants (*Phasianus colchicus*), and mallards (*Anas platyrhynchos*), and found that LC_{50} for these species were in the ranges 311–1869 µg/g for DDT, 37–169 µg/g for dieldrin, and 92–480 µg/g for heptachlor. In field studies Barker (1958) found that poisoning of robins was associated with 53–204 µg/g (wet weight) DDT in earthworms taken as food, while Collett and Harrison (1968) found that levels of DDT in earthworms taken as food that resulted in poisoning of blackbirds (*Turdus merula*) and thrushes (*T. philomelos*) were 13–29 µg/g (wet weight). Beyer and Gish (1980) have pointed out that susceptibility of birds to insecticides varies with species, with age and state of health of individual birds, and with diet, but the data that are available indicate that concentration of organochlorine insecticides in earthworms must be regarded as an important hazard to birds whose diet includes earthworms as a major component.

Shilova *et al.* (1971) measured the concentration in the tissues of moles (*Talpa europea*) of carbaryl, applied to an area of spruce forest in Russia as a wettable powder (85% carbaryl) at a rate of 0.5 g/m². Very high earthworm mortality followed the treatment. Moles decreased in numbers very rapidly after the carbaryl was applied, and this was attributed to the decline in earthworms, which are their principal food. Moles (a total of 66) were caught 10 days after the carbaryl treatment, and analyses showed that the biocide was present in the tissues of > 80% of the moles, with highest concentrations (mean 12.5 µg/g) in the uterus. The sharp decrease in number of moles was attributed to a combination of DDT poisoning and starvation resulting from the high mortality of earthworms.

15
Use of Earthworms for Waste Disposal

Increasing demand for methods of disposal of domestic and industrial wastes that minimize environmental pollution has resulted, in recent years, in increased interest in the possibility of using the organisms that make up the soil "decomposer industry" (Macfadyen 1963), including earthworms, to process large quantities of waste materials, as they have traditionally been used in domestic compost heaps and earth toilets. The major waste material requiring treatment is sewage sludge, the solid end-product of sewage treatment plants. The composition of sludge (Friedman 1978) typically includes large quantities of protein, cellulose, grease and fats; it is high in nitrogen and phosphorus, and contains variable quantities of mineral materials, including a range of heavy metals (p. 330). Farmyard wastes and organic matter-rich industrial wastes also might be handled by the "decomposer industry".

Two main methods of handling disposal have been tested and are now in use. The solid wastes are piled into heaps and treated like compost heaps, or the wastes are spread out over the soil surface, usually on pastures, but sometimes on crops or in forests, for incorporation into the soil. Earthworms are contributors to both of these methods of disposal, and have the added advantage of producing saleable materials from worthless waste in the composting procedure and of favourably influencing soil fertility in the incorporation procedure.

The potential of earthworms in the composting procedure was demonstrated by Fosgate and Babb (1972) who fed animal faeces, mixed with lime to raise the pH, to earthworms which they designated, perhaps incorrectly, as *Lumbricus terrestris*, and obtained 1 kg (fresh weight) of earthworms for each 2 kg (dry weight) of faeces. The earthworm-processed faecal material (casts) was described as a loose, friable humus-type soil, with 3.0% nitrogen;

it was shown to be equal to greenhouse potting mixes for the production of flowering plants, with the added advantage that its weight was only *c.* 50% that of potting soils. The earthworms were dried and converted to a meal, which was found to contain 58% protein and 2.8% fat, and was readily accepted as food by domestic cats.

Sabine (1983) has summarized the potential benefits of large-scale intensive culture of earthworms on organic wastes as follows:
(a) reduction of noxious qualities of a wide variety of organic wastes, e.g., elimination of smell, elimination or reduction of harmful microorganisms, increase in bulk density, greater ease of physical handling;
(b) production from organic wastes of useful and marketable organic fertilizer, compost, or potting media;
(c) protein production for stock food, or even human food, from organic wastes.

I. Suitable Earthworm Species

Many earthworm species would probably be suitable for waste disposal, but attention has focused on a few species that are known to prefer habitats with high concentrations of organic matter and to breed prolifically, especially the lumbricid *Eisenia fetida* and the eudrilid *Eudrilus eugeniae*.

A. *Eisenia fetida*

Graff (1974), Vail (1974), Watanabe and Tsukamoto (1976), Tsukamoto and Watanabe (1977), Hartenstein *et al.* (1979b) and Kaplan *et al.* (1980a) investigated productivity, growth, and population biology of *E. fetida*, when fed on animal dung, compost, or sewage sludge. Hartenstein *et al.* (1979b) found that cocoon production in *E. fetida* began when the earthworms were about four to six weeks old. Worms were kept at 25°C in experimental cultures of 300 cm^3 (78 cm^2 surface area) of horse dung, at populations of 1–16/300 cm^3. Maximum cocoon production was attained when the worms were about 11 weeks old and then declined fairly steadily, regardless of the number of worms per colony; at 16 worms per culture, production ceased after about six months, while at the lowest population densities it ceased after about one year. The overall annual production of cocoons was *c.* 260 for colonies of 3/300 cm^3, rising to a maximum of *c.* 550 for colonies of 8/300 cm^3, then falling with increasing worm numbers to *c.* 420 for colonies of 16/300 cm^3. Growth of *E. fetida* was rapid; for cultures with 4, 8, 12 and

16 individuals the total biomass per culture ranged from 4.7 g to 11.5 g for cultures feeding on horse dung and from 10.0 g to 26.0 g for cultures feeding on activated sewage sludge, i.e., *E. fetida* grew c. 2.5 times faster when fed on activated sewage sludge as when fed on horse dung. Kaplan *et al.* (1980a) found that maximum weight gain occurred at temperatures of 20°C-29°C and moisture contents of 70%-85%, in horse dung or in activated sludge. Optimum pH was found to be c. 7, with all worms dying within one week at pH < 5 or pH > 9. Total soluble salt concentrations > 0.5% were lethal, and ammonium acetate caused 100% mortality at 0.1%; concentrations in dung contaminated by urine may be lethal. The presence of soil, beneath or mixed with activated sludge, produced more rapid growth, and the increase in growth rate was proportional to the inorganic matter content of the soil.

B. Eudrilus eugeniae

This cosmopolitan eudrilid species, originally from Africa, is widely cultured and distributed by earthworm growers in the United States (see Shields 1976), where it is known as the "African night crawler". Its effectiveness for dung and sewage sludge treatment has been studied by Neuhauser *et al.* (1979), using techniques similar to those described above for *Eisenia fetida*. Optimum temperatures for this species were found to be 20°C-28°C, with hatching at 25°C about three weeks after cocoons were deposited, and given unlimited food (either horse dung or activated sludge) and space, it grew to a maximum weight of c. 2.5 g in eight to ten weeks; it then declined in size and suffered increasing mortality from about 12 weeks old, with only 35%-45% of the individuals surviving to 17 weeks. Cultures were established, with 1-16 worms per 1500 cm^3 (324 cm^2 surface area), placed on a 1 cm thick layer of soil. Growth rate was maximal in cultures of 1 worm per 1500 cm^3 and decreased continuously with increasing numbers. Optimum cocoon production was achieved at a density of 8/1500 cm^3, when c. 0.7 cocoons per week were produced per adult between ages seven to 12 weeks, and an average of c. 5 cocoons per worm was produced throughout the life of the culture. This is much lower than the productivity of *E. fetida*, which is about 70 cocoons per worm at densities of 8/300 cm^3 (see above). Converting these figures to a standard volume of substrate, lifetime cocoons production for *E. fetida* is c. 35 times that of *E. eugeniae*. Both *E. fetida* and *E. eugeniae* may have three or more generations per year, so annual production of cocoons by *E. fetida* would also be c. 35 times that of *E. eugeniae*.

On the basis of data on life cycles and reproductive rates presented in

Chapter 4 and from data in this Chapter, it is apparent that *E. fetida*, kept on animal dung or activated sewage sludge at 25°C (optimum temperature), produces cocoons at a rate of c. 3.5 per week. The cocoons hatch in c. three weeks, producing c. two hatchlings per cocoon if provided with food at luxury rates. The worms grow to maturity and begin to produce cocoons c. seven weeks after hatching, with maximum cocoon production attained c. 10 weeks after hatching, and high production rates continuing to c. 27 weeks. Maximum growth rate is attained at c. 11 weeks after hatching. At this age, feeding at 25°C on compost, *E. fetida* weighs c. 650 mg (c. 130 mg dry weight). In optimum conditions there may be four generations each year, so that harvesting the earthworms about 11-12 weeks after hatching, supplying food at luxury rates, and avoiding over-crowding, it would theoretically be possible for a population of *F. fetida* to produce c. 100 times its own weight of tissue per year. It is unlikely that such high production could be maintained in large-scale cultures, but this species has unparalleled possibilities as a producer of protein. For "vermicomposting", *E. fetida* and *E. eugeniae* are both effective species.

C. Other Species

Kaplan *et al.* (1980b) have investigated the capabilities of two pheretimoid (megascolecid) earthworms, *Amynthas hawayana* and *A. rodericensis*, which are cosmopolitan and are readily available in the United States, to process horse dung and activated sewage sludge. Studies on groups of 50-100 adults kept in large boxes of soil, with horse dung or grain supplied to the surface, showed that these two species live in anastomosing burrow systems in the soil, drawing food from the surface down into their burrows in the manner of many lumbricids (anéciques of Bouché, 1972). When kept in activated sludge, alone or overlying a 1 cm thick layer of peat, they rapidly lost weight; with sludge overlying a 1 cm thick layer of soil they gained weight for the first 15 days and then began to lose weight; when kept in a mixed medium of sludge and soil they gained weight. No cocoons were produced in winter, and the species appeared to have an annual generation cycle. Kaplan *et al.* (1980b) concluded that they might have an important role in field conditions in burying and accelerating the decomposition of organic wastes applied to the ground surface, but were not, like *E. fetida* and *E. eugeniae*, suitable species for treatment of concentrations of sewage sludge.

There is some evidence from India (Kale *et al.* 1982) and from the Philippines (Guerrero 1983) that animal manures may be successfully processed by *Perionyx excavatus*, to dispose of waste materials or to yield earthworm protein as an animal feed supplement. *Amynthas asiaticus* is also claimed to be useful for garbage treatment, producing a superior quality potting medium and useful protein for animal food as by-products (W. Tan, pers. comm.).

Satchell (in press) has listed the characters that might be sought or investigated in selecting suitable earthworm species for waste disposal programmes and has discussed the possibility of selective breeding to obtain populations with optimum capability for the purpose. It should be possible, given the large number of species of earthworms and the range of habitats they occupy, to specify the desirable ecological and physiological characters of earthworms for a particular waste disposal problem, and then to identify likely candidate species. At present there is probably insufficient information of the kind required on any but a few lumbricid species, and *Eudrilus eugeniae*, and most of this information has been obtained only very recently by those who have been directly involved in investigations of the possibility of using earthworms for waste disposal projects. There is scope here for much research and for practical application of knowledge of life cycles, phenology, feeding behaviour, digestive capabilities and other aspects of the ecology of earthworms.

There may also be scope for selective breeding from the gene pool represented by the 2500–3000 known species of earthworms; the only information available on this subject derives from a comparison by Graff (1978), who compared *Eisenia fetida* from Germany with the "Tennessee Wiggler", a commercially available stock of *E. fetida* which Graff obtained from the USA. When worms from the two populations were kept in animal dung for 37 weeks, the German worms were larger (mean weight 13% higher than "Tennessee Wigglers") and produced 52% more cocoons than the Tennessee population, although the cocoons were smaller (mean weight 88% of those from the Tennessee worms). When worms from these two populations were kept in compost, the Tennessee population had a mean weigh 32% higher, cocoon production was 44% higher, and mean weight of cocoons was 45% higher than those of the German population. It is likely that the progenitors of the Tennessee population were introduced from England, and have been genetically isolated from German *E. fetida* for a long time, even before their introduction to the USA. It is also likely that the commercial producers have selected for large size and increased cocoon production. The differences that now exist between the two populations indicate that there is considerable variability within *E. fetida* so that breeding and selection for suitability for specific waste disposal problems has potential and should be investigated; other species might reasonably be expected to display variability and should also be studied.

II. Methods of Waste Disposal

Bouché (1979) summarized the ideal objectives of waste disposal, through the use of earthworms, as: (a) to upgrade the value of the original waste materials so that they can be sold as useful materials, (b) to produce the upgraded materials *in situ*, i.e., without having to transport worthless waste materials over long distances, and (c) to obtain a final product free of chemical or biological pollutants.

Much attention has been given to these objectives in recent years.

A. Composting of Municipal and Industrial Wastes

Hartenstein (1978a) described laboratory trials of *"vermicomposting"*, the use of *Eisenia fetida* in the composting of aerobic sludge from waste water (sewage) treatment plants. He compared the end-products, comprising worm casts, that have some value as a fertilizer and potting soil, and earthworms, which are high in protein and could be sold as a protein additive for animal food; with the end-product of other disposal methods. The methods comprise: (a) *composting* with microorganisms, resulting in a material that can be used as a soil amendment but has little value as a fertilizer, (b) *incineration*, which is environmentally undesirable, energy demanding, and leaves a virtually unusable residue, and (c) *burial* which is energy demanding, expensive, and can result in pollution of water supplies.

He recommended a programme of research to investigate the possibilities of large scale "vermicomposting", its economic and environmental potentials, and the value of its end-products. Though the ideas were not new, much research and practical application of "vermicomposting" has followed, especially in the USA, where proposed laws prohibiting the discharge of sewage or other wastes into inland and coastal waters have added urgency to the research.

Mitchell (1978) showed that aerobic sewage sludge ingested and egested as casts by *E. fetida* is decomposed, and thus stabilized (i.e. rendered innocuous) about three times as fast as non-ingested sludge, apparently because of the enhancement of microbial decomposition in the casts; he found that, relative to non-ingested sludge, objectionable odours disappeared more quickly and there was a marked reduction in populations of *Salmonella enteriditis* and other Enterobacteriaceae. Most of the sludge produced in sewage plants is anaerobic, and when fresh is toxic to *E. fetida*. Mitchell *et al.* (1980) linked its toxicity to its Eh, showing than when Eh

> -100 mV *E. fetida* survived and fed on the sludge. Collier (1978) used a large sample of sun-dried anaerobic sludge and showed that it was lethal to *E. fetida* when first moistened, that with daily watering it remained lethal until the tenth day, but had lost its toxicity after 14 days.

Seeds of sunflower, tomatoes and corn were planted in *E. fetida* casts derived from sewage sludge, in unprocessed sludge, ground to a size similar to *E. fetida* casts, and untreated soil, by Collier (1978). All plants growing on unprocessed sludge were dead within two months, while plants in castings and soil thrived and those in the castings weighed four times those in soil. Stark *et al.* (1978) found that worm casts, although they may be rich in plant nutrients, sometimes contained high concentrations of Na^+ and other salts, sufficient to damage plants. He found that harmful soluble salts could be reduced to tolerable levels by leaching the casts with water, while still retaining most of the useful plant nutrient content.

Kaplan (1978) raised the possibility of mixing sewage sludge and other materials, e.g., paper-pulp sludge or other lignin-rich wastes, and composting the mixture with earthworms; maceration and mixing of such materials into earthworm casts might accelerate their decomposition. A variety of possibilities exists for disposal of timber mill wastes and other materials, with simultaneous disposal of municipal sewage sludge and production of useful materials.

B. Practical Application of "Vermicomposting"

At a symposium in Syracuse in 1978 (Hartenstein 1978b), several contributors reported successful establishment of field-scale "vermicomposting" waste treatment projects. On the basis of their experience it was claimed that the method was environmentally more acceptable than engineering solutions to waste disposal problems, such as gas production, incineration and land-filling. An added attraction was that the residue after composting, consisting largely of earthworm casts, was readily accepted and saleable for the manufacture of potting mixes for plants, or as an organic fertilizer and soil conditioner for gardens. Vermicomposting has been recommended particularly for the treatment of municipal garbage and of sewage sludge.

Satchell (in press) warned that the municipal garbage generated in western countries is a particularly unpromising material for vermicomposting, because it consists largely of paper, plastic, glass, metals and various inorganic materials. Satisfactory composting requires screening, to separate non-decomposable materials, and shredding of the remainder. A pilot plant operated in Utah, USA to process 100 t of garbage per day (Prince *et al.* 1981) cost twice as much to operate as to burn the material in a modern combustion unit and from four to six times that to bury the garbage in land fill. Satchell (in press) mentioned vermicomposting operations in Holland and Italy, where screening and shredding operations were already in use, and where vermicomposting of the shredded materials was claimed to be economically viable; data on these operations are not published.

Collier and Livingstone (1981) described successful pilot scale trials of the use of *Eisenia fetida* for the vermicomposting of aerobic and anaerobic sludges from southern Californian municipal sewage plants (USA). Ninety tons of "vermicompost", produced in their trials, were shown to be a favourable medium for plant growth and were readily accepted as a saleable product. However, Satchell (in press) stated that no system involving vermicomposting of sewage sludge had yet been operated successfully on a continuing basis. An operation was established at Lufkin, Texas, (USA) but failed when the earthworms died in a particularly hot summer (Prince *et al.* 1981). Satchell (in press) quoted Bird and Hale (1982), who described a system in Ontario (Canada) using sludge contaminated with heavy metals from industrial sources; vermicompost from the sludge, applied to soils at rates that would provide N at 220 kg ha^{-1} y^{-1}, would raise the concentration of heavy metals in the soil above legally prescribed limits within 45 years.

Perhaps the most promising use of vermicomposting might be in the treatment of domestic and agricultural wastes in non-industrial societies, where heavy metal and biocide contamination is not likely to be high, and wastes contain much less glass, metal or other inorganic material than those in industrial societies. I have seen *E. fetida*, raised in nursery beds of stable manure, used successfully to raise the production of vegetable crops in Shaanxi Province, China. Parallel trenches, *c.* 30 cm deep and 30 cm wide, were dug and filled with mixed organic wastes, with rows of soybean planted between them; cocoons of *E. fetida*, from the nursery beds, were mixed into the organic wastes at *c.* 200/m^2. The result was an increase of *c.* 100% in soybean production relative to crops without added earthworms (Huang Foo Chen, pers. comm.). The input of labour required was large, and would not be economically feasible in highly industrialized societies, but in the context of the Chinese economy, where increased food production is the primary aim, vermicomposting may have an important role as part of the general philosophy of maximum utilization of resources.

C. Toxicity of Earthworm Casts from Sewage Sludge

Indications from earlier work that casts of *E. fetida* feeding on sewage sludge were self-toxic when reingested led Kaplan *et al.* (1980c) to investigate the mechanism of toxicity and the intraspecific and interspecific toxicity of casts between *E. fetida*, *E. eugeniae*, *Amynthas hawayanus* and *A. rodericensis*. Experiments showed that casts of *E. fetida* produced from untreated soil resulted in weight loss when fed to all the species, that the apparent toxicity was due to a water-insoluble component in the *E. fetida* casts, and that interspecific toxicity disappeared and intraspecific toxicity was reduced after two weeks' ageing of the casts at 25°C. Casts derived from activated sludge were self-toxic to both *E. fetida* and *E. eugeniae*, but there was no interspecific toxicity. Investigation of several possible mechanisms led Kaplan *et al.* to the conclusion that toxicity resides in a particular combination of bacteria and/or their metabolic excretions that are present in recently deposited casts. Effects of the toxicity can be overcome by addition of soil or fresh activated sludge, and it was suggested that, to maintain optimum conditions for effective sewage sludge treatment by *E. fetida* or *E. eugeniae*, precautions should be taken to ensure that these species are not exposed for long periods to high concentrations of their own or each other's casts derived from sludge. The validity of the proposition that casts may have toxic properties is questionable, since toxicity has not been detected by any other workers for any species of earthworm (J. E. Satchell, pers. comm.).

D. Application of Organic Wastes and Waste Water to Soils in the Field

The very large concentrations of animal dung that result from intensive farming can not be released into streams and coastal waters, and are commonly mixed with water and used to irrigate pasturs.

Yeates (1976), in New Zealand, studied the earthworm population of a 1.6 ha pasture, irrigated for 270 days (late winter to late autumn) each year with effluent from a milking shed, mixed with water and applied by spray at a rate of 30 000 L/d. The spray irrigators were moved around the field on a 10-15 day cycle, and the annual addition of water was equivalent to *c.* 500 mm additional rainfall. In the irrigated field the mean earthworm population was $1010/m^2$ (range $525-1810/m^2$) with a mean biomass of 308 g/m^2 (range $159-711$ g/m^2); in a control area earthworms range from $300/m^2$ (biomass 95 g/m^2) in summer to $367/m^2$ (biomass 162 g/m^2) in winter. *Aporrectodea caliginosa* and *Lumbricus rubellus* were dominant species, with some *A. longa*. The quantity of animal waste applied to the experimental field was not known, and the principal benefit to plant growth, and hence to earthworm numbers, may have come from the water and not from the included animal wastes. Soil moisture at 0-10 cm was consistently higher in the irrigated than in non-irrigated pasture, pH differed little, organic C was a little higher, and %N was similar. The annual dry matter production of the pasture was 1500–2000 g/m^2, and Yeates (1976) calculated that for each 100 g/m^2 dry-matter production there were about 15.4 g/m^2 mean biomass of earthworms, almost the same ratio as found by Sears (1953) and Sears and Evans (1953) in a non-irrigated pasture 2 km distant. In this case it seems that the animal wastes in the effluent may have had little effect on the soil or the earthworms, but this is not so when wastes are applied at heavy rates.

Curry (1976) investigated the effects on earthworms of application to pastures of slurries of cattle-wastes, pig-wastes and poultry-wastes, for three to four years before sampling, at rates of c. 28 t ha^{-1} y^{-1} and 56 t ha^{-1} y^{-1}, in Ireland. These rates of application corresponded to additions of the order of 98–252 kg/ha nitrogen, 14–56 kg/ha phosphorous and 98–280 kg/ha potassium for cattle slurry, 112–252 kg/ha nitrogen, 56–168 kg/ha phosphorus and 14–84 kg/ha potassium for pig slurry, and 196–560 kg/ha nitrogen, 98–252 kg/ha phosphorus and 56–224 kg/ha potassium for poultry slurry. Earthworm populations on slurry treated plots relative to untreated controls were 41% higher, with 56% increase in biomass on plots with added cattle slurry, 53% higher, with 60% increase in biomass on plots with pig slurry, and 40% higher, with 36% increase in biomass on plots treated with poultry slurry. In further experiments plots were treated with cattle slurry, as follows: (a) a single application of c. 56 t/ha; (b) six applications of c. 112.5 t/ha each, a total of c. 675 t/ha, over a period of 21 months; and (c) two applications, each of c. 560 t/ha, a total of 1120 t/ha 14 months apart.

Earthworm populations, assessed over a two year period following these treatments showed, relative to untreated control plots, an increase of 5% and 31% respectively and increases in biomass of 33% and 95% respectively for treatments (a) and (b). For treatment (c) there was a reduction in population of 56% and biomass was reduced by 37%. *Allolobophora* spp. (especially *A. chlorotica*) were dominant, except in treatment (c), where the near surface-dwelling species *Dendrodrilus rubidus*, *Lumbricus castaneus*, *L. festivus* and *L. rubellus* were very numerous.

A subsequent series of investigations of the effects of moderate and heavy applications of cattle-slurries and pig-slurries on pastures in Ireland was made by Cotton and Curry (1980a, b). In one set of experiments (Cotton and Curry 1980a) cattle slurry and pig slurry were applied at rates of 80–100 t ha^{-1} y^{-1} (c. 9% dry matter content). The cattle slurry contained nitrogen at the rate of c. 530–660 kg/ha, phosphorus at 65–80 kg/ha, and potassium at 366–458 kg/ha, while the pig slurry contained nitrogen at the rate of

595–740 kg/ha, phosphorus at 131–164 kg/ha, potassium at 129–160 kg/ha, and in addition contained copper at 5.6–7.0 kg/ha. As controls, another set of plots was treated with inorganic nitrogen at 125–250 kg ha^{-1} y^{-1}. All treatments had been applied for four years preceding sampling of the earthworms, and additional phosphorus, potassium and calcium had been applied at rates to maintain soil fertility as indicated by soil tests. The mean population counts and biomass of earthworms in all plots were high and not much different between treatments, with 310 m^2 and a biomass of 129.4 g/m^2 wet weight of earthworms in control plots, 388/m^2 and 159 g/m^2 in plots with cattle slurry, and 406/m^2 and 178 g/m^2 in plots with pig slurry. A total of 14 species of lumbricids was recorded, all but one present in each treatment, and there was little difference in species composition between treatments.

In laboratory and field experiments Derby et al. (1982) compared soil pH, and NH$_4^+$, NO$_2^-$ and NO$_3^-$ content of litter and A$_1$ horizons of an acid forest soil from a larch (*Larix*) plantation, treated with pig slurry (surface application), in the presence and in the absence of lumbricid earthworms. Addition of the pig slurry resulted in a pH increase in the A$_0$ horizon from pH 3.8 to 6.0; in the presence of earthworms pH fell within two months to c. 4, while in the absence of earthworms it fell to c. 5. The rapid pH fall with earthworms was accompanied by a rapid decline in NH$_4^+$ and NO$_2^-$ and a corresponding rapid increase in NO$_3^-$, relative to soil without earthworms. It was concluded that the end result of presence of earthworms was to improve the supply of nitrogen (as NO$_3^-$) available to plants. Studies of the food requirements and population densities of lumbricids led Satchell (1967) to conclude that earthworm populations are usually food-limited, especially by the quantity of organic nitrogen available, and it has been demonstrated that addition of nitrogen-rich organic matter, or stimulation of plant growth by addition of inorganic fertilizer nitrogen results in increased populations. The experiments of Curry (1976) and Cotton and Curry (1980a, b) indicate that there is an upper limit to the amount of organic N that will result in population increase, and at very heavy application rates other factors are more important. At very high levels there were changes in earthworm community structure, including a reduction in proportion of immature worms, as well as changes in proportions of the species present. Curry (1976) showed that cattle and pig slurries were highly toxic to earthworms that were confined in small containers in a soil peat moss medium, and that the toxicity declined with time, and suggested that such effects might reduce field populations with very high levels of slurry addition.

Large amounts of copper salts are added to pig rations, so pig slurries have a high copper content, and their application to pastures may result in toxicity to earthworms and other soil organisms. In Holland, van Rhee (1975) compared the earthworm populations of long-established pastures, contrasting those of pastures heavily irrigated with pig slurry, with soil copper content 31.4–64.6 µg/g at 0–5 cm, with those of pastures lightly irrigated with pig slurry, with soil copper content 6.0–11.8 µg/g at 0–5 cm. In the soils with higher copper content earthworm populations were 10–169/m^2, and soil organic matter content was 6.4–12.3% at 0–5 cm, while in soils with low copper content populations were 143–215/m^2 and soil organic matter content was 5.7–7.9% at 0–5 cm. In all cases *Lumbricus rubellus* was the dominant earthworm, followed by *Aporrectodea caliginosa*, and very small numbers of *A. rosea*. It was concluded that copper toxicity had reduced the earthworms and probably other components of the soil biota, thus resulting in accumulation of undecomposed soil organic matter. The pig slurries used by Cotton and Curry (1980a, b) also contained much copper, but soil copper levels in their experiments were always < 20 µg/g, considerably below those recorded by van Rhee. Body copper levels in earthworms at one of the sites of Cotton and Curry (1980b) were 17.5 µg/g on control plots and 18.5 µg/g on plots receiving the highest rates of pig slurry; in a different experiment van Rhee (1977b) measured soil copper contents of 6.7–109.7 µg/g at nine sites treated with pig slurry for > 10 years. There was no correlation between earthworm numbers and soil copper content, but at the two highest levels of soil copper content (98.9, 109.7 µg/g) earthworm populations were respectively 12/m^2 and zero. There was a good correlation between soil copper and earthworm copper content (r^2 = 0.91, P < 0.001).

The significance of copper toxicity to earthworms, and any consequent effects on organic matter decomposition, are not clear. It is known that earthworms are absent in soils with very high levels of copper resulting from long-continued use of copper-rich fungicides (p. 232); it might be expected that long continued application of copper-rich pig slurry would have similar effects, but any such effects might be expected to vary with variations in climate, soils, drainage, earthworm species, and perhaps many other factors. Uptake of copper by earthworms from pig slurry applied to pastures, whether or not the copper is significantly toxic to the earthworms, raises questions of environmental contamination and transfer along food chains of copper and other heavy metals by earthworms from many kinds of organic wastes. Helmke *et al.* (1979) measured the concentrations of 29 elements in earthworms from fields treated with various quantities of anaerobically digested sewage sludge, and suggested that analyses of earthworms might be used to monitor levels of environmental contamination with a number of heavy metals. Various aspects of the role of earthworms in circulation of heavy metals in the environment are discussed on pp. 234-241.

Dindal *et al.* (1977) compared the effects on earthworms of municipal waste-water irrigation, applied to grassland, a mixed herbaceous/white spruce community, and a mixed oak/hardwood forest. Earthworm numbers and biomass increased in all treatments, but some species of earthworms were shown to be advantaged or disadvantaged, and these effects were not consistent between sites.

The Niimi process of waste water disposal, devised in Japan and used in rural areas, involves filtration of waste water through a trench containing a prepared bed of soil overlying a layer of stones, broken earthenware pipes, or other porous material. Nakamura (1982) has shown that *Eisenia fetida* and some microdrilids, mainly *Enchytraeus* spp., have an important role in the successful operation of the process. *Eisenia fetida* attains high populations in the overlying soil layer, feeding on suspended organic matter, especially hydrocarbons and fatty protein analogues. Presence of *E. fetida* prevents the soil layer from clogging, so allowing the maintenance of aerobic conditions, which are necessary for successful water treatment.

16
Earthworms as a Protein Source

Large scale processing of wastes by the composting method has as a corollary the production of large numbers of earthworms. The early experiments of Fosgate and Babb (1972) showed that 1 kg of earthworms (fresh weight) could be produced from 2 kg (dry weight) of animal dung consumed by *Lumbricus terrestris*.

Lawrence and Millar (1945) found the protein content of an earthworm they designated as *L. terrestris* was 62.0%-71.5% of dry weight, and quoted previous investigations that demonstrated the presence in earthworm tissue of "all the usual amino-acids produced by hydrolysis of mammalian tissue". Bouché (1967) found the protein content of *L. terrestris* was 66.25%, of *Aporrectodea longa* 68.12% and of *Octolasion cyaneum* 60.00% of the dry weight. Sabine (1978) cultured *Eisenia fetida* on sheep manure, harvested the worms, freeze-dried or air-dried them, and fed the product to chickens or to pigs. The dry-matter content of the worms was 20%-25% of fresh weight, and on a dry weight basis they contained 60%-61% protein, 7%-10% fat, 8%-10% ash, including 0.55% calcium and 1.0% phosphorus. Sabine (1978) concluded that earthworms could possibly be a significant source of high grade protein for intensive livestock industries, for pet food, and perhaps for human food.

I. Nutrient Content and Food Quality of Earthworms

The potential of earthworms as a food source, and their nutrient and protein content have been reviewed recently by Sabine (1983). The favoured species for protein production must be *Eisenia fetida* because of its short generation time, extraordinarily high productivity (pp. 70–71) and its preference for life in organic matter-rich habitats such as compost and dung heaps. *Eudrilus eugeniae* might also be a candidate species for protein production, and may be more suited than *E. fetida* to tropical conditions.

The protein content and amino-acid content of protein meal made from *E. fetida* or *E. eugeniae* closely resemble those of meat meal and fish meal commonly used as protein sources in stock foods prepared for the intensive raising of livestock (Sabine 1983). Weight gains of livestock fed earthworm meal are comparable to those when other generally accepted protein sources are used. There would seem to be no reason to reject earthworm protein-meal on a nutritional basis, but there are economic problems and possible hazards of disease, and toxicity from heavy metals and biocides in livestock that may not be easily overcome.

II. Economic Problems

Sabine (1983) quoted Mohr and Littleton (1978) who calculated, on the basis that meat meal for stock food manufacture costs $177/t in Australian, earthworm meal, on the basis of its proven food value, is an economic substitute in poultry food at prices up to $236/t, and may be economic up to *c.* $300/t; this price is far below that at present available to producers of *E. fetida* for fish bait, so earthworm protein production for stock food would not be commercially attractive.

Satchell (in press) discussed the costs of production of earthworm protein. He quoted Tomlin (1983), who calculated the price of *E. fetida* in Canada as $21/kg ($21 000/t) and that of dressed tenderloin beef as $16/kg ($16 000/t). Satchell (in press) quoted calculations, based on the amino-acid, mineral, energy and water content of earthworms, that in England earthworm meal (1983 prices) was worth £230/t (dry weight), compared with £160/t for meat and bone meal and £230/t for fish meal; the value of earthworm meal (wet weight) on this basis would be £53/t, and Satchell compared this with a current price of £7700/t for *E. fetida* asked by the largest producer of earthworms in Britain.

It is apparent that economically viable production of protein from earthworms is not without problems.

III. Disease Hazards

A number of common mammalian and avian parasites, especially some nematodes and protozoans, have earthworms as secondary hosts (p. 151). If earthworm meal is to be used for stock food it would be necessary to sterilize the product to ensure that it did not contain viable cysts or eggs of such parasites, or to ensure that earthworms that fed on waste materials derived from one group of susceptible animals were used to provide protein for another group of animals whose earthworm-hosted parasites differed from those used to provide the culture medium.

There is a further possibility that microorganisms that are human pathogens may survive in earthworm meal prepared from earthworms fed on sewage sludge, so that use of earthworm meal as a pet food or of casts as a medium for plant growth in a domestic environment could present a health hazard to the community (Satchell, in press).

IV. Heavy Metal and Biocide Toxicity

Earthworms are known to accumulate some heavy metal pollutants (pp. 234–239) and biocides (p. 310) from the environment. In industrialized countries, where these pollutants are already sometimes present in the environment in hazardous or near-hazardous concentrations, the possibility exists for their further concentration in earthworms fed on composted wastes. Graff (1982) examined the heavy metal content of *Eisenia fetida* and *Eudrilus eugeniae*, before and after feeding on compost made from municipal garbage in Germany. In *E. fetida* concentrations (on a dry weight basis) of copper increased from 4 µg/g to 29 µg/g, of zinc from 140 µg/g to 640 µg/g, of lead from 3 µg/g to 14 µg/g, of cadmium from 2 µg/g to 9 µg/g, and of mercury from 0.1 µg/g to 13 µg/g; corresponding increases in *E. eugeniae*, obtained for the experiments from Nigeria, were from 17 µg/g to 55 µg/g for copper, 165 µg/g to 360 µg/g for zinc, 10 µg/g to 72 µg/g for lead, 4 µg/g to 6 µg/g for cadmium, and 1 µg/g to 15 µg/g for mercury. If composted municipal and industrial wastes were to be widely used to produce earthworms as a protein source, it would be necessary to monitor the heavy metal content of the wastes and the earthworms, as concentrations of the order demonstrated by Graff (1982) would not be tolerable in protein meal for stock food.

Similarly, it would be necessary to monitor food sources and earthworms produced for biocide concentrations, which might be particularly hazardous as contaminants in earthworms produced from organic wastes in horticultural and intensive agricultural farming regions.

Appendix
Field Sampling Methods

Earthworms present some difficult problems for quantitative sampling of their abundance and biomass. Their distribution is often patchy, with numbers varying greatly over small distances; they are concealed from sight beneath the surface, and are active at depths in soil that vary between species, between life-stages within species, and also vary seasonally and sometimes diurnally within a population; many live in extensive burrow systems and can respond to the disturbance of digging by moving quickly away from the centre of disturbance; some species pass part of the year deep in the soil, in resting stages; small species and juveniles of larger species are easily overlooked, and frequently behave differently from adults. Some knowledge of the life cycles and seasonal variation in life stages and behaviour of the target species will help to appreciate possible phenological and behavioural factors that could affect the efficiency of sampling. Unit sampling areas and sample depths must be suited to the population density of the target species. Because of the shortcomings of all methods of extraction and collection, counts of field populations are most likely to be underestimates; some methods for correcting the raw data from field collections are discussed later in this Section.

The sampling methods most commonly used are of two kinds: (a) *passive*, in which the earthworms are physically sorted from soil, litter, and other habitats, and (b) *behavioural*, in which they are stimulated to move out or are captured when they voluntarily move out from cover.

Combinations of passive and behavioural techniques are used with considerable success. Less commonly used are *mark-recapture*, and indirect methods, e.g., counting surface casts.

Some of the methods discussed here have been reviewed by Bouché (1972, pp. 35-42) and Edwards and Lofty (1977, pp. 118-123) and their publications should be consulted for references to much of the literature up to about 1972. Bouché and Gardner (1984) have provided a more recent review of some sampling methods and their relative efficiencies.

Nearly all the methods have been developed for field sampling of Lumbricidae which live mainly in the soil, and little has been done on the sampling of non-lumbricid species, or of species that are not soil inhabitants.

In planning field studies of abundance, biomass or both, some preliminary sampling, preferably using several methods, should be done to assess the relative efficiency of the methods in the particular circumstances and to match sample area and depth to the behaviour, population density and vertical distribution in the soil profile of the species to be sampled. It is usually necessary to compromise between numbers and size of samples that are desirable for statistical purposes and the practical limitations of time and labour available to sort and count the numbers collected. This is a problem common to most population studies; it is necessary to appreciate the limitations that it may impose on the reliability of the results.

For population studies that involve repeated sampling, sets of samples should be taken according to a predetermined plan that obviates the possibility of resampling from a previously sampled point; a minimum distance of 5 m between sampling points should avoid disturbance between adjacent samples (Bouché 1972).

I. Passive Methods

A. Hand Sorting

Soil samples of a standard surface area are dug out, and the earthworms are sorted from them by hand. Edwards and Lofty (1977) traced this method back to Bretscher (1896), and it is commonly used to this day. It is important that sampling depth is matched to the burrowing habits of the species present, and the method is not practicable if a substantial proportion of the population lives in very deep burrows. Sampling depth must often be varied seasonally, as most earthworms that are active near the surface retreat deeper into the soil and may become inactive during hot, dry or very cold periods of the year. For example, Persson and Lohm (1977), in Sweden, found that in the warmer months of the year sampling to 30 cm depth was adequate, while in the winter it was necessary to sample to 60 cm.

Sampling is most commonly done by digging with a spade, first cutting around the edge of the sample, or digging a trench to isolate the block of soil to be sampled. Zicsi (1958b) used a square sampling tool, like four sharpened spades welded together, pushed into the soil to cut and isolate a sample 25 cm × 25 cm × 20 cm deep before it was dug out, enclosed in the square sampling tool. Svendsen (1955) used a semi-circular spade-like tool to cut a circular sample c. 19 cm in diameter and 20 cm deep.

Samples should be sorted agrainst a pale-coloured background to increase the probability of detecting earthworms; white or blue trays or plastic sheet are probably most effective.

The number of samples necessary and the size of individual samples vary with population density. Abrahamsen (1969), sampling Enchytraeidae in Norwegian forest soils, found that the coefficient of variation of his sample means increased very rapidly as the means fell below 40 per soil sample, but the coefficient was not greatly improved when the mean was > 40. Zicsi (1958b) compared samples of 25 × 25, 50 × 50, 50 × 100 and 100 × 100 cm surface area from a Hungarian pasture; he took 50 samples of each size and found that the estimated mean numbers per square metre decreased with increasing sample size, and were respectively 84.5/m^2, 77.7/m^2, 62.8/m^2 and 62.0/m^2. He attributed the decrease to the increasing tedium of sorting large samples and concluded that samples of 25 × 25 cm surface area were preferable. The mean number of earthworms beneath this surface area (1/16 m^2) would have been only a little more than five, and this is rather too small a number for accurate population estimation, in view of the variability of numbers that might reasonably be expected. On the basis of his investigation Zicsi recommended taking 16 samples, each 25 × 25 cm and 20 cm deep. Reinecke and Ljungström (1969) took four samples, each 25 × 25 cm and 16 cm deep, when sampling populations in a South African pasture, and estimated population densities varying between 72/m^2 and 112/m^2 (4.5–6.95 per sample) through the year. Rundgren (1975) also took four samples, each 35 × 35 cm (c. 1/8 m^2) and 100 cm deep, in Swedish forest and pasture lands.

Samples are usually divided horizontally into layers and each layer is sorted separately to determine vertical distribution.

Raw (1960) tested the accuracy of hand sorting as a method of estimating lumbricid populations in a wet grassland soil where the worms were confined mainly to a thick turf mat. After hand sorting the soil and roots were shaken in a 2 mm sieve, suspended in a 0.5 mm sieve, with the sieves immersed in a bowl of water. Careful searching of the sieves showed that the hand sorting had recovered 52% of the total numbers, but 84% of the total biomass. Small species and immature specimens of larger species were most commonly missed in hand sorting. For one small species, *Eiseniella tetraedra*, hand sorting recovered only 38% of the numbers and 57% of the biomass; hand sorting indicated that this species made up 49% of the total earthworm population and 18% of the biomass, while subsequent sieving showed that *E. tetraedra* constituted 68% of the population and 26% of the biomass. Applying the same tests to soil from a pasture with no root mat, Raw (1960) recovered 89% of the total population and 95% of the biomass by hand sorting, while in an old arable field with a poorly structured clay soil he recovered 59% of the population and 90% of the biomass. Nelson and Satchell (1962) took undisturbed soil samples from a grassland, killed the earthworms in them by freezing, and then introduced known numbers of a mixture of species of various sizes and life stages. Subsequent hand sorting recovered 93% of the total numbers and 99% of the biomass added, but for small immature *Allolobophora chlorotica* and *Lumbricus castaneus* the recovery rates were only 80% and 74% respectively of the numbers introduced. They concluded that the method is satisfactory for specimens of live weight $>$ 200 mg. It is reasonably accurate for biomass determinations, but is unlikely to give reliable results for the lower age classes in population studies. Persson and Lohm (1977) also found hand sorting unsatisfactory for specimens $<$ 200 mg live weight, while Axelsson *et al.* (1971) set the lower limit at 160 mg, and Reynolds (1973) found that specimens $<$ 2 cm in length tended to escape detection.

Lavelle (1978) experimented with several methods for sampling a mixed population of Megascolecidae and Eudrilidae in deep sandy soils of savannas and gallery forests in the Ivory Coast. The populations included some species that inhabit near-surface layers and some from deeper layers of the soil, and included a size range from Megascolecidae up to 70 cm long, 30 g in weight, and numbering *c.* $1/m^2$ to immature Eudrilidae *c.* 1 mg in weight, numbering several hundred per square metre. The most satisfactory method was found to require two operations: hand sorting of large samples, 1 m^2 surface area, 50 cm deep, subdivided into 10 cm-thick layers, supplemented by washing and sieving of 10 cm-thick layers from smaller samples, 20 $\times$ 20 $\times$ 50 cm deep, taken beside the larger samples, to recover small specimens and cocoons. Twelve of each kind of samples were taken at each (monthly) sampling occasion.

Lavelle (1978) found it necessary to apply corrections to his raw population data, to allow for inadequacies in the collection techniques. He treated large species and small species separately. For the large *Millsonia* spp. and *Agastrodrilus* spp. (Megascolecidae), numbers collected by direct hand sorting were adequate and did not require correction, even for juveniles. The populations at Lamto included three species of Eudrilidae whose fresh weights are usually < 200 mg when mature, and one *Dichogaster agilis* (Megascolecidae), usually < 500 mg. Over a period of 16 months of collection, using direct hand sorting of large samples and washing and sieving of smaller samples (576 of each sample type), as described above, Lavelle found that the relationship between numbers detected by the hand sorting and washing and sieving methods was expressed by the equation

$$y = 0.67x + 0.12 \quad (P < 0.01)$$

where y is number from washing/sieving; x is number from hand sorting. For individuals whose fresh weight was > 150 mg hand sorting was adequate; for the smallest size class considered (< 10 mg fresh weight) hand sorting resulted in collection of only 6.7% of those collected by washing and sieving. Lavelle's results give some indication of the levels of accuracy that might be expected using hand sorting in populations elsewhere that include small species or small juveniles of larger species.

In summary, hand sorting is a reasonably reliable method for biomass estimates, provided the soil is friable enough that samples can be readily dug out, that there is not a thick turf mat at the surface, and that deep-burrowing or very small species are not numerous. It needs to be supplemented by a washing and sieving technique to recover small earthworms and cocoons. Given these provisoes, it has the advantages that it requires only the simplest equipment and that it recovers cocoons and also earthworms that are dormant. Its greatest disadvantages are that it is labour-intensive and tedious. The number of samples necessary at a sampling occasion will vary with the population density, the number of species present, the level of species differentiation required and the input of labour available. Samples should ideally be taken at not less than 5 m spacings.

B. Washing and Sieving

The advantages of combining a soil-washing and sieving technique with hand sorting have been discussed. Some techniques employ soil washing/sieving wihout preliminary hand sorting. Selection of suitable sieve sizes ensures collection of small immature earthworms, cocoons, and also aestivating or torpid specimens, but if the sieving is done by hand it is very laborious and Raw (1960) concluded that it was probably not worth the additional labour, compared with hand sorting, for most population studies.

Edwards *et al.* (1970) described a mechanized method of sieving and washing, in which soil samples were placed in rotating vertical drums, with sieves fitted to their bottoms, and subjected to high pressure jets of water. This achieved results similar to those with hand sieving, with a greatly reduced input of labour.

A further development of the washing/sieving technique has been described by Bouché (1969, 1972) and Bouché and Beugnot (1972). Soil cores or blocks of soil, taken in an undisturbed state from the field, are first soaked for two days in a solution containing sodium hexametaphosphate (2%), to disperse clays, and formaldehyde (4%) to kill and fix earthworms and other soil animals. The soil samples are then placed in a water-filled trough, where they are broken up by paddles that are rotated mechanically on a horizontal shaft, and simultaneously flushed with water and sieved. The sieved material includes plant roots, leaves, macroarthropods and the coarser soil particles, as well as earthworms and their cocoons and though it is necessary to sort the earthworms from these materials, much manual labour is avoided. Bouché has successfully used this method for many years; it has the additional advantage of extracting macroarthropods, so it can be used for studies of total soil macrofauna.

In summary, washing/sieving methods are probably the most effective available for extracting earthworms from soil. They make it possible to extract small specimens, cocoons, aestivating or dormant earthworms, and with the reduced labour input that results from the mechanical aids developed by Bouché and his coworkers they have wide application to population studies. Soils that have very high clay content and are difficult to disperse, or are stony, present problems for separating earthworms by any method, but washing and sieving, with addition of dispersants to clay soils, is probably the most effective method available.

C. Flotation

Flotation methods involve initial hand sorting or washing and sieving, and then the residues of soil and plant material, containing small earthworms and cocoons are stirred in containers through which a solution of magnesium sulphate (specific gravity 1.16–1.20) is pumped; the $MgSO_4$ solution is stirred and allowed to overflow into sieves; plant material, earthworms and their cocoons and other soil animals float off and are collected in the sieves; further separation of the earthworms and other animals from plant material is done by flotation at a kerosene–water interface. Martin (1976) has described a successful flotation technique. His apparatus and technique are based on those originally developed by Salt and Hollick (1944) for extracting soil arthropods. Raw (1960) and Gerard (1967) used $MgSO_4$ solutions, with less complicated apparatus than Martin's and found the method very effective. Martin's apparatus perhaps has the advantage that it requires less input of labour and he found it possible to treat large numbers of samples. It should, however, be noted that Bouché (1975a) was unable to satisfactorily float off earthworms with $MgSO_4$ solutions, and even using solutions of potassium bromide, with a maximum density of 1.28, did not succeed in floating off all earthworms from soil samples.

The flotation method perhaps has some advantages where soil samples contain large numbers of plant roots, but it involves time-consuming work, additional to hand sorting or washing and sieving. Bouché (1975a) found that his mechanized washing/sieving method gave more satisfactory results for population studies.

II. Behavioural Methods

A. Chemical Repellants

These methods involve saturation of the soil with solutions of irritant chemicals to induce earthworms to abandon their burrows and emerge on the soil surface, where they can be collected. The chemicals that have been used are mercuric chloride ($HgCl_2$), mowrah meal (p. 306), potassium permanganate ($KMnO_4$) and formalin (HCHO, methanal).

Mercuric chloride (c. 0.1% concentration) was used by Eaton and Chandler (1942) for studies of earthworms of forest soils, and mowrah meal by Jefferson (1955) for studies on the earthworms of sports fields, but these materials have not been widely adopted.

Evans and Guild (1947) first adopted potassium permanganate solution (1.5 g/L, applied to the ground surface at 6.8 L/m^2) to bring earthworms to the surface, and subsequently used this method in population studies of earthworms in the United Kingdom (see Edwards and Lofty 1977). Potassium permanganate is a powerful oxidising agent that attacks the tissues of earthworms; earthworms that emerge when it is applied must be quickly washed in large volumes of water to keep them in reasonable condition, and many are killed before they can reach the surface.

Svendsen (1955) compared the permanganate method with hand sorting. He took ten samples, applying potassium permanganate to 1 m^2, and beside each of these samples he took five soil cores, each c. 20 cm diameter and 20 cm deep, and hand sorted these for earthworms. The mean earthworm population density shown by the permanganate method was 35/m^2, and by the hand sorting method 447.6/m^2. Comparison between the two methods showed great variability between species, e.g. permanganate extracted 33.5%-56.6% of *Aporrectodea longa* and 4.6%-20.1% of *A. rosea* numbers collected by hand sorting.

Raw (1959) demonstrated that formalin (about 70 mL of 40% formalin solution in 12.7 L of water (0.22% concentration) applied per square metre of soil surface) was effective in bringing lumbricid earthworms to the surface.

The formalin method has been widely used for population studies, but it has serious limitations. Raw (1959) found the method most effective for *Lumbricus terrestris*, which lives in more or less vertical burrows that open at the surface, but less useful for other lumbricid species that live in more horizontal or in branching burrow systems. Baker (1983) found that, of 15 lumbricid species collected from a reclaimed peat soil in Ireland, *L. terrestris*, *Dendrobaena octaedra* and *Satchellius mammalis* were most efficiently sampled by the formalin method, while others, especially *Aporrectodea* spp., were very inefficiently sampled compared with parallel sampling by hand sorting. Satchell (1969) used three concentrations of formalin (0.165%, 0.275%, 0.550%), applied in three lots of 9 L each at 10-minute intervals to areas of 0.5 m^2, and found the three concentrations equally effective for *Lumbricus* spp. and *Allolobophora* s.l. Four further applications to the same areas showed that 95% of the total catch was expelled by the first three applications. He found that, relative to numbers obtained by hand sorting, the catch with the formalin method varied with soil temperature, with a maximum when the temperature at 10 cm depth was 10.6°C, and with soil moisture, increasing linearly to 38% moisture content and decreasing for soil moisture contents of 44% or more. From his data he calculated a correction factor, as follows:

Estimated population =
observed population $\times$ e[0.0075 (T − 10.6)] $\times$ e[−0.0214 (M − 40)]

where T is soil temperature at 10 cm depth, and M is soil moisture percentage at 0-5 cm depth. The method, with the correction applied, was found to be most satisfactory for *L. terrestris* and less satisfactory for other species. The temperature effect apparently relates to changes in activity of *L. terrestris*, which Satchell (1967) found to be at a maximum at *c.* 10°C. It should, however, be noted that temperature optima vary within earthworm species and cannot be assumed to be the same in all environmental conditions (p. 45). It is likely that penetration of solutions into the soil is progressively impeded with increasing soil moisture; penetration is also dependent on soil structure, porosity and the presence or absence of compacted soil horizons.

The effectiveness of the formalin method for extracting lumbricids has been tested by many others including Bouché (1969, 1975a), Lakhani and Satchell (1970), Nordström and Rundgren (1972), Barnes and Ellis (1979) and it is generally agreed that it is most suitable for *L. terrestris*, and less suitable for other spcies. Bouché (1975a) compared six methods for sampling populations; these included digging followed by hand sorting, soil-washing, washing and sieving, flotation, and extraction with potassium permanganate or formalin solutions. For a site at Cîteaux he calculated correction coefficients to be applied to the raw data, and separately assessed the efficiency of the methods for sampling épigées, anéciques and endogées, with reference to a very carefully sampled reference population at the same site. He concluded that realistic corrections were possible, but it was essential to have data from similar habitats, with similar proportions of the ecological categories (épigées, anéciques, endogées), collected in equivalent weather conditions, and at the same season, if corrections were to be applied at more than one site. Bouché's paper provides much useful information on the relative efficiencies of the methods tested and the importance of behavioural and ecological variations between and within species that are important in affecting the raw data. Barnes and Ellis (1979) compared the efficiencies of formalin extraction of earthworm (lumbricid) populations in ploughed and direct-drilled wheat and barley fields, and found formalin more effective in direct-drilled fields, where earthworm burrows are little disturbed and open directly to the surface, than in ploughed fields, where many burrows are broken up or blocked.

Applications of formalin to the surface may be useful to bring subsoil-inhabiting, surface-feeding species (endogées) up towards the surface, where they are more likely to be taken by hand sorting techniques (Bouché 1969, St Remy and Daynard 1982). Similarly, application of formalin to the bottom of pits from which soil samples have been taken for hand sorting can be used to collect deeper-dwelling species (Martin 1976, Barnes and Ellis 1979).

For earthworms other than Lumbricidae the formalin sampling method may be successful or may be totally unsuccessful. Block and Banage (1968) and Madge (1969) used it successfully to estimate mixed populations, including Glossoscolecidae and Eudrilidae, in Uganda and Nigeria, but Lavelle (1978) found it a total failure for sampling Megascolecidae and Eudrilidae in the soils of savannas of the Ivory Coast. Reynolds (1976) found that formalin was not satisfactory for sampling North American *Diplocardia* spp.

In summary, the formalin method is the most satisfactory of a rather poor choice among chemical repellant techniques. Its one advantage is that it requires a small input of labour. However, it has been shown to be unreliable for nearly all species of earthworms for which it has been tested. The use of formalin should be confined to species that inhabit vertical burrows that open at the soil surface, and with suitable corrections it might be used especially for *L. terrestris*.

The soil surface should be scraped bare of vegetation and litter before formalin is applied, so that the earthworms that emerge can be readily seen and collected. The pathways taken by the formalin solution beneath the surface will depend upon the directions taken by burrows or other large soil pores, and it may not be possible to relate the volume of soil samples to the surface area to which formalin is applied.

The method is capable of sampling only those earthworms that are active at the time, is unsatisfactory for earthworms that are aestivating or are torpid, and is incapable of sampling cocoons. It can be recommended for comparing the levels of activity of susceptible species through the year at a particular site, or between sites if soils, soil temperature and moisture, and the ecological categories of earthworms present and their susceptibility to the method are closely similar, and if sufficient data are available to correct the results for these variables. It cannot be recommended as a substitute for hand sorting or washing and sieving for studies that require accurate counts of total earthworm populations, but it may be a useful check to apply formalin to the bottom of pits dug to obtain samples for hand sorting or washing and sieving, to detect susceptible deep-dwelling species not detected in the primary sampling.

B. Heat Extraction

Satchell (1969) described a method based on that of Baermann (1917), originally developed for the extraction of free-living soil nematodes and subsequently modified for extraction of Enchytraeidae from soil samples. Blocks of soil 20 × 20 × 10 cm deep were placed on wire gauze fixed 5 cm from the bottom of a plastic bath 55 × 45 cm; the bath was then filled with water to a level 5 cm below the soil surface and heat was applied for three hours from 14 light bulbs (each 60 W) suspended 2 cm above the soil surface. The earthworms retreated from the heated soil into the water in the bath. Satchell reported that this method was superior to hand sorting or the formalin method for extracting the small lumbricid *Allolobophora eiseni* from the dense root mat of *Deschampsia flexuosa*, and it may be a useful technique for extracting earthworms from root mats of other plants or plant communities. Bouché (1972) reported that this technique was effective for extracting *Microscolex phosphoreus* from surface soil horizons.

Abrahamsen (1972) used the Baermann funnel technique, taking soil cores 9.1 cm diameter, 10 cm deep (24 or 30 per sampling site), and dividing them into 5 cm depths for extraction. Careful hand sorting of the samples after the heat extraction indicated that the heat extraction method yielded about 95% of the total population. Formalin extraction of adjacent sampling sites was used simultaneously, and showed that the heat extraction method was usually more effective, but was less effective than the formalin method for rare species (abundance $< 5-7/m^2$); in pine forests that they sampled, *Aporrectodea rosea* and *Lumbricus terrestris* were below the limits satisfactorily sampled by the heat extraction method. At least for *L. terrestris* it is likely that 0-10 cm samples would not be deep enough to satisfactorily sample the population; for rare species the total surface area of samples is also important — this was 0.195 m^2 (30 samples) or 0.156 m^2 (24 samples) for the heat extraction method, compared with 3 m^2 for the formalin method.

C. Electrical Method

It is well known that some earthworms will come to the surface close to an electrode pushed into the soil when an alternating current is discharged. Satchell (1955b) tested the method and found that earthworms in a pasture soil at Rothamsted emerged over a convenient area most readily when current output from the electrode was 2-4 A at 360 V AC (50 cycles/s). He showed that the method was superior to potassium permanganate extraction, which was the most commonly used method at the time, for three out of the seven lumbricid species present, but was inferior for the remaining four species.

The volume of soil affected by the current flow is dependent on its conductivity, which is dependent principally on soil moisture, the electrolyte content of the soil water, and the temperature; these properties of soils are variable from place to place and from time to time, and vertically depending on the nature of the soil profile at any one place. It is therefore not possible to make any generally applicable estimate of the volume of soil sampled, so the method has little practical value for quantitative population studies.

I have tested the method in garden soils in Adelaide, Australia, where it caused a small proportion of the lumbricids to emerge at the surface. In further tests in montane soils at Mt Kosciusko (New South Wales) I found the method totally ineffective for the numerous megascolecids that could be collected by digging and hand sorting. In these wet montane soils the method was dangerous to the operator and it was not possible to touch the soil surface within several metres of the electrodes without receiving a severe electric shock. In carefully selected circumstances, where the soil is not excessively wet and where there are earthworms that are susceptible to this method of expulsion from the soil, the electrical method may be useful for qualitative, but not for quantitative studies of some species. It may be particularly applicable to situations where sampling must be done without physically disturbing the soil, but it is likely that the formalin method would give as good or better results.

D. Mechanical Vibration

Some earthworms can be induced to leave their burrows and come to the surface by vibrating the soil, and this method is well known to fishermen, especially in some parts of the USA; a spade or a flexible rod is pushed into the ground and then violently vibrated. Reynolds (1973) reported that this method does not seem to work for Lumbricidae, but induces some *Acanthodrilidae (e.g., Diplocardia mississipiensis, D. floridana)* and Megascolecidae (e.g., *Amynthas diffringens*) to come to the surface in large numbers.

The method may be useful for obtaining specimens of susceptible species, but it has no value for quantitative studies of earthworm populations.

E. Trapping Methods

The use of pitfall traps in behavioural studies to sample those earthworms that are active on the surface, especially at night, was reviewed by Bouché (1972). The traps must contain a fixative, to kill and preserve the earthworms and other animals that fall into them, and to prevent earthworms from escaping or being eaten by predators. Formalin has been used, but the fumes are repellent and Bouché found a saturated solution of picric acid more suitable. Bouché (1972) developed a large pitfall trap, 107 mm in diameter, and containing 900 mL of picric acid solution, with provision for drainage of rain water that fell into the trap, and with a removable inner container to collect the earthworms.

The method, like all pitfall trapping methods, assumes that the target animals will not actively avoid falling into the trap; for elongate animals like earthworms, which can easily withdraw the anterior portion of the body from an unfavourable position, it cannot be assumed that all species that encounter a trap will be sampled in proportion to their true population density on the soil surface. The method is best regarded as a qualitative and perhaps semi-quantitative technique for sampling surface-active earthworms, and for comparing the incidence of surface activity of species on a diurnal or seasonal basis, or in relation to weather patterns. Bouché (1972) has successfully used the method for this purpose.

Satchell (1980c) used a baited trap method for qualitative sampling of earthworms that were present in very low numbers ($< 1/m^2$) in *Calluna* heathland in northern England. Earthenware seed trays, 20 cm diameter and 5 cm deep, with holes drilled at the bottom for drainage, were filled to a depth of 4 cm with *Calluna* raw humus covered with a 1 cm deep layer of milled cow dung, and were set into the ground level with the soil surface. The traps were searched at intervals of 1–2 months and were found to attract earthworms from the surrounding soil.

F. Other Behavioural Methods

Sugi and Tanaka (1978a) made use of an unusual behavioural pattern to determine population density of the megascolecid *Amynthas sieboldi*. *Amynthas sieboldi* is a large species, up to 40 g (wet weight) for mature individuals; it is found only in upland forest areas of Kyushu Island, southwestern Japan, and its population density is of the order of 1–2/100 m^2 surface area. From about April to November it migrates to sloping ground, where it lives in the surface litter layer, and lies with the anterior portion of the body horizontal along the upper surface of the H-layer (see Figure 4) and the posterior portion in the H-layer. If the overlying litter is removed the worms emerge completely onto the surface of the H-layer and move rapidly down-slope. Sugi and Tanaka used rakes to remove the overlying litter and collected the earthworms as they emerged and moved away. I know of no other species that behaves in this way, but the method illustrates how a thorough knowledge of species' behaviour may sometimes be used to advantage in population studies, especially for species like *A. sieboldi*, whose population density is normally very low.

III. Mark–Recapture Methods

It is not easy to mark earthworms so that they can be recognized in later collections.

Meinhardt (1976) described a technique for staining earthworms with a water-soluble non-toxic green dye, which stains the bodies of earthworms and persists for several months. By collecting large numbers of earthworms, staining them, returning them to the soil and re-collecting later it is possible to estimate populations. The method has been used with considerable success by Mazaud and Bouché (1980) to study earthworm dispersal and mortality rates in a pasture soil in France.

Richter (1973) used a freeze-branding method for labelling slugs, which might work well for earthworms, especially for the larger species. He made a branding tool from copper wire (1 mm diameter), bent at the tip to make a right angle, and mounted on a plastic handle. The wire was dipped in liquid nitrogen for several minutes, and the tip was then placed on the dorsal surface of a slug to make a burn-mark that was recognizable for several months. By orienting the right angle at eight predetermined intervals of 45° to the dorsal midline Richter was able to number the slugs and recognize them individually. The branding did not appear to affect mortality rates compared with unbranded individuals in field experiments.

Radioactive isotopes, introduced into the coelomic cavity, have been used to label earthworms for studies of behaviour, and may in some circumstances be useful for small-scale studies of populations. Joyner and Harmon (1961) injected labelled gold (^{198}Au) into the coelom of *L. terrestris* and were able to study the behaviour of the earthworms in burrows; Gerard (1963) inserted small pieces of ^{182}Ta-labelled (tantalum) wire into the coelom of *L. terrestris* through small slits in the body wall. It is possible that the presence of radioactive materials in the body would themselves affect the survival and behaviour of earthworms, and the cost and technical difficulty of labelling large numbers of specimens would outweight any advantage over the more simple staining or branding methods described above for population studies. Use of radioactive isotopes is more appropriate in studies of assimilation of food, biochemical pathways in metabolism or water relationships than in population studies.

IV. Counting Casts

It is doubtful that this method could be used for quantitative estimates of abundance, but for species that cast at the surface, levels of activity can be estimated. Seasonal or shorter term variations in activity can be assessed and compared between sites by collecting and counting numbers of casts at suitable time intervals. Evans and Guild (1947), for example, showed a close correlation between numbers of casts deposited and numbers of *Aporrectodea longa* and *A. caliginosa* extracted with potassium permanganate, which preferentially extracts active species with burrows opening at the surface. The method must be subject to variable errors, as casting rate varies greatly with temperature and soil moisture.

V. Summary of Sampling Methods

For most practical purposes, hand sorting or washing and sieving methods give the most accurate results and should be preferred. The mechanized washing and sieving method developed by Bouché (see Bouché and Buegnot 1972, Bouché 1972) is perhaps the most successful available at present. Formalin extraction is the best of the available behavioural methods, mainly because it requires little labour, but it is satisfactory only for some species, mostly surface-feeding Lumbricidae with surface-opening burrows, and should not be used unless its effectiveness has first been checked against hand sorting or washing and sieving in the soils to be sampled. Some of the other methods described here may prove useful in particular circumstances. The potassium permangate extraction method is particularly ineffective, and should not be used, while the electrical method is inaccurate and may be dangerous.

VI. Biomass Estimation

The weight of a fully hydrated earthworm is generally > 80% water (p. 33). Its state of hydration depends on the availability of water in its immediate environment and, especially for soil-inhabiting species, this is variable on a seasonal and often on a diurnal time scale. Earthworms can survive water loss from their tissues, at levels as high as 80% of their water content when fully hydrated (p. 36). Over a wide range of water contents many species of earthworms show little or no sign of stress. It is therefore essential that earthworm biomass should be expressed in terms of dry weight, or of weight at a standard level of hydration; the weight of gut contents should be excluded since, especially for species that ingest soil, the weight of gut contents may be a significant proportion of the total, and its contribution to the apparent biomass varies with feeding or burrowing behaviour at the time of sampling. Corrections may also be necessary for changes in weight of earthworms in preservative solutions.

A. Corrections to Biomass Measurements

For each investigation, it is necessary to determine experimentally the relationships between dry weight and fresh or preserved weight of the species present, if a high standard of accuracy is required. Some indication of the relationships can be derived from results of previous investigations.

Lakhani and Satchell (1970) found that for *L. terrestris* in an English woodland soil, 1 g dry weight (for worms free of gut contents) corresponded to 6.37 g fresh weight, i.e., dry weight equalled 15.7% fresh weight. Lavelle (1978) distinguished species of Megascolecidae and Eudrilidae from the savanna lands of Ivory Coast on the basis of their burrowing and feeding behaviour; he found that the dry weight of litter-feeding species was about 15% of fresh weight, while for soil-feeding species dry weight was 7.1%–8.6% of wet weight. For *Agastrodrilus* spp. dry weight was about 21% of fresh weight, but it seems likely that these earthworms are predatory on small earthworms (p. 29) and probably ingest little soil.

Earthworms preserved in formalin commonly lose a significant proportion of their fresh body weight. Piearce (1972b) found that *Aporrectodea caliginosa*, preserved in 5% formalin, had lost 3.79% of their body weight overnight, 3.46% after one week, while *Lumbricus rubellus* had lost 4.74% overnight and 5.09% after one week. Losses of this kind should be taken into account if accurate biomass determinations are to be made. Persson and Lohm (1977) preserved specimens of Lumbricidae from forest soils in Sweden in 6% formalin solution. Dissection and weighing of the gut contents showed that gut contents accounted for 11.8% of the initial dry weight for *Aporrectodea rosea*, 15% for *A. caliginosa*, and 10% for *Dendrobaena* spp. and *Lumbricus* spp. When placed in formalin *A. rosea* exuded much mucus, and this was found to correspond to 11.6% of the initial dry weight, almost exactly compensating for the mean weight of gut contents.

The use of body volume or body length to calculate biomass was advocated by Bouché (1966b, 1967) and by Nordström and Rundgren (1972). Relationships between body length and body weight have been calculated by Edwards (1967), Reynolds (1972), Mazanceva (1976) for some lumbricid species and by Madge (1969) for some non-lumbricid species from Nigeria. Reynolds (1972) calculated regressions relating body length to weight of lumbricids from Indiana, as follows:

$$\log D = 0.4055 + 2.3900 \log L$$

and

$$\log W = 0.2800 + 2.3362 \log L$$

where D is dry weight (mg), W is fresh weight (mg), L is length (cm); the correlation coefficient for both these equations was 0.91. Sugi and Tanaka (1978b) calculated regressions to relate dry weight and fresh weight to body volume of a *Pheretima* sp. from a Japanese woodland soil, as follows:

$$D = 0.0695\,V + 1.3304$$

$$W = 1.03338\,V - 2.1669$$

where D is dry weight (mg), W is fresh weight (mg), V is body volume (mm³). Body volume was calculated from length and diameter.

Abrahamsen (1973) proposed that body surface area was more meaningful as a parameter for measuring the significance of earthworms in ecosystems than was the abundance or biomass, because metabolic activity is usually proportional to surface area (see Macfadyen [1963], Ganong [1969]). Abrahamsen calculated the body surface area of earthworms from the equation

$$A = 1.9\,(\pi.W.L)^{1/2}$$

where A is body surface area, W is live weight, L is length of earthworm preserved in 5% formalin. He determined the mean density of earthworms (gut empty) was 1.04, and from this could calculate live weight from length and mean diameter. In a brown forest soil in Sweden, where the biomasses of enchytraeids and lumbricids were respectively 3.3 g/m² and 27.5 g/m², Abrahamsen calculated that the mean body surface areas of the two groups were respectively 5.2 dm²/m² and 4.1 dm²/m². Assuming that metabolic activity is proportional to body surface area, the enchytraeids would be more important in energy consumption than the lumbricids, since the enchytraeids are much smaller and thus have a much higher surface/volume ratio.

To empty the gut of soil the usual method has been to keep the earthworms alive on filter paper (Joest 1897) or in cellulose powder (Doeksen and Couperus 1962) until all soil is voided, or to kill and fix the worms and carefully dissect and wash out the gut contents. Bouché (1966a) described a method using a large water-filled syringe, with a plastic catheter inserted into the oesophagus, to flush soil from the gut of some lumbricid species. The method works well for some species but Lavelle (1978) found it impracticable for very small species and for some large species from the Ivory Coast.

References

Abdel-Fattah, R. F. (1954). The chloragogen tissue of earthworms and its relation to urea metabolism. *Proc. Egypt. Acad. Sci.* **10**, 36-50.

Abrahamsen, G. (1969). Sampling design in studies of population densities in Enchytraeidae (Oligochaeta). *Oikos* **20**, 54-66.

Abrahamsen, G. (1972). Ecological study of Lumbricidae (Oligochaeta) in Norwegian coniferous forest soils. *Pedobiologia* **12**, 267-281.

Abrahamsen, G. (1973). Biomass and body-surface area of populations of Enchytraeidae and Lumbricidae (Oligochaeta) in Norwegian coniferous forest soils. *Pedobiologia* **13**, 28-39.

Agarwal, G. S., Rao, K. S. K., and Negi, L. S. (1958). Influence of certain species of earthworms on the structure of some hill soils. *Curr. Sci. (Bangalore)* **27**, 213.

Aichberger, R. von (1914). Untersuchungen über die Ernährung des Regenwurmes. *Kleinwelt* **6**, 53-88

Aldag, R. and Graff, O. (1975a). Einfluss der Regenwurmtätigkeit auf Proteingehalt und Proteinqualität junger Haferpflanzen. *Landwirtschaftliche Forschung* (1975), 31/II, pp. 277-284.

Aldag, R. and Graff, O. (1975b). N-Fraktionen in Regenwurmlosung und deren Ursprungsboden. *Pedobiologia* **15**, 151-153.

Andersen, C. (1979). Cadmium, lead and calcium content, number and biomass, in earthworms (Lumbricidae) from sewage sludge treated soil. *Pedobiologia* **19**, 309-319.

Andersen, C. and Laursen, J. (1982). Distribution of heavy metals in *Lumbricus terrestris*, *Aporrectodea longa* and *A. rosea* measured by atomic absorption and X-ray fluorescence spectrometry. *Pedobiologia* **24**, 347-356.

Arthur, D. R. (1965). Form and function in the interpretation of feeding in lumbricid worms. *In* "Viewpoints in Biology" (J. D. Carthy and C. L. Duddington, eds), Vol. 4, pp. 204-251. Butterworths, London.

Ash, C. P. J. and Lee, D. L. (1980). Lead, cadmium, copper and iron in earthworms from roadside sites. *Environ. Pollut.* **A22**, 59-67.

Atlavinyte, O. (1974a). Effect of earthworms on the biological productivity of barley. *Liet TSR Mokslu. Akad. Darb., Ser. C, Biol. Mokslai* **1**, 69-79.

Atlavinyte, O. (1974b). Earthworms (Lumbricidae). *In* "The Effect of Agronomical Practices on Soil Invertebrates". pp. 41-68. Acad. Sci. Lithuanian SSR, Inst. Zool. Parasitol. Vilnius, 1974.

Atlavinyte, O. (1975). The effect of chemical substances on the activity of Lumbricidae in the process of straw disintegration. *In* "Progress in Soil Zoology". (J. Vanek, ed.). Proc. 5th Intl. Colloq. Soil Zool., Prague, 1973, pp. 515-519. Junk, The Hague/Academia, Prague.

Atlavinyte, O. and Daciulyte, J. (1969). The effect of earthworms on the accumulation of vitamin B_{12} in soil. *Pedobiologia* **9**, 165-170.

Atlavinyte, O. and Payarskaite, A. I. (1962). The effect of erosion on earthworms (Lumbricidae) during the growing season. *Zool. Zh.* **41**, 1631-1636.

352 References

Atlavinyte, O. and Pociene, C. (1973). The effect of earthworms and their activity on the amount of algae in the soil. *Pedobiologia* **13**, 445-455.

Atlavinyte, O. and Vanagas, J. (1973). Mobility of nutritive substances in relation to earthworm numbers in the soil. *Pedobiologia* **13**, 344-352.

Atlavinyte, O. and Vanagas, J. (1982). The effect of earthworms on the quality of barley and rye grain. *Pedobiologia* **23**, 256-262.

Atlavinyte, O., Bagdonaviciene, Z. and Budviciene, L. (1968). The effect of Lumbricidae on the barley crops in various soils. *Pedobiologia* **8**, 415-423.

Atlavinyte, O., Daciulyte, J. and Lugauskas, A. (1971). Correlations between the number of earthworms, microorganisms and vitamin B_{12} in soil fertilised with straw. *Liet. TSR Mokslu. Akad. Darb., Ser. B.,* **3**, 43-56.

Atlavinyte, O., Kuginyte, Z. and Pileckis, S. (1974). Erosion effect on soil fauna under different crops. *Pedobiologia* **14**, 35-40.

Atlavinyte, O., Galvelis, A., Daciulyte, J. and Lugauskas, A. (1982). Effects of entobacterin on earthworm activity. *Pedobiologia* **23**, 372-379.

Avel, M. (1959). Classe des annélides oligochètes. *In* "Traité de Zoologie", (P. P. Grassé, ed.), Vol. 5, pp. 224-270. Masson, Paris.

Axelsson, B., Gardefors, D., Lohm, U., Persson, T. and Tenow, O. (1971). Reliability of estimating standing crop of earthworms by hand sorting. *Pedobiologia* **11**, 338-340.

Baermann, G. (1917). Ein einfach Methode zur Auffindung von *Ankylostomum* (Nematoden) Larven in Erdproben. *Meded. Geneesk. Lab. Weltevr.* 1917, 41-47.

Bahl, K. N. (1919). On a new type of nephridia found in Indian earthworms of the genus *Pheretima. Q. J. Microsc. Sci.* **64**, 67-119.

Bahl, K. N. (1947). Excretion in the Oligochaeta. *Biol. Rev. Camb. Philos. Soc.* **22**, 109-147.

Baker, G. H. (1983). Distribution, abundance and species associations of earthworms (Lumbricidae) in a reclaimed peat soil in Ireland. *Holarct. Ecol.* **6**, 74-80.

Baltzer, R. (1956). Die Regenwürmer Westfalens. Eine tiergeographische, ökologische und sinnesphysiologische Untersuchung. *Zool. Jahrb. Syst.* **84**, 355-414.

Barker, R. J. (1958). Notes on some ecological effects of DDT sprayed on elms. *J. Wildl. Manage.* **22**, 269-274.

Barley, K. P. (1959a). The influence of earthworms on soil fertility. I. Earthworm populations found in agricultural land near Adelaide. *Aust. J. Agric. Res.* **10**, 171-178.

Barley, K. P. (1959b). The influence of earthworms on soil fertility. II. Consumption of soil and organic matter by the earthworm *Allolobophora caliginosa* (Savigny). *Aust. J. Agric. Res.* **10**, 179-185.

Barley, K. P. (1959c). Earthworms and soil fertility. IV. The influence of earthworms on the physical properties of a red-brown earth. *Aust. J. Agric. Res.* **10**, 371-376.

Barley, K. P. (1961). The abundance of earthworms in agricultural land and their possible significance in agriculture. *Adv. Agron.* **13**, 249-268.

Barley, K. P. and Jennings, A. C. (1959). Earthworms and soil fertility. III. The influence of earthworms on the availability of nitrogen. *Aust. J. Agric. Res.* **10**, 364-370.

Barley, K. P. and Kleinig, C. R. (1964). The occupation of newly irrigated lands by earthworms. *Aust. J. Sci.* **26**, 290-291.

Barnes, B. T. and Ellis, F. B. (1979). Effects of different methods of cultivation and direct drilling and disposal of straw residues, on populations of earthworms. *J. Soil Sci.* **30**, 669-679.

Bassalik, K. (1913). Über Silikatersetzung durch Bodenbakterien. *Z. GärPhysiol.* **2**, 1-32.

Bauer, K. (1964). Studien über Nebenwirkungen von Pflantzenschutzmitteln auf die Bodenfauna. *Mitt. Biol. Budensanst. Land- Forstwirtsch. Berl- Dahlem* **112**, 1-42.

Baweja, K. D. (1939). Studies of the soil fauna with special reference to the recolonization of sterilised soil. *J. Anim. Ecol.* **8**, 120-161.
Beadle, L. C. (1933). Scientific results of the Cambridge Expedition to the East African swamps, 1930-31. 13. Adaptation to aerial respiration in *Alma emini* Mich., an oligochaete from East African swamps. *J. Linn. Soc. Lond. Zool.* **38**, 347-350.
Beadle, L. C. (1957). Respiration in the African swampworm *Alma emini* Mich. *J. Exp. Biol.* **34**, 1-10.
Beaugé, C. (1912). Les vers de terre et la fertilité du sol. *J. Agr. Prat. (Paris)* n.s. **23**, 506-507.
Bengtson, S.-A., Nilsson, A., Nordström, S. and Rundgren, S. (1975). Habitat selection of lumbricids in Iceland. *Oikos* **26**, 253-263.
Bengtson, S.-A., Nilsson, A., Nordström, S. and Rundgren, S. (1976). Effect of bird predation on lumbricid populations. *Oikos* **27**, 9-12.
Bengston, S.-A., Nilsson, A., Nordström, S. and Rundgren, S. (1979). Short-term colonization success of lumbricid founder populations. *Oikos* **33**, 308-315.
Benham, W. B. (1903). On some new species of aquatic Oligochaeta from New Zealand. *Proc. Zool. Soc., Lond.* 1903(2), 202-232.
Benham, W. B. (1905). Earthworms from the Kermadecs. *Trans. N.Z. Inst.* **37**, 299-300.
Benham, W. B. (1915). Oligochaeta from the Kermadec Islands. *Trans. N.Z. Inst.* **47**, 174-185.
Benz, G. and Altwegg, A. (1975). Safety of *Bacillus thuringiensis* for earthworms. *J. Invertebr. Pathol.* **26**, 125-126.
Beyer, W. N. and Gish, C. D. (1980). Persistence in earthworms and potential hazards to birds of soil applied DDT, dieldrin and heptachlor. *J. Appl. Ecol.* **17**, 295-307.
Bhandari, G. S., Randhawa, N. S. and Maskin, M. S. (1967). On the polysaccharide content of earthworm casts. *Curr. Sci. (Bangalore)* **36**, 519-520.
Bhatnagar, T. (1975). Lombriciens et humification: Un aspect nouveau de l'incorporation microbienne d'azote induite par les vers de terre. *In* "Biodégradation et Humification". (G. Kilbertus, O. Reisinger, A. Mourey, J. A. Cancela de Fonseca, eds), pp. 169-182. Pierron, Sarreguemines.
Bird, S. J. G. and Hale, I. M. (1982). Development of vermicomposting operations at a prototype plant. Consultants' Report to A. E. Burgess, Waste Control Division, Environment Canada, Ottawa, Ontario. 175 pp.
Bishop, S. H. and Campbell, J. W. (1965). Arginase and urea biosynthesis in the earthworm *Lumbricus terrestris*. *Comp. Biochem. Physiol.* **15**, 51-71.
Black, W. M. and Neely, D. (1975). Effect of soil-injected benomyl on resident earthworm populations. *Pestic. Sci.* **6**, 543-545.
Blancke, E. and Giesecke, F. (1924). The effect of earthworms on the physical and biological properties of soil. *Z. PflanzErnähr. Dung. Bodenkund.* **3B**, 198-210.
Block, W. and Banage, W. B. (1968). Population density and biomass of earthworms in some Uganda soils. *Rev. Ecol. Biol. Sol.* **5**, 515-521.
Bocock, K. L. (1963). The digestion and assimilation of food by *Glomeris*. *In* "Soil Organisms" (J. Doeksen and J. van der Drift, eds), pp. 85-91. North Holland Publishing Company, Amsterdam.
Bodenheimer, F. S. (1935). Soil conditions, which limit earthworm distribution. *Zoogeographica* **2**, 572-578.
Bolton, P. J. and Phillipson, J. (1976). Burrowing, feeding, egestion and energy budgets of *Allolobophora rosea* (Savigny) (Lumbricidae). *Oecologia (Berl.)* **23**, 225-245.
Bornebusch, C. H. (1953). Laboratory experiments on the biology of worms. *Dansk. Skovforen. Tidsskr.* **38**, 557-579.

Bosse, I. (1967). Wiederbelebung biologisch verarmter Weinbergsböden, dargestellt am Beispiel des Regenwurmbesatzes. *In* "Progress in Soil Biology" (O. Graff and J. E. Satchell, eds), pp. 299-309. North Holland Publishing Company, Amsterdam.

Bouché, M. B. (1966a). Sur un nouveau procédé d'obtentien de la vacuité artificielle du tube digestif des lombricides. *Rev. Ecol. Biol. Sol.* **3**, 479-482.

Bouché, M. B. (1966b). Application de la volumétrie à l'évaluation quantitative de la faune endogée. *Rev. Ecol. Biol. Sol.* **3**, 19-30.

Bouché, M. B. (1967). Etablissement et comparaison de diverses bioquantités pour trois espèces de Lumbricidae. *In* "Progress in Soil Biology" (O. Graff and J. E. Satchell, eds), pp. 595-600. North Holland Publishing Company, Amsterdam.

Bouché, M. B. (1969). Comparaison critique de méthodes d'évaluation des populations de Lombricidés. *Pedobiologia* **9**, 26-34.

Bouché, M. B. (1970). Observations sur les Lombricidés, VII. Une adaptation à la vie amphibie: La respiration caudale aquatique. *Rev. Ecol. Biol. Sol.* **7**, 533-536.

Bouché, M. B. (1971). Relations entre les structures spatiales et fonctionelles des écosystèmes, illustrées par le rôle pédobiologique des vers de terre. *In* "La Vie dans les Sols" (C. Delamere Debouttevelle, ed.), pp. 189-209. Gauthiers Villars, Paris.

Bouché, M. B. (1972). "Lombriciens de France. Ecologie et Systématique". INRA Publ. 72-2. Institut National des Recherches Agiculturelles, Paris.

Bouché, M. B. (1975a). Fonctions des lombriciens. IV. Corrections et utilisations des distorsions causées par les méthodes de capture. *In* "Progress in Soil Zoology" (J. Vanek, ed.), pp. 571-582. Proc. 5th Intl. Colloq. Soil Zool., Prague, 1973. Junk, The Hague/Academia, Prague.

Bouché, M. B. (1975b). Action de la faune sur les états de la matière organique dans les écosystèmes. *In* "Biodégradation et Humification" (K. Gilbertus, O. Reisinger, A. Mourey and J. A. Cancela da Fonseca, eds), pp. 157-168. Pierron, Sarruguemines.

Bouché, M. B. (1975c). Fonctions des lombriciens. III. Premières estimations quantitatives des stations francaises du PBI. 1er Colloq. de la Sócieté d'Ecologie. 1973. *Rev. Ecol. Biol. Sol.* **12**, 12-25.

Bouché, M. B. (1975d). La reproduction de *Spermophorodrilus albanianus* nov. gen., nov. sp. (Lumbricidae), explique-t-elle la fonction des spermatophores? *Zool. Jb. Syst.* **102**, 1-11.

Bouché, M. B. (1976). Etude de l'activité des invertébrés épigés prairiaux. I. Résultats généraux et géodrilogiques (Lumbricidae: Oligochaeta). *Rev. Ecol. Biol. Sol.* **13**, 261-281.

Bouché, M. B. (1977a). Fauna in the grassland system. Role of earthworms. *In* "Department of Zoology. Orientations and Means; Presentation of Some Research Projects". Publ. Institut National des Recherches Agriculturelles, France, pp. 21-23.

Bouché, M. B. (1977b). Stratégies lombriciennes. *In* "Soil Organisms as Components of Ecosystems" (U. Lohm and T. Persson, eds), *Biol. Bull. (Stockholm)* **25**, 122-132.

Bouché, M. B. (1978). Fonctions des lombriciens. I. Recherches francaises et résultats d'un programme forestier coopératif (R.C.P.40). *Bull. Scientif. Bourgogne* **30**, 139-228.

Bouché, M. B. (1979). Valorisation des déchets organiques par les lombriciens. *In* "La Documentation Francaise, Collection Recherche et Environnement" 11, 384.

Bouché, M. B. (1980). Objectifs, compartimentation et faisabilité du modèle R.E.A.L. (Rôle Ecologique et Agronomique des Lombriciens). *Pedobiologia* **20**, 197-211.

Bouché, M. B. (1981). Contribution des lombriciens aux migrations d'éléments dans les sols tempérés. Colloques Internationaux du Centre National des Recherches Scientifiques, no. 303, 145-153.

Bouché, M. B. (1982). Ecosystème prairial. 4.3 Un exemple d'activité animale: Le rôle des lombriciens. *Acta Oecol., Oecol. Gen.* **3**, 127-154.

Bouché, M. B. (1983). Ecophysiologie des lombriciens: Acquis récents et perspectives. *In* "New Trends in Soil Biology" (Ph. Lebrun, H. M. André, A. de Medts, C. Grégoire-Wibo and G. Wauthy, eds), pp. 321-333. Proc. 8th Intl. Colloq. Soil Zool., Louvain-la-Neuve, 1982. Dieu-Brichart, Ottignies-Louvain-la-Neuve.

Bouché, M. B. (1984). Ecotoxicologie des lombriciens. I. Ecotoxicologie contrôlée. *Acta Oecologica* **5**, 271-287.

Bouché, M. B. and Beugnot, M. (1972). Contribution à l'approche méthodologique de l'étude des biocénoses. II. L'extraction des macroéléments du sol par lavage-tamisage. *Ann. Zool. Ecol. Anim.* **4**, 537-544.

Bouché, M. B. and Gardner, R. H. (1984). Earthworm functions. VIII. Population estimation techniques. *Rev. Ecol. Biol. Sol.* **21**, 37-63.

Bouché, M. B. and Kretzschmar, A. (1974). Fonctions des lombriciens. II. Recherches méthodologiques pour l'analyse du sol ingéré (étude du peuplement de la station R.C.P.-165/P.B.I.). *Rev. Ecol. Biol. Sol.* **11**, 127-139.

Bouché, M. B. and Kretzchmar, A. (1977). REAL: Un modèle du rôle écologique et agronomique des lombriciens. *In* "Soil Organisms as Components of Ecosystems" (U. Lohm and T. Persson, eds), *Ecol. Bull. (Stockholm)* **25**, 402-408.

Brauns, A. (1955). Angewandte Bodenbiologie, waldbauliche Probleme. *Neues Archiv. für Neidersachsen*, Part 1/3, 1-18.

Bremeyer, A. (1974). Structure of the tropical grassland ecosystem in Panama. *In* Proc. IBP Symposium cum Synthesis Meeting, Benares Hindu University, Varanasi, India.

Bretscher, K. (1896). The Oligochaeta of Zurich. *Rev. Suisse Zool.* **3**, 499-532.

Breznak, J. A., Brill, W. J., Mertins, J. W. and Coppel, H. C. (1973). Nitrogen fixation in termites. *Nature (Lond.)* **244**, 577-580.

Brinkhurst, R. O. and Jamieson, B. G. M. (1971). "Aquatic Oligochaeta of the World". Oliver and Boyd, Edinburgh.

Brookfield, H. C. and Hart, D. (1971). "Melanesia. A Geographical Interpretation of an Island World." Methuen, London.

Brown, A. W. A. (1978). "Ecology of Pesticides". John Wiley and Sons, New York.

Brown, B. A. and Mitchell, M. J. (1981). Role of the earthworm, *Eisenia foetida*, in affecting survival of *Salmonella enteritidis* ser. *typhimurium*. *Pedobiologia* **22**, 434-438.

Brüsewitz, G. (1959). Investigations on the effects of earthworms on the population, species distribution and activity of microorganisms in soil. *Arch. Mikrobiol* **33**, 52-82.

Bull, K. R., Roberts, R. D., Inskip, M. J. and Goodman, G. T. (1977). Mercury concentrations in soil, grass, earthworms and small mammals near an industrial emission source. *Environ. Pollut.* **12**, 135-140.

Buntley, G. J. and Papendick, R. I. (1960). Worm-worked soils of eastern South Dakota, their morphology and classification, *Soil Sci. Soc. Am. Proc.* **24**, 128-132.

Byzova, J. B. (1965a). Comparative rate of respiration in some earthworms (Lumbricidae, Oligochaeta). *Rev. Ecol. Biol. Sol.* **2**, 207-216.

Byzova, J. B. (1965b). Oxygen consumption dependence on mode of life and body size, exemplified by earthworms. *J. Gen. Biol. Moscow* **26**, 555-562.

Byzova, J. B. (1974). The dynamics of some blood indices in earthworms (Oligochaeta, Lumbricidae). *Rev. Ecol. Biol. Sol.* **11**, 325-332.

Byzova, J. B. (1975). Specific haemoglobin content as an ecological characteristic in earthworms (Lumbricidae). *In* "Progress in Soil Zoology" (J. Vanek, ed.), pp. 559-562. Proc. 5th Intl. Colloq. Soil Zool., Prague, 1973. Junk, The Hague/Academia, Prague.

Byzova, J. B. (1977). Haemoglobin content of *Allolobophora caliginosa* (Sav.) (Lumbricidae, Oligochaeta) during aestivation. *Doklady Biological Sciences* **236**, 763-765. (English translation).

Calow, P. (1977). Ecology, evolution and energetics: A study in metabolic adaptation. *Adv. Ecol. Res.* **10**, 1-62.

Carley, W. W. (1975). Effects of brain removal on integumental water permeability and ion content of the earthworm *Lumbricus terrestris* L. *Gen. Comp. Endocrinal.* **27**, 509-516.

Carley, W. W. (1978). Water economy of the earthworm *Lumbricus terrestris* L.: Coping with the terrestrial environment. *J. Exp. Zool.* **205**, 71-78.

Carter, A., Heinonen, J. and de Vries, J. (1982). Earthworms and water movement. *Pedobiologia* **23**, 395-397.

Carter, A., Kenney, E. A., Guthrie, T. F. and Timmenga, H. (1983). Heavy metals in earthworms in non-contaminated and contaminated agricultural soil from near Vancouver, Canada. *In* "Earthworm Ecology" (J. E. Satchell, ed.), pp. 267-274. Chapman and Hall, London.

Carter, G. S. and Beadle, L. C. (1931). The fauna of the swamps of the Paraguayan Chaco in relation to its environment. III. Respiratory adaptations in the Oligochaeta. *J. Linn. Soc. Lond. Zool.* **37**, 379-386.

Caseley, J. C. and Eno, C. F. (1966). Survival and reproduction of two species of earthworm and a rotifer following herbicide treatment. *Soil. Sci. Soc. Am. Proc.* **30**, 346-350.

Cernosvitov, L. (1945). Oligochaeta from Windermere and the Lake District. *Proc. Zool. Soc., Lond.* **114**, 523-548.

Chapron, C. (1971). Etude de rôle de la glande de Morren des lombriciens dans la régulation de l'eau. *J. Microsc. (Paris).* **10**, 351-356.

Chen, C. M. and Liu, C. L. (1963). Dynamics of the populations and communities of rice insect pests in the bank of Fung-Ting Lake region, Hunan. *Acta Entomol. Sinica* **12**, 649-657.

Citernesi, U., Neglia, R., Seritti, A., Lepidi, A. A., Filippi, C., Bagnoli, G., Nuti, M. P. and Galluzzi, R. (1977). Nitrogen fixation in the gastro-enteric cavity of soil animals. *Soil Biol. Biochem.* **9**, 71-72.

Cleland, W. (1973). The kinetics of enzyme-catalyzed reactions with two or more substrates or products. *Biochim. Biophys. Acta.* **67**, 104-137.

Clements, R. O. (1982). Some consequences of large and frequent pesticide applications to grassland. *In* Proc. 3rd Australasian Conf. Grassl. Invert. Ecol. (K. E. Lee, ed.), pp. 393-396. South Australian Government Printer, Adelaide.

Coleman, D. C. and Sasson, A. (1978). Decomposer subsystem. *In* "Grasslands, Systems Analysis and Man" (A. J. Breymeyer and G. M. van Dyne, eds), I.B.P. 19, pp. 609-655. Cambridge University Press, Cambridge.

Coleman, D. C., Andrews, R., Ellis, J. E. and Singh, J. S. (1976). Energy flow and partitioning in selected man-managed and natural ecosystems. *Agro-Ecosystems* **3**, 45-54.

Coleman, D. C., Reid, C. P. P. and Cole, C. V. (1983). Biological strategies of nutrient cycling in soil systems. *Adv. Ecol. Res.* **13**, 1-55.

Coles, G. C. (1970). Some biochemical adaptations of the swamp worm *Alma emini* to low oxygen levels in tropical swamps. *Comp. Biochem. Physiol.* **34**, 481-489.

Collett, N. and Harrison, D. L. (1968). Some observations on the effects of using organochlorine sprays in an orchard. *N.Z. J. Sci.* **11**, 371-379.

Collier, J. (1978). Use of earthworms in sludge lagoons. *In* "Utilization of Soil Organisms in Sludge Management" (R. Hartenstein, ed.), pp. 131-133. Natl Tech. Inf. Services, PB286932. Springfield, Virginia.

Collier, J. and Livingstone, D. (1981). Conversion of municipal wastewater treatment plant residual sludges into earthworm castings for use as a fertile topsoil. Report to U.S. National Science Foundation, Appropriate Technology Program. Grant No. ENV77-16832 A01.

Collins, N. M. (1980). The distribution of soil macrofauna on the west ridge of Gunung (Mount) Mulu, Sarawak. *Oecologia (Berl.)* **44**, 263–275.

Combault, A. (1907). Quelques expériences pour déterminer le rôle des glandes calcifères des Lombrics. *C.R. Séances Soc. Biol. (Paris)* **62**, 440–442.

Combault, A. (1909). Contribution à l'étude de la respiration et de la circulation de lombriciens. *J. Anat. (Paris)* **45**, 358–399.

Contreras, E. (1980). Studies on the intestinal actinomycete flora of *Eisenia lucens* (Annelida, Oligochaeta). *Pedobiologia* **20**, 411–416.

Cook, A. G., Critchley, B. R., Critchley, U., Perfect, T. J. and Yeadon, R. (1980). Effects of cultivation and DDT on earthworm activity in a forest soil in the sub-humid tropics. *J. Appl. Ecol.* **17**, 21–29.

Cook, M. E. and Swait, A. A. J. (1975). Effects of some fungicide treatments on earthworm populations and leaf removal in apple orchards. *J. Hortic. Sci.* **50**, 495–499.

Cooke, A. (1983). The effects of fungi on food selection by *Lumbricus terrestris* L. In "Earthworm Ecology" (J. E. Satchell, ed.), pp. 365–373. Chapman and Hall, London.

Cooke, A. and Luxton, M. (1980). Effect of microbes on food selection by *Lumbricus terrestris*. *Rev. Ecol. Biol. Sol.* **17**, 365–370.

Cosgrove, W. B. and Schwartz, J. B. (1965). The properties and functions of the blood pigment of the earthworm, *Lumbricus terrestris*. *Physiol. Zool.* **38**, 206–212.

Cotton, D. C. F. and Curry, J. P. (1980a). The effects of cattle and pig slurry fertilizers on earthworms (Oligochaeta, Lumbricidae) in grassland managed for silage production *Pedobiologia* **20**, 181–188.

Cotton, D. C. F. and Curry, J. P. (1980b). The response of earthworm populations (Oligochaeta, Lumbricidae) to high applications of pig slurry. *Pedobiologia* **20**, 189–196.

Critchley, B. R., Cook, A. G., Critchley, U., Perfect, T. J., Russell-Smith, A. and Yeadon, R. (1979). Effects of bush clearing and soil cultivation on the invertebrate fauna of a forest soil in the humid tropics. *Pedobiologia* **19**, 425–438.

Crossley, D. A., Reichle, D. E. and Edwards, C. A. (1971). Intake and turnover of radioactive cesium by earthworms (Lumbricidae). *Pedobiologia* **11**, 71–76.

Cuendet, G. (1977). Etude du comportement alimentaire des Mouettes rieuses (*Larus ridibundus*) et de leur influence sur les populations de vers de terre. *Ornithol. Beob.* **2**, 87–88.

Cuendet, G. (1983). Predation on earthworms by the black-headed gull (*Larus ridibundus* L.). In "Earthworm Ecology" (J. E. Satchell, ed.), pp. 415–424. Chapman and Hall, London.

Cuénot, L. (1898). Etudes physiologiques sur les oligochètes. *Arch. Biol.* **15**, 79–124.

Curry, J. P. (1976). Some effects of animal manures on earthworms in grassland. *Pedobiologia* **16**, 425–438.

Czerwinski, Z., Jakubczyk, H. and Nowak, E. (1974). Analysis of a sheep pasture ecosystem in the Pieniny Mountains (The Carpathians). XII. The effect of earthworms on the pasture soil. *Ekol. Pol.* **22**, 635–650.

Dales, R. P. (1978). Defence mechanisms. In "Physiology of Annelids" (P. J. Mill, ed.), pp. 479–507. Academic Press, London.

Darwin, C. R. (1837). On the formation of mould. *Trans. Geol. Soc.* **5** (ser. 2), 505–509.

Darwin, C. R. (1881). "The Formation of Vegetable Mould through the Action of Worms, with Observations on their Habits". Murray, London.

Dash, M. C. and Patra, U. C. (1977). Density, biomass and energy budget of a tropical earthworm population from a grassland site in Orissa, India. *Rev. Ecol. Biol. Sol.* **14**, 461-471.

Dash, M. C. and Patra, U. C. (1979). Wormcast production and nitrogen contribution to soil by a tropical earthworm population from a grassland site in Orissa, India. *Rev. Ecol. Biol. Sol.* **16**, 79-83.

Dash, M. C. and Senapati, B. K. (1980). Cocoon morphology, hatching and emergence pattern in tropical earthworms. *Pedobiologia* **20**, 316-324.

Dash, M. C. and Senapati, B. K. (1982). Environmental regulation of oligochaete reproduction in tropical pastures. *Pedobiologia* **23**, 270-271.

Dash, M. C., Patra, U. C. and Thambi, A. V. (1974). Comparison of primary production of plant material and secondary production of oligochaetes in a tropical grassland of Orissa, India. *Trop. Ecol.* **15**, 16-21.

Dash, M. C., Senapati, B. K. and Mishra, C. C. (1980). Nematode feeding by tropical earthworms. *Oikos* **34**, 322-325.

Davis, B. N. K. (1971). Laboratory studies on the uptake of dieldrin and DDT by earthworms. *Soil Biol. Biochem.* **3**, 221-233.

Davis, J. B. and Slater, W. K. (1928). The anaerobic metabolism of the earthworm (*Lumbricus terrestris*). *Biochem. J.* **22**, 338-343.

Day, G. M. (1950). The influence of earthworms on soil microorganisms. *Soil Sci.* **69**, 175-184.

Derby, J. M., Houssiau, M., Lemasson-Florenville, M., Wauthy, G. and Lebrun, Ph. (1982). Impact de populations lombriciennes introduites sur le pH et sur la dynamique de l'azote dans un sol traité avec du lisier de porcs. *Pedobiologia* **23**, 157-171.

Dexter, A. R. (1978). Tunnelling in soil by earthworms. *Soil Biol. Biochem.* **10**, 447-449.

Dhennin, Léone, Heim de Balsac, H., Verge, J. and Dhennin, Louis (1963). Recherches sur le rôle éventuel de *Lumbricus terrestris* dans la transmission expérimentale du virus de la fièvre aphteuse. *Bull. Acad. Vet. Fr.* **36**, 153-155.

Diaz Cosin, D. J., Jesus, B. J. and Moreno, A. G. (1981). Earthworms taxocoenosis of Chelva (Valencia, Spain). *Pedobiologia* **21**, 125-131.

Dietz, S. (1979). Etude de l'incorporation de la litière en système herbacé à l'aide de matériel végétal marqué au ^{14}C. Thèse de 3° cycle écol. terr. Univ. Sci. Tech., Languedoc, Montpellier.

Dietz, S. and Bottner, P. (1981). Etude par autoradiographie de l'enfouissement d'une litière marquée au ^{14}C en milieu herbacé. Colloques internationaux du Centre National des Recherches Scientifiques, no. 303, 125-132.

Dietz, T. H. and Alvarado, R. H. (1970). Osmotic and ionic regulation in *Lumbricus terrestris* L. *Biol. Bull. (Woods Hole)* **138**, 247-261.

Dindal, D. L., Schwert, D. P., Moreau, J.-P. and Theoret, L. (1977). Earthworm communities and soil nutrient levels as affected by municipal wastewater irrigation. *In* "Soil Organisms as Components of Ecosystems" (U. Lohm and T. Persson, eds), *Ecol. Bull. (Stockholm)* **25**, 284-290.

Doane, C. C. (1962). Effects of certain insecticides on earthworms. *J. Econ. Ent.* **55**, 416-418.

Dobrovol'skii, G. V. and Titkova, N. F. (1960). Characteristics of the structure of soils of bottomland oak stands. *Pochvovedenie* 1960, no. 1, 15-25.

Dobzansky, T. (1950). Evolution in the tropics. *Am. Sci.* **38**, 209-221.

Doeksen, J. (1964). Notes on the activity of earthworms. III. The conditioning effects of earthworms on the surrounding soil. *Jaarb. Inst. Biol. Scheikd.* 1964, 187-191.

Doeksen, J. and Couperus, H. (1962). Het vaststellen van groei bij regenwormen. *Inst.*

Biol. Scheikd. Onderz. Landbouwegewassen Wageningen Jaarversl. 1964, no. 256, 187-191.
Dommergues, Y. (1968). Dégagement tellurique de CO_2. Mesure et signification. *Ann. Inst. Pasteur (Paris).* **115**, 627-656.
Dorsett, D. A. (1978). Organization of the nerve cord. *In* "Physiology of Annelids" (P. J. Mill, ed.), pp. 115-160. Academic Press, London.
Douglas, J. T., Goss, M. J. and Hill, D. (1989). Measurements of pore characteristics in a clay soil under ploughing and direct drilling, including use of a radioactive tracer (^{144}Ce) technique. *Soil and Tillage Res.* **1**, 11-18.
Drift, J. van der (1963). The influence of biocides on the soil fauna. *Neth. J. Plant. Pathol.* **69**, 188-199.
Drummond, H. (1888). The white ant. A theory. Chapter 6 *In*: "Tropical Africa" London.
Dubash, P. J. and Ganti, S. S. (1963). Preliminary studies on earthworms in relation to soil. *In* "Symposium on Ecological Problems in the Tropics". *Proc. Nat. Acad. Sci. India Sect. B (Biol. Sci.)* **33**, 193-197.
Dubash, P. J. and Ganti, S. S. (1964). Earthworms and amino acids in soil. *Curr. Sci. (Bangalore)* **33**, 219-220.
Duff, H. A. (1958). Pasture improvement in low altitude tussock and shrub-land. *N.Z. Grassl. Ass. Proc.* **20**, 70-79.
Dzangaliev, A. D. and Belousova, N. K. (1969). Earthworm populations in irrigated orchards under various soil treatments. *Pedobiologia* **9**, 103-105.
Easton, E. G. (1979). A revision of the "acaecate" earthworms of the *Pheretima* group (Megascolecidae: Oligochaeta): *Archipheretima, Metapheretima, Planapheretima, Pleionogaster,* and *Polypheretima. Bull. Brit. Mus. Nat. Hist. Zool.* **35**, 1-126.
Easton, E. G. (1983). A guide to the valid names of Lumbricidae (Oligochaeta). *In* "Earth Ecology", (J. E. Satchell, ed.), pp. 475-487. Chapman and Hall, London.
Eaton, T. H. and Chandler, R. F. (1942). The fauna of forest-humus layers in New York. *Mem. Cornell Univ. Agric. Exp. Stn* 247.
Edwards, C. A. (1965). Effects of pesticide residues on soil invertebrates and plants. *In* "Ecology and the Industrial Society" (G. T. Goodman, R. W. Edwards and J. M. Lambert, eds), pp. 239-261. *5th Symp. Brit. Ecol. Soc.*, Blackwell, London.
Edwards, C. A. (1967). Relationships between weights, volumes and numbers of soil animals. *In* "Progress in Soil Biology" (O. Graff and J. E. Satchell, eds), pp. 585-594. North Holland Publishing Company, Amsterdam.
Edwards, C. A. (1974). Macroarthropods. *In* "Biology of Plant Litter Decomposition" (C. H. Dickinson and G. J. F. Pugh, eds), Vol. 2, pp. 533-554. Academic Press, New York.
Edwards, C. A. and Heath, G. W. (1975). Studies in leaf litter breakdown. III. The influence of leaf age. *Pedobiologia* **15**, 348-354.
Edwards, C. A. and Lofty, J. R. (1977). "Biology of Earthworms". Second edition. Chapman and Hall, London.
Edwards, C. A. and Lofty, J. R. (1978). The influence of arthropods and earthworms upon root growth of direct drilled cereals. *J. Appl. Ecol.* **15**, 789-795.
Edwards, C. A. and Lofty, J. R. (1979). The effects of straw residues and their disposal on the soil fauna. *In* "Straw Decay and Its Effect on Dispersal and Utilization" (E. Grossbard, ed.), pp. 37-44. Wiley, New York.
Edwards, C. A. and Lofty, J. R. (1980). Effects of earthworm inoculation upon the root growth of direct drilled cereals. *J. Appl. Ecol.* **17**, 533-543.
Edwards, C. A. and Lofty, J. R. (1982a). Nitrogenous fertilizers and earthworm populations in agricultural soils. *Soil Biol. Biochem.* **14**, 515-521.
Edwards, C. A. and Lofty, J. R. (1982b). The effect of direct drilling and minimal cultivation on earthworm populations. *J. Appl. Ecol.* **19**, 723-734.

Edwards, C. A. and Thompson, A. R. (1973). Pesticides and the soil fauna. *Residue Rev.* **45**, 1-79.

Edwards, C. A., Thompson, A. R. and Beynon, K. I. (1968). Some effects of chlorfenvinphos, an organophosphorus insecticide, on populations of soil animals. *Rev. Ecol. Biol. Sol.* **5**, 199-224.

Edwards, C. A., Whiting, A. E. and Heath, G. W. (1970). A mechanised washing method for separation of invertebrates from soil. *Pedobiologia* **10**, 141-148.

Edwards, P. J. and Brown, S. M. (1982). Use of grassland plots to study the effect of pesticides on earthworms. *Pedobiologia* **24**, 145-150.

Edwards, W. M., van der Ploeg, R. R. and Ehlers, W. (1979). A numerical study of the effects of noncapillary sized pores upon infiltration. *Soil. Sci. Soc. Amer. J.* **43**, 851-856.

Ehlers, W. (1975). Observations on earthworm channels and infiltration on tilled and untilled loess soil. *Soil Sci.* **119**, 242-249.

Eitminaviciute, I., Bagdanaviciene, Z., Budaviciene, I., Atlavinyte, O., Liepinis, A., Strazdiene, V., Sukackiene, I. and Slepetiene, J. (1971). Untersuchungen der Beziehungen zwischen Gruppen von Wirbellosen lebewesen und Mikroorganismen sowie der B-Vitamingruppe in unterschiedlichen Böden. *In* "IV Colloquium Pedobiologiae" (J. d'Aguilar, ed.), pp. 93-97. Institut National des Recherches Agriculturelles Publ. 71-7, Paris.

Ellenby, C. (1945). Influence of earthworms on larval emergence in the potato root eelworm *Heterodera rostochiensis* Wollenweber. *Ann. Appl. Biol.* **31**, 332-339.

Emerson, W. W. (1954). The determination of the stability of soil crumbs. *J. Soil Sci.* **5**, 233-250.

Emiliani, F., Ljungström, P. O., Priano, L. Gutiérrez, T. and Calamante, R. (1971). Sobre la ecologia de la *Eukerria halophila* (Oligochaeta, Ocnerodrilidae). *Bol. R. Soc. Espan. Hist. Nat. (Biol.)* **69**, 19-22.

Eno, C. F. (1966). The effect of gamma radiation on soil microorganisms, their metabolic processes, and the fertility of the soil. *Rep. Fla. Agric. Exp. Stns 1966*, 175-176.

Enwezor, W. P. and Moore, A. W. (1966). Phosphorus status of some Nigerian soils. *Soil Sci.* **102**, 322-326.

Evans, A. C. (1948). Studies on relationships between earthworms and soil fertility. II. Some effects of earthworms on soil structure. *Ann. Appl. Biol.* **35**, 1-13.

Evans, A. C. and Guild, W. J. McL. (1947). Studies on the relationships between earthworms and soil fertility. I. Biological studies in the field. *Ann. Appl. Biol.* **34**, 307-330.

Evans, A. C. and Guild, W. J. McL. (1948). Studies on the relationships between earthworms and soil fertility. IV. On the life cycles of some British Lumbricidae. *Ann. Appl. Biol.* **35**, 471-484.

Ferrière, G. (1980). Fonctions des lombriciens. VII. Une méthode d'analyse de la matière organique végétale ingérée. *Pedobiologia* **20**, 263-273.

Finck, A. (1952). Ökologische und Bodenkundliche Studien über die Leistungen der Regenwürmer für die Boden fruchtbarkeit. *Z. PflErnähr. Düng. Bodenk.* **58**, 120-145.

Finlayson, D. G., Campbell, C. J. and Roberts, H. A. (1975). Herbicides and insecticides: Their compatibility and effects on weeds, insects and earthworms in the minicauliflower crop. *Ann. Appl. Biol.* **79**, 95-108.

Fleming, T. P. and Richards, K. S. (1981a). A technique to quantify surface adsorption of heavy metals by soft-bodied invertebrates. *Comp. Biochem. Physiol.* **69C**, 391-394.

Fleming, T. P. and Richards, K. S. (1981b). Uptake and surface adsorption of zinc by the freshwater tubificed oligochaete *Tubifex tubifex*. *Comp. Biochem. Physiol.* **71C**, 69-75.

Fleming, T. P. and Richards, K. S. (1982). Localization of adsorbed heavy metals on the earthworm body surface and their retrieval by chelation. *Pedobiologia* **23**, 415-418.

Florkin, M. (1969). Nitrogen metabolism. *In* "Chemical Zoology. Vol. 4. Annelida, Echiura, Sipuncula" (M. Florkin and B. T. Scheer, eds), pp. 147-162. Academic Press, New York.

Fosgate, O. T. and Babb, M. R. (1972). Biodegradation of animal waste by *Lumbricus terrestris. J. Dairy Sci.* **55**, 870-872.

Franz, H. and Leitenberger, L. (1948). Biologisch-chemische Untersuchungen über Humusbildung durch Bodentiere. *Öst Zool. Z.* **1**, 498-518.

French, J. R. J., Turner, G. L. and Bradbury, J. F. (1976). Nitrogen fixation by bacteria from the hindgut of termites. *J. Gen. Microbiol.* **95**, 202-206.

Friedmann, A. A. (1978). Fundamental research needs from the engineer's point of view. *In* "Utilization of Soil Organisms in Sludge Management" (R. Hartenstein, ed.), pp. 9-25. Natl. Tech. Inf. Services, PB286932. Springfield, Virginia.

Ganong, W. F. (1969). "Review of Medical Physiology". Blackwell Scientific Publications, Oxford.

Gansen-Semal, P. van (1959). Structure des glandes calciques d'*Eisenia foetida* Sav. *Bull. Biol. Fr. Belg.* **93**, 38-63.

Gardner, C. R. (1976). The neuronal control of locomotion in the earthworm. *Biol. Rev. Camb. Philos. Soc.* **51**, 25-52.

Gates, G. E. (1941). Ecology of some earthworms with special reference to seasonal activity. *Am. Midl. Nat.* **66**, 61-86.

Gates, G. E. (1948). On segment formation in normal and regenerative growth of earthworms. *Growth* **12**, 165-180.

Gates, G. E. (1959). Earthworms. *In* "The Natural History of Rennell Island, British Solomon Islands". Vol. 2, pp. 7-23. Danish Science Press, Copenhagen.

Gates, G. E. (1961). On some species of the Oriental earthworm genus *Pheretima* Kinberg, 1867. *Zool. Meded. (Leiden)* **37**, 293-312.

Gates, G. E. (1963). Earthworm. *In* "Encyclopaedia Britannica". Chicago.

Gates, G. E. (1970). Miscellanea Megadrilogica. VIII. *Megadrilogica* **1**(2), 1-14. Index Oligochaetarum, Bangor, Maine.

Gates, G. E. (1972). Burmese earthworms. An introduction to the systematics and biology of megadrile oligochaetes with special reference to Southeast Asia. *Trans. Am. Philosoph. Soc. N.S.* **62**(7), 1-326.

Gates, G. E. (1976). More on oligochaete distribution in North America. *Megadrilogica* **2**, 1-8.

Gates, G. E. (1979). South Dakota does have earthworms! *Megadrilogica* **3**, 165-166.

Gavrilov, K. (1939). Sur la réproduction de *Eiseniella tetraedra* (Sav.) forma *typica. Acta Zool. Acad. Sci. Hung.* **20**, 439-464.

Gavrilov, K. (1948). Reproduccion uni y biparental de los oligoquetos. *Acta Zool. Lilloana* **5**, 221-311.

Gavrilov, K. (1962). Role of earthworms in the enrichment of soil by biologically active substances. *Voprosy Ekologii Vysshaya Shkola Moscow* **7**, 34.

Gavrilov, K. (1963). Earthworms, producers of biologically active substances. *Zh. Obshch. Biol.* **24**, 149-154.

Gerard, B. M. (1960). "The biology of certain British earthworms in relation to environmental conditions" Ph.D. Thesis, University of London.

Gerard, B. M. (1963). The activities of some species of Lumbricidae in pasture land. *In* "Soil Organisms" (J. Doeksen and J. van der Drift, eds). pp. 49-54. North Holland Publishing Company, Amsterdam.

Gerard, B. M. (1967). Factors affecting earthworms in pastures. *J. Anim. Ecol.* **36**, 235-252.

Gerard, B. M. and Hay, R. K. M. (1979). The effect on earthworms of ploughing, tined cultivation, direct drilling and nitrogen in a barley monoculture system. *J. Agric. Sci. Cambridge* **93**, 147-155.

Germann, P. and Beven, K. (1981a). Water flow in soil macropores. I. An experimental approach. *J. Soil Sci.* **32**, 1-13.

Germann, P. and Beven, K. (1981b). Water flow in soil macropores. II. A combined flow model. *J. Soil Sci.* **32**, 15-29.

Germann, P. and Beven, K. (1981c). Water flow in soil macropores. III. A statistical approach. *J. Soil Sci.* **32**, 31-39.

Ghabbour, S. I. (1966). Earthworms in agriculture. A modern evaluation. *Rev. Ecol. Biol. Sol.* **3**, 259-271.

Ghabbour, S. I. (1975). Ecology of water relations in Oligochaeta. I. Survival in various relative humidities. *Bull. Zool. Soc. Egypt* **27**, 1-10.

Ghabbour, S. I. (1976). Ecology of water relations in Oligochaeta. II. The role of body lipids and minerals. *Bull. Zool. Soc. Egypt* **28**, 1-7.

Ghabbour, S. I. and Imam, M. (1967). The effect of five herbicides on three oligochaete species. *Rev. Ecol. Biol. Sol.* **4**, 119-122.

Ghabbour, S. I. and Shakir, S. H. (1982). Population density and biomass of earthworms in agro-ecosystems of the Mariut coastal desert region. *Pedobiologia* **23**, 189-198.

Ghilarov, M. S. (1965). Some practical problems of soil zoology. *Pedobiologia* **5**, 189-204.

Ghilarov, M. S. and Mamajev, B. M. (1967). Das Einsetzen von Regenwürmem (Oligochaeta Lumbricidae) in die Böden artesisch bewässerter Oasen als Massnahme zur Steigerung der biologischen Aktivitat. *In* "Progress in Soil Biology" (O. Graff, and J. E. Satchell, eds), pp. 275-281. North Holland Publishing Company, Amsterdam.

Gish, C. D. and Christensen, R. E. (1973). Cadmium, nickel, lead and zinc in earthworms from roadside soil. *Environ. Sci. Technol.* **7**, 1060-1062.

Gissel-Nielsen, M. and Gissel-Neilsen, G. (1975). Selenium in soil-animal relationships. *Pedobiologia* **15**, 65-67.

Glaessner, M. F., Priess, W. V. and Walter, H. R. (1969). Precambrian columnar stromatolites in Australia: Morphological and stratigraphic analysis. *Science* **164**, 1056-1058.

Goh, B. S. (1978). Robust stability concepts for ecosystem models. *In* "Theoretical Systems Ecology" (E. Halfon, ed.). Academic Press, New York.

Gotwald, W. W. (1974). Foraging behaviour of *Anomma* driver ants in Ghana cocoa farms (Hymenoptera: Formicidae). *Bull. Inst. Fondam. Afr. Noire Ser. A. Sci. Nat.* **36**, 705-713.

Graff, O. (1953a). Bodenzoologische Untersuchungen mit besonderer Berucksichtigung der terrikolen Oligochaeten. *Z. Pflanzern. Dung. Bodenk.* **61**(106), 72-77.

Graff, O. (1953b). Die Regenwürmer Deutschlands. *Shcrift. Forsch. Land. Braunschweig-Volkenrode* 7.

Graff, O. (1957). De lumbricidis quibusdam in Lusitania habitantibus. *Agronom. Lusit.* **19**, 299-306.

Graff, O. (1961). Die Regenwürmer (Oligochaeta, Lumbricidae) auf dem Gelände der Forschungsanstalt für Landwirtchaft. *Landbauforsch. Völkenrode* **2**, 19-22.

Graff, O. (1969). Regenwurmtaetigkeit im Ackerboden unter verschiedenem Redeckungsmaterial, gemessen an der Losungsablage. *Pedobiologia* **9**, 120-127.

Graff, O. (1970). Effect of different mulching materials on the nutrient content of earthworm tunnels in the subsoil. *Pedobiologia* **10**, 305-319.

Graff, O. (1971). Stickstoff, Phosphor und Kalium in der Regenwurmlosung auf der Wiesenversuchsfläche des Sollingprojektes. *In* "IV Colloquium Pedobiologiae" (J. D'Aguilar, ed.), pp. 503-511. Institut National des Recherches Agriculturelles Publ. 71-7, Paris.

Graff, O. (1974). Gewinnung von Biomasse aus Abfallstoffen durch Kultur des

Kompostregenwurms *Eisenia foetida* (Savigny 1826). *Landbauforsch Völkenrode* **2**, 137-142.

Graff, O. (1978). Physiologische Rassen bei *Eisenia foetida*? Ein Beitrage zur Frage der Domestikation dieser Art. *Rev. Ecol. Biol. Sol.* **15**, 251-263.

Graff, O. (1982). Vergleich der Regenwurmarten *Eisenia foetida* und *Eudrilus eugeniae* hinsichtlich ihrer Eignung zur Proteingewinnung aus Abfallstoffen. *Pedobiologia* **23**, 277-282.

Graff, O. (1983). Darwin on earthworms — the contemporary background and what the critics thought. *In* "Earthworm Ecology" (J. E. Satchell, ed.), pp. 5-18. Chapman and Hall, London.

Graff, O. and Makeschin, F. (1980). Beeinflussung des Ertrags von Weidelgras (*Lolium multiflorum*) durch Ausscheidungen von Regenwürmen dreier verschiedener Arten. *Pedobiologia* **20**, 176-180.

Grant, J. D. (1983). The activities of earthworms and the fates of seeds. *In* "Earthworm Ecology" (J. E. Satchell, ed.), pp. 107-122. Chapman and Hall, London.

Grant, W. C. (1955a). Studies on moisture relationships in earthworms. *Ecology* **36**, 400-407.

Grant, W. C. (1955b). Temperature relationships in the megascolecid earthworm *Pheretima hupeiensis*. *Ecology* **36**, 412-417.

Greacen, E. L., Farrell, D. A. and Forrest, J. A. (1967). The measurement of density patterns in soil. *J. Agric. Eng. Res.* **12**, 311-313.

Greenslade, P. J. M. (1972). Evolution in the staphylinid genus *Priochirus* (Coleoptera). *Evolution* **26**, 203-220.

Greenslade, P. J. M. (1982). Selection processes in arid Australia. *In* "Evolution of the Flora and Fauna of Arid Australia" (W. R. Barker and P. J. M. Greenslade, eds), pp. 125-130. Peacock Publications, Adelaide.

Greenslade, P. J. M. (1983). Adversity selection and the habitat templet. *Am. Nat.* **122**, 352-365.

Greenslade, P. J. M. and Greenslade, P. (1983). Ecology of soil invertebrates. *In* "Soils: An Australian viewpoint", Division of Soils, CSIRO, pp. 645-669, Commonwealth Scientific Industrial Research Organization, Melbourne and Academic Press, London.

Griffiths, D. C., Raw, F. and Lofty, J. R. (1967). Effect on soil fauna of insecticides tested against wireworms in wheat. *Ann. Appl. Biol.* **60**, 479-490.

Griffiths, M. (1978). "The Biology of the Monotremes". Academic Press, New York.

Grime, J. P. (1974). Vegetation classification by reference to strategies. *Nature (Lond.)* **250**, 26-31.

Grime, J. P. (1977). Evidence for the existence of three primary strategies in plants and its relevance to ecological and evolutionary theory. *Am. Nat.* **111**, 1169-1194.

Grime, J. P. (1979). "Plant Strategies and Vegetation Processes". John Wiley, Chichester, U.K.

Gruia, L. (1969). Répartition quantitative des Lombricidés dans les sols de Roumanie. *Pedobiologia* **9**, 99-102.

Gruner, B. and Zebe, E. (1978). Studies on the anaerobic metabolism of earthworms. *Comp. Biochem. Physiol.* **60B**, 441-445.

Guardabassi, A. (1957). Le ghiandole calcifere (ghindole di Morren) di *Eisenia foetida*. Studio isto e citochimico. *Z. Zellforsch. Mikrosk. Anat.* **46**, 619-634.

Guerrero, R. D. (1983). The culture and use of *Perionyx excavatus* as a protein resource in the Philippines. *In* "Earthworm Ecology" (J. E. Satchell, ed.), pp. 309-313. Chapman and Hall, London.

Guild, W. J. McL. (1948). Effect of soil type on populations. *Ann. Appl. Biol.* **35**, 181-192.

Guild, W. J. McL. (1951). The distribution and population density of earthworms (Lumbricidae) in Scottish pasture fields. *J. Anim. Ecol.* **20**, 88-97.
Guild, W. J. McL. (1952). Variation in earthworm numbers within field populations. *J. Anim. Ecol.* **21**, 169-181.
Guild, W. J. McL. (1955). Earthworms and soil structure. *In* "Soil Zoology" (D. K. McE. Kevan, ed.), pp. 83-98. Butterworth, London.
Gupta, M. L. and Sakal, R. (1967). The role of earthworms in the availability of nutrients in garden and cultivated soils. *J. Indian Soc. Soil Sci.* **15**, 149-151.
Haantjens, H. A. (1965). Morphology and origin of patterned ground in a humid tropical lowland area, New Guinea. *Aust. J. Soil Res.* **3**, 111-129.
Haantjens, H. A. (1969). Fire and wind erosion, or earthworms as the cause of microrelief in the Lower Sepik Plains, New Guinea. *Aust. J. Sci.* **32**, 52-54.
Hall, F. G. (1922). The vital limit of desiccation of certain animals. *Biol. Bull. (Woods Hole)* **42**, 31-51.
Handley, W. R. C. (1954). Mull and mor formation in relation to forest soils. *Bull. For. Comm. Lond.* **23**, 1-115.
Handreck, K. A. (1978). "Earthworms for Gardeners and Fishermen". Discovering Soils No. 5. Division of Soils, Commonwealth Scientific and Industrial Research Organization, Adelaide, Australia.
Hanschko, D. (1958). The earthworm faunas of different plant communities. *Forsch. Beratung.* **3**, 16-29.
Hansen, R. P. and Czochanska, Z. (1974). Earthworm lipids: Occurrence of phytanic, pristanic and 4, 8, 12-trimethyl-tridecanoic acids. *Lipids* **9**, 825-827.
Hansen, R. P. and Czochanska, Z. (1975). The fatty acid composition of the lipids of earthworms. *J. Sci. Food Agric.* **26**, 961-971.
Harding, D. J. L. and Studdard, R. A. (1974). Microarthropods. *In* "Biology of Plant Litter Decomposition" (C. H. Dickinson and G. J. F. Pugh, eds), Vol. 2, pp. 489-532. Academic Press, New York.
Harker, W. (1955). Appendix B. Pasture agronomist report no. 34. *In* Water Resources Survey of Uganda. A. Gibb and Partners, 1955. Uganda Government Publication.
Hartenstein, R. (1973). Characteristics of a peroxidase from the freshwater crayfish *Cambarus bartoni*. *Comp. Biochem. Physiol.* **45B**, 749-762.
Hartenstein, R. (1978a). The most important problem in sludge management as seen by a biologist. *In* "Utilization of Soil Organisms in Sludge Management" (R. Hartenstein, ed.), pp. 2-8. Natl. Tech. Inf. Services, PB286932. Springfield, Virginia.
Hartenstein, R. (Ed.) (1978b). "Utilization of Soil Organisms in Sludge Management". Natl Tech. Inf. Services, PB286932. Springfield, Virginia.
Hartenstein, R. (1982). Effect of aromatic compounds, humic acids and lignins on growth of the earthworm *Eisenia foetida*. *Soil Biol. Biochem.* **14**, 595-599.
Hartenstein, R., Neuhauser, E. F. and Kaplan, D. L. (1979a). A progress report on the potential use of earthworms in sludge management. Proc. 8th National Sludge Conf., pp. 238-241. Information Transfer, Silver Spring, Maryland.
Hartenstein, R., Neuhauser, E. F. and Kaplan, D. L. (1979b). Reproductive potential of the earthworm *Eisenia foetida*. *Oecologia (Berl.)* **43**, 329-340.
Heath, G. W. (1964). Biology of forest soils. *For. Comm. Rep. For. Res.* 1963, 94-96.
Heath, G. W. and King, H. G. C. (1964). Litter breakdown in deciduous forest soils. Proc. 4th Int. Congr. Soil Sci., Bucharest, 1964, 3, 979-987.
Helmke, P. A., Robarge, W. P., Korotev, R. L. and Schomberg, P. J. (1979). Effects of soil-applied sewage sludge on concentrations of elements in earthworms. *J. Environ. Qual.* **8**, 322-327.

Hess, W. N. (1924). Reactions to light in the earthworm, *Lumbricus terrestris. J. Morphol.* **39**, 515-542.
Hess, W. N. (1925). Photoreceptors of *Lumbricus terrestris* with special reference to their distribution, structure and function. *J. Morphol.* **41**, 63-95.
Heungens, A. (1966). Control of earthworms in spruce litter and in vitro. *Meded. Rijksfac. Landbouwwet Gent* **31**, 329-432.
Heungens, A. (1969). The physical decomposition of pine litter by earthworms. *Plant Soil* **31**, 22-30.
Hill, E. F., Heath, R. G., Spann, J. W. and Williams, J. D. (1975). Lethal dietary toxicities of environmental pollutants to birds. Special Scientific Report — Wildlife 191. United States Department of the Interior, Fish and Wildlife Service, Washington.
Hoeksema, K. J. and Jongerius, A. (1959) On the influence of earthworms on the soil structure in mulched orchards. *Proc. Int. Symp. Soil Struct. Ghent 1958*, pp. 188-194.
Hoffmann, J. A. and Purdy, L. H. (1964). Germination of dwarf bunt (*Tilletia controversa*) teliospores after ingestion by earthworms. *Phytopathology* **54**, 878-879.
Hogben, L. and Kirk, R. L. (1944). Studies on temperature regulation. 1. The Pulmonata and Oligochaeta. *Proc. Roy. Soc. Lond. B. Biol. Sci.* **132**, 239-252.
Holter, P. (1979). Effect of dung-beetles (*Aphodius* spp.) and earthworms on the disappearance of cattle dung. *Oikos* **32**, 393-402.
Holter, P. (1983). Effect of earthworms on the disappearance rate of cattle droppings. In "Earthworm Ecology" (J. E. Satchell, ed.), pp. 49-57. Chapman and Hall, London.
Hoogerkamp, M., Rogaar, H. and Eijsackers, H. J. P. (1983). Effect of earthworms on grassland on recently reclaimed polder soils in the Netherlands. In "Earthworm Ecology" (J. E. Satchell, ed.), pp. 85-105. Chapman and Hall, London.
Hopkins, A. R. and Kirk, V. M. (1957). Effects of several insecticides on the English red worm. *J. Econ. Entomol.* **50**, 699-700.
Hopp, H. (1946). Earthworms fight erosion too. *Soil Conserv. U.S. Dept Agric.* **11**, 252-254.
Hopp, H. (1973). "What every gardener should know about earthworms". Garden Way Publishing Company, Vermont, USA.
Hopp, H. and Hopkins, H. T. (1946). The effect of cropping systems on the winter populations of earthworms. *J. Soil Water Conserv.* **1**, 85-88.
Hopp, H. and Linder, P. J. (1947). A principle for maintaining earthworms in farm soils. *Science* **105**, 663-664.
Hopp, H. and Slater, C. S. (1948). Influence of earthworms on soil productivity. *Soil Sci.* **66**, 421-428,
Hopp, H. and Slater, C. S. (1949a). The effect of earthworms on the productivity of agricultural soil. *J. Agric. Res.* **78**, 325-339.
Hopp, H. and Slater, C. S. (1949b). A principal for maintaining structure in clean-cultivated soils. *J. Agric. Res.* **78**, 347-352.
Howell, C. D. (1939). The responses to light in the earthworm *Pheretima agrestis* Goto and Hatai, with special reference to the function of the nervous system. *J. Exp. Zool.* **81**, 231-259.
Hughes, M. K., Lepp, N. W. and Phipps, D. A. (1980). Aerial heavy metal pollution and terrestrial ecosystems. *Adv. Ecol. Res.* **11**, 218-237.
Huhta, V. (1979). Effects of liming and deciduous litter on earthworm (Lumbricidae) populations of a spruce forest, with an inoculation experiment on *Allolobophora caliginosa. Pedobiologia* **19**, 340-345.
Huhta, V., Karpinnen, E., Nurminen, M. and Valpas, A. (1967). Effect of silvicultural practices upon arthropod, annelid and nematode populations in coniferous forest soil. *Ann. Zool. Fenn.* **4**, 87-143.

Hurwitz, S. H. (1910). The reactions of earthworms to acids. *Proc. Am. Acad. Arts and Sci.* **46**, 67-81.

Hutchinson, S. A. and Kamel, M. (1956). The effect of earthworms on the dispersal of soil fungi. *J. Soil Sci.* **7**, 213-218.

Hyche, L. L. (1956). Control of mites infesting earthworm beds. *J. Econ. Ent.* **49**, 409-410.

Inoue, T. and Kondo, K. (1962). Susceptibility of *Branchiura sowerbyi*, *Limnodrilus socialis* and *L. willeyi* for several agricultural chemicals. *Botyu-Kagaku (Japan)* **27**, 97-99.

Ireland, M. P. (1975a). Metal content of *Dendrobaena rubida* (Oligochaeta) in a base metal mining area. *Oikos* **26**, 74-79.

Ireland, M. P. (1975b). The effect of the earthworm *Dendrobaena rubida* on the solubility of lead, zinc and calcium in heavy metal contaminated soil in Wales. *J. Soil Sci.* **26**, 313-318.

Ireland, M. P. (1977). Heavy worms. *New Sci.* **76**, No. 1079, 486-487.

Ireland, M. P. (1979). Metal accumulation by the earthworms *Lumbricus rubellus*, *Dendrobaena veneta* and *Eiseniella tetraedra* living in heavy metal polluted sites. *Environ. Pollut.* **13**, 201-206.

Ireland, M. P. (1983). Heavy metal uptake and tissue distribution in earthworms. *In* "Earthworm Ecology (J. E. Satchell, ed.), pp. 247-265. Chapman and Hall, London.

Ireland, M. P. and Richards, K. S. (1977). The occurrence and localisation of heavy metals and glycogen in the earthworms *Lumbricus rubellus* and *Dendrobaena rubida* from a heavy metal site. *Histochemistry* **51**, 153-166.

Ireland, M. P. and Wooton, R. J. (1976). Variations in the lead, zinc and calcium content of *Dendrobaena rubida* (Oligochaeta) in a base metal mining area. *Environ. Pollut.* **10**, 201-208.

Jackson, R. B. (1983). Pesticide residues in soils. *In* "Soils: An Australian viewpoint", pp. 825-842. Commonwealth Scientific and Industrial Organization, Melbourne and Academic Press, London.

Jaenike, J., Ausabel, S. and Grimaldi, D. A. (1982). On the evolution of clonal diversity in parthenogenetic earthworms. *Pedobiologia* **23**, 304-310.

Jamieson, B. G. M. (1978). Phylogenetic and phenetic systematics of the opisthoporous Oligochaeta (Annelida: Clitellata). *Evol. Theory* **3**, 195-233.

Jamieson, B. G. M. (1981). "The Ultrastructure of the Oligochaeta". Academic Press, London,

Jeanson, C. (1960a). Etude expérimentale de l'action de *Lumbricus herculeus* (Savigny) (Oligochète Lumbricide) sur la stabilité structurale des terres. *C. R. Hébd. Séances Acad. Sci.* **250**, 3041-3043.

Jeanson, C. (1960b). Evolution de la matière organique du sol sous l'action de *Lumbricus herculeus* Savigny (Oligochète Lumbricidae). *C. R. Hébd. Séances Acad. Sci.* **250**, 3500-3502.

Jeanson, C. (1961). Sur une méthode d'étude du comportement de la faune du sol et de sa contribution à la pédogenèse. *C. R. Hébd. Séances Acad. Sci.* **253**, 2571-2573.

Jeanson, C. (1964). Micromorphology and experimental soil zoology; contribution to the study, by means of giant-sized thin sections, of earthworm-produced artificial soil structure. *In* "Soil Micromorphology". (A. Jongerius, ed). Proc. 2nd Int. Wk Mtg Soil Micromorph. Arnhem, pp. 47-55.

Jeanson, C. (1971). Structure d'une galerie de lombric à la microsonde électronique. *In* "IV Colloquium Pedobiologiae" (J. d'Aguilar, ed.), pp. 513-525. Institut National des Recherches Agriculturelles Publ. 71-7, Paris.

Jefferies, D. J. (1974). Earthworms in the diet of the red fox *Vulpes vulpes*. *J. Zool. Lond.* **173**, 251-252.

Jefferson, P. (1955). Studies on the earthworms of turf. A. The earthworms of experimental turf plots. *J. Sports Turf Res. Inst.* **9**, 6-27.
Jefferson, P. (1958). Studies on the earthworms of turf. C. Earthworms and casting. *J. Sports Turf Res. Inst.* **9**, 437-452.
Jeuniaux, C. (1969). Nutrition and digestion. *In* "Chemical Zoology, Vol. 4 Annelida, Echiura, Sipuncula" (M. Florkin and B. T. Scheer, eds), pp. 69-91. Academic Press, New York.
Joest, E. (1897). Transplantationsversuche an Lumbriciden. *Arch. Entwicklungsmech. Org. (Wilhelm Roux)* **5**, 419.
Johansen, K. and Martin, A. W. (1965). Circulation in a giant earthworm, *Glossoscolex giganteus*. I. Contractile processes and pressure gradients in the large blood vessels. *J. Exp. Biol.* **43**, 333-347.
Johansen, K. and Martin, A. W. (1966). Circulation in a giant earthworm, *Glossoscolex giganteus*. II. Respiratory properties of the blood and some patterns of gas exchange. *J. Exp. Biol.* **45**, 165-172.
Johnson, M. L. (1942). The respiratory function of the haemoglobin of the earthworm. *J. Exp. Biol.* **18**, 266-277.
Jordan, G. A., Reynolds, J. W. and Burnett, A. J. (1976). Computer plotting and analysis of earthworm population distribution in Prince Edward Island. *Megadrilogica* **2**, 1-7.
Josens, G. (1974). Les termites de la savane de Lamto. *In* "Analyse d'un Ecosystème Tropical Humide: La Savane de Lamto (Côte d'Ivoire)," Bull. de Liaison des Chercheurs de Lamto, No. Spéc. 1974, **5**, 91-131.
Joshi, N. V. and Kelkar, B. V. (1952). The role of earthworms in soil fertility. *Indian J. Agric. Sci.* **22**, 189-196.
Joyner, J. W. and Harmon, N. P. (1961). Burrows and oscillative behaviour therein of *Lumbricus terrestris*. *Proc. Indiana Acad. Sci.* **71**, 378-384.
Kale, R. D. and Krishnamoorthy, R. V. (1980). The calcium content of the body tissues and castings of *Pontoscolex corethrurus* (Annelida, Oligochaeta). *Pedobiologia* **20**, 309-315.
Kale, R. D. and Krishnamoorthy, R. V. (1981). What affects the abundance and diversity of earthworms in soils? *Proc. Indian Acad. Sci. (Anim. Sci.)* **90**, 117-121.
Kale, R. D., Bano, K. and Krishnamoorthy, R. V. (1982). Potential of *Perionyx excavatus* for utilizing organic wastes. *Pedobiologia* **23**, 419-425.
Kaplan, D. L. (1978). The biochemistry of sludge decomposition, sludge stabilization and humification. *In* "Utilization of Soil Organisms in Sludge Management" (R. Hartenstein, ed.), pp. 78-86. Natl Tech. Inf. Services, PB286932. Springfield, Virginia.
Kaplan, D. and Hartenstein, R. (1977). Absence of nitrogenase and nitrate reductase in soil macroinvertebrates. *Soil Sci.* **124**, 328-331.
Kaplan, D. L., Hartenstein, R., Neuhauser, E. F. and Malecki, M. R. (1980a). Physicochemical requirements in the environment of the earthworm *Eisenia foetida*. *Soil Biol. Biochem.* **12**, 347-352.
Kaplan, D. L., Hartenstein, R., Malecki, M. R. and Neuhauser, E. F. (1980b). Role of the pheretimoid worms, *Amynthas hawayana* and *A. rodericensis* in soils and recycling. *Rev. Ecol. Biol. Sol* **17**, 165-171.
Kaplan, D. L., Hartenstein, R. and Neuhauser, E. F. (1980c). Coprophagic relations among the earthworms *Eisenia foetida*, *Eudrilus eugeniae* and *Amythas* spp. *Pedobiologia* **20**, , 74-84.
Kelsey, J. M. and Arlidge, E. Z. (1968). Effects of isobenzan on soil fauna and soil structure. *N.Z. J. Agric. Res.* **11**, 245-260.
Keogh, R. G. (1979). Lumbricid earthworm activities and nutrient cycling in pasture ecosystems. *In* Proc. 2nd Australasian Conf. Grassl. Invert. Ecol. (T. K. Crosby and R. P. Pottinger, eds), pp. 49-51. Government Printer, Wellington.

Keogh, R. G. and Christensen, M. J. (1976). Influence of passage through *Lumbricus rubellus* Hoffmeister earthworms on viability of *Pithomyces chartarum* (Berk. and Curt.) M.B. Ellis spores. *N.Z. J. Agric. Res.* **19**, 255-256.

Keogh, R. G. and Whitehead, P. H. (1975). Observations on some effects of pasture spraying with benomyl and carbendazim on earthworm activity and litter removal from pasture. *N.Z. J. Exp. Agric.* **3**, 103-104.

Khalaf El-Duweini, A. and Ghabbour, S. I. (1964). Effect of pH and of electrolytes on earthworms. *Bull. Zool. Soc. Egypt.* **19**, 89-100.

Khalaf El-Duweini, A. and Ghabbour, S. I. (1965a). Population density and biomass of earthworms in different types of Egyptian soils. *J. Appl. Ecol.* **2**, 271-287.

Khalaf El-Duweini, A. and Ghabbour, S. I. (1965b). Temperature relations of three Egyptian oligochaete species. *Oikos* **16**, 9-15.

Khalaf El-Duweini, A. and Ghabbour, S. I. (1968). Nephridial systems and water balance of three oligochaete genera. *Oikos* **19**, 61-70.

Khalaf El-Duweini, A. and Ghabbour, S. I. (1971). Nitrogen contribution by live earthworms to the soil. *In* "IV Colloquium Pedobiologiae" (J. d'Aguilar, ed.), pp. 495-501. Institut National des Recherches Agriculturelles Publ. 71-7, Paris.

Khambata, S. R. and Bhat, J. V. (1957). A contribution to the study of the intestinal microflora of Indian earthworms. *Arch. Mikrobiol.* **28**, 69-80.

Khan, A. W. (1966). Earthworms of West Pakistan and their utility in soil improvement. *Agric. Puk.* **17**, 415-434.

Khan, N., Mohammad Ashraf and Karimullah. (1967) Studies on earthworms with special reference to populations and their cast production. *Sci. Ind. (Karachi)* **5**, 413-417.

King, H. G. C. and Heath, G. W. (1967). The chemical analysis of small samples of leaf material and the relationship between the disappearance and composition of leaves. *Pedobiologia* **7**, 192-197.

Kitazawa, Y. (1971). Biological regionality of the soil fauna and its function in forest ecosystem types. *In* "Productivity of Forest Ecosystems". (P. Duvigneaud, ed.), Proc. Brussels, Symp., 1969. Ecology and Conservation **4**, 485-498.

Kollmannsperger, F. (1934). "Die Oligochaeten des Bellinchengebietes, eine Okologische, Ethologische und Tiergeographische Untersuchung". Ph.D. Thesis, Köln.

Kollmannsperger, F. (1952). Über die Bedeutung des Regenwurmes für die Fruchtbarkeit des Bodens. *Decheniana* **105/106**, 165.

Kollmannsperger, F. (1956a). Lumbricidae of humid and arid regions and their effect on soil fertility. *6th Congr. Int. Sci. Sol Rapp. C*, pp. 293-297.

Kollmannsperger, F. (1956b). Über die Bedeutung der Regenwürmer für die Fruchtbarkeit der Bergsavannen Kameruns. *Zool. Anz.* **157**, 216-219.

Kozlovskaya, L. S. and Zhdannikova, E. N. (1957). Joint action of earthworms and microflora in forest soils. *Doklady Akad. Nauk SSSR* **139**, 470-473.

Kretzschmar, A. (1978). Quantification écologique des galeries de lombriciens. Techniques et premières estimations. *Pedobiologia* **18**, 31-38.

Kretzschmar, A. (1982). Description des galeries des vers de terre et variation saisonnière des réseaux (observations en conditions naturelles). *Rev. Ecol. Biol. Sol* **19**, 579-591.

Krivolutsky, D. A., Tichomirova, A. L. and Turcaninova, V. A. (1972). Strukturänderungen des Tierbesatzes (Land- und Bodenwirbellose) unter dem Einfluss der Kontamination des Bodens mit Sr^{90}. *Pedobiologia* **12**, 374-380.

Krivolutsky, D., Turcaninova, V. and Mikhaltsova, Z. (1982). Earthworms as bioindicators of radioactive soil pollution. *Pedobiologia* **23**, 263-265.

Kubiena, W. L. (1953). "The Soils of Europe". Murby, London.

Kudrjaseva, I. V. (1976). The role of large soil invertebrates in the oak woods of the steppe forest zones of the Soviet Union. *Pedobiologia* **16**, 18-26.

Kühle, J. C. (1983). Adaptation of earthworm populations to different soil treatments in apple orchard. *In* "New Trends in Soil Biology" (Ph. Lebrun, ed.), pp. 487-501. Proc. 8th Intl Colloq. Soil Zool., Louvain-la-Neuve, 1982. Dieu-Brichart, Ottignies-Louvain-la-Neuve.

Lakhani, K. H. and Satchell, J. E. (1970). Production by *Lumbricus terrestris* L. *J. Anim. Ecol.* **39**, 473-492.

Lal, R. (1974). No-tillage effects on soil properties and maize (*Zea mays* L.) production in western Nigeria. *Plant Soil* **40**, 321-331.

Lan, H. an der and Aspöck, H. (1962). The effect of Sevin on earthworms. *Anz. Schädlingsk.* **35**, 180-182.

Langdon, F. E. (1895). The sense organs of *Lumbricus agricola* Hoffm. *J. Morphol.* **11**, 193-234.

Langmaid, K. K. (1964). Some effects of earthworm invasion in virgin podzols. *Canad. J. Soil Sci.* **44**, 34-37.

Lavelle, P. (1971a). Etude préliminaire de la nutrition d'un ver de terre africain *Millsonia anomala* (Acanthodrilidae, Oligochètes). *In* "IV Colloquium Pedobiologiae" (J. d'Aguilar, ed.), pp. 133-145. Institut National des Recherches Agriculturelles Publ. 71-7, Paris.

Lavelle, P. (1971b). Recherches écologiques dans la savane de Lamto (Côte d'Ivoire): Production annuelle d'un ver de terre *Millsonia anomala* Omodeo. *Terre Vie* **2**, 240-254.

Lavelle, P. (1971c). Recherches sue la démographie d'un ver de terre d'Afrique; *Millsonia anomala* Omodeo (Oligochètes Acanthodrilidae). *Bull. Soc. Ecol.* **2**, 302-312.

Lavelle, P. (1974). Les vers de terre de la savane de Lamto. *In* "Analyse d'un Ecosystème Tropical Humide: La Savane de Lamto (Côte d'Ivoire)". Bull. de Liaison des Chercheurs de Lamto, No. Spéc. 1974, **5**, 133-136.

Lavelle, P. (1978). Les vers de terre de la savane de Lamto (Côte d'Ivoire). Peuplements, populations et fonctions de l'écosystème. *Publ. Lab. Zool. E.N.S.* **12**, 1-301.

Lavelle, P. (1979). Relations entre types écologiques et profils démographiques chez les vers de terre de la savane de Lamto (Côte d'Ivoire). *Rev. Ecol. Biol. Sol* **16**, 85-101.

Lavelle, P. (1981a). Stratégies de reproduction chez les vers de terre. *Acta Oecol. Oecol. Gener.* **2**, 117-133.

Lavelle, P. (1981b). Un ver de terre carnivore des savanes de la moyenne Côte d'Ivoire: *Agastrodrilus dominicae* nov. sp. (Oligochètes-Megascolecidae). *Rev. Ecol. Biol. Sol* **18**, 253-258.

Lavelle, P. (1983a). The soil fauna of tropical savannas. II. The earthworms. *In* "Ecosystems of the World. 13. Tropical Savannas", pp. 485-504. Elsevier, Amsterdam.

Lavelle, P. (1983b). The structure of earthworm communities. *In* "Earthworm Ecology" (J. E. Satchell, ed.), pp. 449-466. Chapman and Hall, London.

Lavelle, P. (1983c). *Agastrodrilus* Omodeo and Vaillaud, a genus of carnivorous earthworms from the Ivory Coast. *In* "Earthworm Ecology" (J. E. Satchell, ed.), pp. 425-429. Chapman and Hall, London.

Lavelle, P. and Meyer, J. A. (1977). Modélisation et simulation de la dynamique, de la production et de la consommation des populations du ver de terre géophage *Millsonia anomala* Omodeo (Oligochètes — Acanthodrilidae) dans la savane de Lamto (Côte d'Ivoire). *In* "Soil Organisms as Components of Ecosystems", (U. Lohm and T. Persson, eds), *Ecol. Bull. (Stockholm)* **25**, 420-430.

Lavelle, P. and Meyer, J. A. (1983). ALLEZ-LES-VERS, a simulation model of dynamics and effect on soil of populations of *Millsonia anomala* (Oligochaeta-Megascolecidae). *In* "New Trends in Soil Biology", (Ph. Lebrun, H. M. André, A. de Medts, C. Grégoire-Wibo and G. Wauthy, eds), pp. 503-517. Proc. 8th Intl Colloquium Soil Zool., Louvain-la-Neuve, 1982. Dieu-Brichart, Ottignies-Louvain-la-Neuve.

Lavelle, P., Sow, B. and Schaefer, R. (1980). The geophagous earthworms community in the Lamto savanna (Ivory Cost): Niche partitioning and utilization of soil nutritive resources. *In* "Soil Biology as Related to Land Use Practices" (D. L. Dindal, ed.), pp. 653-672. Proc. 7th Intl Soil Zool. Colloquium, Syracuse, 1979. Environmental Protection Agency, Washington D.C.

Lavelle, P., Rangel, P. and Kanyonyo, J. (1983a). Intestinal mucus production by two species of tropical earthworms: *Millsonia lamtoiana* and *Pontoscolex corethrurus* (Glossoscolecidae). *In* "New Trends in Soil Biology", (Ph. Lebrun, H. M. André, A. de Medts, C. Grégoire-Wibo and G. Wauthy, eds), pp. 405-410. Proc. 8th Intl Colloquium Soil Zool., Louvain-la-Neuve, 1982. Dieu-Brichart, Ottignies-Louvain-la-Neuve.

Lavelle, P., Zaidi, Z. and Schaefer, R. (1983b). Interactions between earthworms, soil organic matter and microflora in an African savanna soil. *In* "New Trends in Soil Biology", (Ph. Lebrun, H. M. André, A. de Medts, C. Grégoire-Wibo and G. Wauthy, eds), pp. 253-259. Proc. 8th Intl Colloquium Soil Zool., Louvain-la-Neuve, 1982. Dieu-Brichart, Ottignies-Louvain-la-Neuve.

Laverack, M. S. (1960). Tactile and chemical perception in earthworms — I. Responses to touch, sodium chloride, quinine and sugars. *Comp. Biochem. Physiol.* **1**, 155-163.

Laverack, M. S. (1961). Tactile and chemical perception in earthworms — II. Responses to acid pH solutions. *Comp. Biochem. Physiol.* **2**, 22-34.

Laverack, M. S. (1963). "The Physiology of Earthworms". International Ser. Monographs on Pure and Appl. Biol., Zool. no. 15. Pergamon, Oxford.

Lawrence, R. D. and Millar, H. R. (1945). Protein content of earthworms. *Nature (Lond.)* **155**, 517.

Lee, K. E. (1953). A note on the earthworm fauna of the Kermadec Islands. *Trans. R. Soc. N.Z.* **81**, 49-51.

Lee, K. E. (1958). Biological studies of some tussock grassland soils. X. Earthworms. *N.Z. J. Agric. Res.* **1**, 998-1002.

Lee, K. E. (1959). The earthworm fauna of New Zealand. *N.Z. Dept Sci. Industr. Res. Bull.* 130.

Lee, K. E. (1961). Interactions between native and introduced earthworms. *Proc. N.Z. Ecol. Soc.* **8**, 60-62.

Lee, K. E. (1967). Microrelief features in a humid tropical lowland area, New Guinea, and their relation to earthworm activity. *Aust. J. Soil Res.* **5**, 263-274.

Lee, K. E. (1968). Oligochaeta from subantarctic islands. *B.A.N.Z. Antarctic Res. Exped. Rep.* **B8**, 149-165.

Lee, K. E. (1969). Earthworms of the British Solomon Islands Protectorate. *Philos. Trans. R. Soc. Lond. B Biol. Sci.* **255**, 345-354.

Lee, K. E. (1975a). Introductory remarks. *In* "A Discussion on the Results of the 1971 Royal Society — Percy Sladen Expedition to the New Hebrides" (E. J. H. Corner and K. E. Lee, eds), *Philos. Trans. R. Soc. Lond. B Biol. Sci.* **272**, 269-276.

Lee, K. E. (1975b). Conclusions. *In* "A Discussion on the Results of the 1971 Royal Society — Percy Sladen Expedition to the New Hebrides" (E. J. H. Corner and K. E. Lee, eds), *Philos. Trans. R. Soc. Lond. B Biol. Sci.* **272**, 477-486.

Lee, K. E. (1981). Earthworms (Annelida: Oligochaeta) of Vanua Tu (New Hebrides Islands). *Aust. J. Zool.* **29**, 535-572.

Lee, K. E. (1983). The influence of earthworms and termites on soil nitrogen cycling. *In* "New Trends in Soil Biology", (Ph. Lebrun, H. M. André, A. de Medts, C. Grégoire-Wibo and G. Wauthy, eds), pp. 35-48. Proc. 8th Intl Colloquium Soil Zool., Louvain-la-Neuve, 1982. Dieu-Brichart, Ottignies-Louvain-la-Neuve.

Lee, K. E. and Correll, R. (1978). Litter fall and its relationship to nutrient cycling in a South Australian dry sclerophyll forest. *Aust. J. Ecol.* **3**, 243-252.

Lee, K. E. and Ladd, J. N. (1984). Some recent advances in soil biology and biochemistry. *Proc. Natl. Soils Conf., Brisbane, 1984,* pp. 83-103. Australian Society of Soil Science, Brisbane.

Lee, K. E. and Wood, T. G. (1971a). Physical and chemical effects on soils of some Australian termites and their pedological significance. *Pedobiologia* **11**, 376-409.

Lee, K. E. and Wood, T. G. (1971b). "Termites and Soils". Academic Press, London.

Legg, D. C. (1969). Comparison of various worm-killing chemicals. *J. Sports Turf Res. Inst. no.* **44**, 47-48.

Lesser, E. J. and Taschenberg, E. W. (1908). Über Fermente des Regenwurms. *Z. Biol.* **50**, 446-470.

Liebmann, E. (1946). On trephocytes and trephocytosis: A study on the role of leucocytes in nutrition and growth. *Growth* **10**, 291-330.

Ljungström, P.-O. (1968). Spermatophores in three species of Swedish lumbricids. *Zool. Anz.* **181**, 53-60.

Ljungström, P.-O. (1970). Introduction to the study of earthworm taxonomy. *Pedobiologia* **10**, 265-285.

Ljungström, P.-O. (1972). Introduced earthworms of South Africa. On their taxonomy, distribution, history of introduction and on the extermination of endemic earthworms. *Zool. Jahrb. Syst.* **99**, 1-81.

Ljungström, P.-O. and Emiliani, F. (1971). Contribución al conocimiento de la ecología y distributión geográfica de las lombrices de tierra (oligoquetos) de la Prov. de Santa Fe (Argentina). *Idia (Buenos Aires)* August 1971, pp. 19-32.

Ljungström, P.-O. and Reinecke, A. J. (1969). Ecology and natural history of the microchaetid earthworms of South Africa. 4. Studies on influence of earthworms upon the soil and the parasitological question. *Pedobiologia* **9**, 152-157.

Ljungström, P.-O., Orellana, J. A. de, and Priano, L. J. J. (1973). Influence of some edaphic factors on earthworm distribution in Sante Fe Province, Argentina. *Pedobiologia* **13**, 236-247.

Loquet, M. (1978). The study of respiratory and enzymatic activities of earthworms-made pedological structures in a grassland soil at Citeaux, France. *Scientific Proc. R. Dublin Soc. Ser.* **A6**, 207-214.

Low, A. J. (1972). The effect of cultivation on the structure and other physical characteristics of grassland and arable soils (1945-1970). *J. Soil Sci.* **23**, 363-380.

Low, A. J. (1976). Effects of long periods under grass on soils under British conditions. *J. Sci. Food Agric.* **27**, 571-582.

Lunt, H. A. and Jacobson, G. M. (1944). The chemical composition of earthworm casts. *Soil Sci.* **58**, 367-375.

Ma, Wei-Chun (1982). The influence of soil properties and worm-related factors on the concentration of heavy metals in earthworms. *Pedobiologia* **24**, 109-119.

MacArthur, R. H. and Wilson, E. O. (1967a). Some generalized theorems of natural selection. *Proc. Nat. Acad. Sci. U.S.A.* **48**, 1893-1897.

MacArthur, R. H. and Wilson, E. O. (1967b). "The Theory of Island Biogeography". Princeton University Press, Princeton.

McCalla, T. M. (1953). Microbiology studies of stubble mulching. *Nebr. Agric. Exp. Stn Bull.* 417.

McColl, H. P. (1982). Interactions between the earthworm *Allolobophora caliginosa* and ryegrass (*Lolium perenne*) in subsoil after topsoil stripping. *In* "Proc. 3rd Australasian Conf. Grassl. Invert. Ecol." (K. E. Lee, ed.), pp. 321-330. South Australian Government Printer, Adelaide.

McColl, H. P. and Lautour, M.-L. de (1978). Earthworms and topsoil mining at Judgeford. *N.Z. Soil News* **26**, 148-152.

Macdonald, D. W. (1980). The red fox, *Vulpes vulpes*, as a predator upon earthworms, *Lumbricus terrestris. Z. Tierpsychol.* **52**, 171-200.

Macdonald, D. W. (1983). Predation on earthworms by terrestrial vertebrates. *In* "Earthworm Ecology" (J. E. Satchell, ed.), pp. 393-414. Chapman and Hall, London.

Macfadyen, A. (1963). The contribution of the fauna to total soil metabolism. *In* "Soil Organisms" (J. Doeksen and J. van der Drift, eds), pp. 3-17. North Holland Publishing Company, Amsterdam.

McIntyre, A. D. and Mills, C. F. (Eds) (1975). "Ecological Toxicity Research". Plenum Press, New York.

Mackay, A. D., Syers, J. K., Springett, J. A. and Gregg, P. H. (1982). Plant availability of phosphorus in superphosphate and a phosphate rock as influenced by earthworms. *Soil Biol. Biochem.* **14**, 281-287.

Mackay, A. D., Springett, J. A., Syers, J. K. and Gregg, R. E. H. (1983). Origin of the effect of earthworms on the availability of phosphorus in a phosphate rock. *Soil Biol. Biochem.* **15**, 63-73.

McRill, M. (1974). The ingestion of weed seed by earthworms. Proc. 12th British Weed Control Conf., 2, 519-524.

McRill, M. and Sagar, G. R. (1973). Earthworms and seeds. *Nature (Lond.)* **243**, 482.

Madge, D. S. (1965). Leaf fall and disappearance in a tropical forest. *Pedobiologia* **5**, 273-288.

Madge, D. S. (1969). Field and laboratory studies on the activities of two species of tropical earthworms. *Pedobiologia* **9**, 188-214.

Maldague, M. E. (1979). "Rôle des Animaux Edaphiques dans la Fertilité des Sols Forestiers". *Publ. Inst. Nat. Etude Agron. Congo Ser. Sci.* No. 112.

Maldague, M. and Couture, G. (1971). Utilisation de litières radioactives par *Lumbricus terrestris. In* "IV Colloquium Pedobiologie" (J. d'Aguilar, ed.), pp. 147-154. Institut National des Recherches Agriculturelles Publ. 71-7, Paris.

Malecki, M. R., Neuhauser, E. F. and Loehr, R. C. (1982). The effect of metals on the growth and reproduction of *Eisenia foetida* (Oligochaeta, Lumbricidae). *Pedobiologia* **24**, 129-137.

Malone, C. E. and Reichle, D. E. (1973). Chemical manipulation of soil biota in a fescue meadow. *Soil Biol. Biochem.* **5**, 629-639.

Mangum, C. P. (1978). Temperature adaptation. *In* "Physiology of Annelids" (P. J. Mill, ed.), pp. 447-478. Academic Press, London.

Marialigeti, K. (1979). On the community-structure of the gut-microbiota of *Eisenia lucens* (Annelida, Oligochaeta). *Pedobiologia* **19**, 213-220.

Marshall, V. G. (1971). Effects of soil arthropods and earthworms on the growth of black spruce. *In* "IV Colloquium Pedobiologiae" (J. d'Aguilar, ed.), pp. 109-117. Institut National des Recherches Agriculturelles Publ. 71-7, Paris.

Martin, L. W. and Wiggans, S. C. (1959). The tolerance of earthworms to certain insecticides, herbicides and fertilizers. *Okla. Agric. Exp. Stn Process Ser.* p. 344.

Martin, M. H. and Coughtrey, P. J. (1976). Comparisons between the levels of lead, zinc and cadmium within a contaminated environment. *Chemosphere* **5**, 313-318.

Martin, N. A. (1976). Effect of four insecticides on the pasture ecosystem. V. Earthworms (Oligochaeta: Lumbricidae) and Arthropoda extracted by wet sieving and salt flotation. *N.Z. J. Agric. Res.* **19**, 111-115.

Martin, N. A. (1978). Aspects of the biology of some lumbricid earthworms in New Zealand pastures. *N.Z. J. Ecol.* **1**, 175-176.

Martin, N. A. (1982a). The interaction between organic matter in soil and the burrowing activity of three species of earthworms (Oligochaeta: Lumbricidae). *Pedobiologia* **24**, 185-190.

Martin, N. A. (1982b). The effects of herbicides used on asparagus on the growth rate of the earthworm *Allolobophora caliginosa*. *Proc. 35th N.Z. Weed and Pest Control Conf.* pp. 328-331.

Martin, N. A. and Charles, J. C. (1979). Lumbricid earthworms and cattle dung in New Zealand pastures. *In* "Proc. 2nd Australasian Conf. Grassl. Invert. Ecol." (T. K. Crosby and R. P. Pottinger, eds), pp. 52-54. Government Printer, Wellington.

May, R. M. (1979). The structure and dynamics of ecological communities. *In* "Population Dynamics" (R. M. Anderson, B. D. Turner and L. R. Taylor, eds). 20th Symp. Brit. Ecol. Soc., pp. 385-407. Blackwell, Oxford.

Mazanceva, G. P. (1976). Estimation of the body weight of some soil invertebrates on the basis of their linear size measurements. *Pedobiologia* **16**, 44-50.

Mazantseva, G. P. (1982). Growth patterns in the earthworm *Nicodrilus caliginosus* (Oligochaeta: Lumbricidae) during the first year of life. *Pedobiologia* **23**, 272-276.

Mazaud, D. (1979). "Evaluation de méthodes de marquage permettant le repérage des lombriciens au terrain: premières applications". Thèse Docteur-ingénieur, Sciences Agronomiques. Institut National Agronomique, Paris-Grignon.

Mazaud, D. and Bouché, M. B. (1980). Introductions en surpopulation et migrations de lombriciens marqués. *In* "Soil Biology as Related to Land Use Practices" (D. L. Dindal, ed.), pp. 687-701. Proc. 7th Intl Soil Zool. Colloquium, Syracuse, 1979. Environmental Protection Agency, Washington D.C.

Mba, C. C. (1978). Influence of different mulch treatments on growth rate and activity of the earthworm *Eudrilus eugeniae* (Kinberg). *Z. Pflanzenernaehr. Bodenkd.* **141**, 453-468.

M'Dowall, J. (1926). Preliminary work towards a morphological and physiological study of the calciferous glands of the earthworm. *Proc. Phys. Soc. Edinb.* **21**, 65-72.

Meijer, J. (1972). An isolated earthworm population in the recently reclaimed Lauwerszeepolder. *Pedobiologia* **12**, 409-411.

Meinhardt, U. (1976). Dauerhafte Markierung von Regenwürmern durch ihre Lebendfärbung. *Nachrichtenbl. Dtsch. Pflanzenschutdienstes (Braunschw.)* **28**, 84-86.

Merker, E. and Braunig, G. (1972). Die Empfindlichkeit feuchthäutiger Tiere im Lichte. 3. Die Atemnot feuchthäutiger Tiere in Licht der Quarzquecksiblerlampe. *Zool. Jahrb.* **43**, 277-338.

Meyer, L. (1943). Experimenteller Beitrag zu makrobiologischen Wirkungen auf Humus- und Bodenbildung. *A. Pflanzenernaehr. Dung. Bodenk.* **29**, 119-140.

Michaelsen, W. (1895). Oligochaeten. *In* "Handbuch zer Zoologie Vol. 8. Oligochaeten" (W. G. Kukenthal and T. Krumbach, eds), p. 32. Walter de Gruyter, Berlin.

Michaelsen, W. (1900). "Das Tierreich Vol. 10 Oligochaeta". (J. W. Spengel, ed.), Friedländer, Berlin.

Michel, C. and de Villez, E. J. (1978). Digestion. *In* "Physiology of Annelids" (P. J. Mill, ed.), pp. 509-553. Academic Press, London.

Michon, J. (1957). Contribution expérimentale à l'étude de la biologie des Lumbricidae. Les variations pondérales au cours des différentes modalités du développement postembryonnaire. *Ann. Biol.* **33**, 367-376.

Miles, H. B. (1963a). Soil protozoa and earthworm nutrition. *Soil Sci.* **95**, 407-409.

Miles, H. B. (1963b). Heat death temperature in *Allolobophora terrestris* (Sav.) forma *longa* (Ude) and *Eisenia foetida* (Sav.). *Nature (Lond.)* **199**, 826.

Mill, P. J. (1978). Sense organs and sensory pathways. *In* "Physiology of Annelids" (P. J. Mill, ed.), pp. 63-114. Academic Press, London.

Miller, R. B., Stout, J. D. and Lee, K. E. (1955). Biological and chemical changes following scrub burning on a New Zealand hill soil. *N.Z. J. Sci. Tech.* **37**, 290-313.

Mitchell, M. J. (1978). Role of invertebrates and microorganisms in sludge decomposition.

In "Utilization of Soil Organisms in Sludge Management" (R. Hartenstein, ed.), pp. 35-50. Natl Tech. Inf. Services, PB286932, Springfield, Virginia.

Mitchell, M. J. (1983). A simulation model of earthworm growth and population dynamics: Application to organic waste conversion. *In* "Earthworm Ecology" (J. E. Satchell, ed.), pp. 339-349. Chapman and Hall, London.

Mitchell, M. J., Hornor, S. G. and Abrams, B. I. (1980). Decomposition of sewage sludge in drying beds and the potential role of the earthworm, *Eisenia foetida*. *J. Environ. Qual.* **9**, 373-378.

Miyasaka, M. (1959). The formation of a granular structure by the earthworms. (In Japanese — abstract in Soils in Fert. Bibliog.). *J. Agric. Eng. Soc. Japan* **26**, 190-194.

Moeed, A. (1976). Birds and their food resources at Christchurch International Airport, New Zealand. *N.Z. J. Zool.* **3**, 373-390.

Mohr, G. and Littleton, I. (1978). Preliminary economic evaluation of some exotic feedstuffs. *Proc. 2nd Australian Poult. Stockfeed Conv.*, Sydney, pp. 226-231.

Mountford, M. D. (1962). An index of similarity and its application to classificatory problems. *In* "Progress in Soil Zoology" (P. W. Murphy, ed.), pp. 43-50. Butterworth, London.

Müller, P. E. (1878). Studier over Skovjord. I. Om Bögemuld od Bögemor paa Sand og Ler. *Tidsskr. Skogbruk* **3**, 1-124.

Murchie, W. R. (1960). Biology of the oligochaete *Bimastos zeteki* Smith and Gittins (Lumbricidae) in Northern Michigan. *Am. Midl. Nat.* **64**, 194-215.

Muskett, C. J. and Jones, M. P. (1980). The dispersal of lead, cadmium and nickel from motor vehicles and effects on roadside invertebrate macrofauna. *Environ. Pollut.* **A23**, 231-242.

Musso, H. (1967). Phenol coupling. *In* "Oxidative Coupling of Phenols" (W. I. Taylor and A. R. Battersby, eds), pp. 1-94. Marcel Dekker, New York.

Nakahara, H. and Bevelander, G. (1969). An electron microscope and autoradiographic study of the calciferous glands of the earthworm, *Lumbricus terrestris*. *Calcif. Tissue Res.* **4**, 193-201.

Nakamura, Y. (1967). Population density and biomass of terrestrial earthworms in grasslands of three different soil types near Sapporo. *Jpn J. Appl. Entomol. Zool.* **11**, 164-168.

Nakamura, Y. (1968). Studies on the ecology of terrestrial Oligochaeta. I. Seasonal variation in the population density of earthworms in alluvial soil grassland in Sapporo, Hokkaido. *Appl. Entomol. Zool.* **3**, 89-95.

Nakamura, Y. (1975). Decomposition of organic materials and soil fauna in pasture. 3. Disappearance of cow dung and associated soil macrofauna succession. *Pedobiologia* **15**, 210-221.

Nakamura, Y. (1982). Colonization by earthworms of NIIMI waste water treatment trenches. *Pedobiologia* **23**, 399-402.

Nakatsugawa, T. and Nelson, P. A. (1972). Insecticide detoxication in invertebrates. Enzymological approach to the problem of biological magnification. *In* "Environmental Toxicology of Pesticides" (F. Matsumara, G. M. Boush and T. Misato, eds), pp. 501-524. Academic Press, New York.

Neal, E. G. (1977). "Badgers". Blandford Press, Poole.

Needham, A. E. (1957). Components of nitrogenous excreta in the earthworms *Lumbricus terrestris* L. and *Eisenia foetida* (Savigny). *J. Exp. Biol.* **34**, 425-446.

Nelson, J. M. and Satchell, J. E. (1962). The extraction of Lumbricidae from soil with special reference to the hand-sorting method. *In* "Progress in Soil Zoology" (P. W. Murphy, ed.), pp. 294-299. Butterworths, London.

Németh, A. (1981). Estudio ecológico de las lombrices de tierra (Oligochaeta) en ecosistemas de bosque humedo tropical en San Carlos de Río Negro, Territorio Federal Amazonas. Ph.D. Thesis, Universidad Central de Venezuela.

Németh, A. and Herrera, R. (1982). Earthworm populations in a Venezuelan tropical rain forest. *Pedobiologia* **23**, 437-443.

Neuhauser, E. F. and Hartenstein, R. (1978). Reactivity of macroinvertebrate peroxidases with lignins and lignin model compounds. *Soil Biol. Biochem.* **10**, 341-342.

Neuhauser, E. F., Hartenstein, R. and Connors, W. J. (1978). Soil invertebrates and the degradation of vanillin, cinnamic acid, and lignins. *Soil Biol. Biochem.* **10**, 431-435.

Neuhauser, F., Kaplan, D. L. and Hartenstein, R. (1979). Life history of the earthworm *Eudrilus eugeniae*. *Rev. Ecol. Biol. Sol* **16**, 525-534.

Newell, P. F. (1967). Mollusca. *In* "Soil Biology" (A. Burges and F. Raw, eds), pp. 413-434. Academic Press, London.

Nielsen, C. O. (1962). Carbohydrases in soil and litter invertebrates. *Oikos* **13**, 200-215.

Nielsen, G. A. and Hole, F. E. (1964). Earthworms and the development of coprogenous A_1 horizons in forest soils of Wisconsin. *Soil Sci. Soc. Amer. Proc.* **28**, 426-430.

Nielson, R. L. (1951a). Earthworms and soil fertility. *Proc. 13th Conf. N.Z. Grassl. Assoc. New Plymouth*, pp. 158-167.

Nielson, R. L. (1951b). Effect of soil minerals on earthworms. *N.Z. J. Agric.* **83**, 433-435.

Nielson, R. L. (1953). Earthworms (recent research work). *N.Z. J. Agric.* **86**, 374.

Nielson, R. L. (1965). Presence of plant growth substances in earthworms demonstrated by paper chromatography and the Went pea test. *Nature (Lond.)* **208**, 1113-1114.

Nishioka, M. and Kirita, H. (1978). Litterfall. *In* "Biological Production in a Warm-temperate Evergreen Oak Forest of Japan" (T. Kira, Y. Ono and T. Hosokawa, eds). J.I.B.P. Synthesis Vol. 18, pp. 231-238. University of Tokyo Press, Tokyo.

Noble, J. C., Gordon, W. T. and Kleinig, C. R. (1970). The influence of earthworms on the development of mats of organic matter under irrigated pastures in southern Australia. *Proc. 11th Intl Grassl. Conf. Brisbane 1970*, pp. 465-468.

Nomura, S. and Usuki, I. (1951). Oxidation-reduction potential and pH in the soil of the habitat of earthworms. *Sci. Rep. Tohoku Univ., Ser. 4 (Biol.)* **19**, 104-112.

Nordström, S. (1975). Seasonal activity of lumbricids in southern Sweden. *Oikos* **26**, 307-315.

Nordström, S. and Rundgren, S. (1972). Methods of sampling lumbricids. *Oikos* **23**, 344-352.

Nordström, S. and Rundgren, S. (1973). Associations of lumbricids in southern Sweden. *Pedobiologia* **13**, 301-326.

Nordström, S. and Rundgren, S. (1974). Environmental factors and lumbricid associations in southern Sweden, *Pedobiologia* **14**, 1-27.

Nowak, E. (1975). Population density of earthworms and some elements of their production in several grassland environments. *Ekol. Pol.* **23**, 459-491.

Nye, P. H. (1955). Some soil-forming processes in the humid tropics. IV. The action of the soil fauna. *J. Soil Sci.* **6**, 73-83.

Oglesby, L. C. (1969). Inorganic components and metabolism; ionic and osmotic regulation: Annelida, Sipuncula, and Echiura. *In* "Chemical Zoology. Vol. 4. Annelida, Echiura, Sipuncula" (M. Florkin and B. T. Scheer, eds), pp. 211-310. Academic Press, New York.

Oglesby, L. C. (1978). Salt and water balance. *In* "Physiology of Annelids" (P. J. Mill, ed.), pp. 555-658. Academic Press, London.

Olive, P. J. W. and Clark, R. B. (1978). Physiology of reproduction. *In* "Physiology of Annelids" (P. J. Mill, ed), pp. 271-368. Academic Press, London.

Oliver, J. H. (1962). A mite parasitic in the cocoons of earthworms. *J. Parasitol.* **48**, 120-123.
Omodeo, P. (1952). Cariologia dei Lumbricidae. *Caryologia* **4**, 173-275.
Omodeo, P. (1955). Cariologia dei Lumbricidae. II. *Caryologia* **8**, 135-178.
Omodeo, P. (1963). Distribution of the terricolous oligochaetes on the two shores of the Atlantic. *In* "North Atlantic Biota and their History" (A. Löve and D. Löve, eds), pp. 127-141. Pergamon Press, New York.
Otanes, F. G. and Sison, P. L. (1947). Pests of rice. *Philippines J. Agric.* **13**, 36-88.
Parle, J. N. (1963a). Micro-organisms in the intestines of earthworms. *J. Gen. Microbiol.* **31**, 1-11.
Parle, J. N. (1963b). A microbiological study of earthworm casts. *J. Gen. Microbiol.* **31**, 13-22.
Patel, H. K. and Patel, R. M. (1959). Preliminary observations on the control of earthworms by soapdust (*Sapindus laurifolius* Vahl) extract. *Indian J. Entomol.* **21**, 251-255.
Peredel'skii, A. A., Poryadkova, N. A. and Rodionova, L. Z. (1957). The role of earthworms in purification of soil contaminated with radioactive isotopes. *Dokl. Akad. Nauk. SSSR* **115**, 809-812.
Peredel'skii, A. A., Bogatyrev, I. O. and Karavyanskii, N. S. (1960a). Effect of earthworms and wireworms on absorption by plants of the radioactive isotopes Ca^{45} and Sr^{90} from soil. *Dokl. Akad. Nauk SSSR* **134**, 1450-1452.
Peredel'skii, A. A., Shain, S. S. and Karavyanskii, N. S. (1960b). Redistribution of radioisotopes in soil by earthworms (Lumbricidae). *Dokl. Akad. Nauk SSSR* **135**, 185-188.
Perel', T. S. (1977). Differences in lumbricid organization connected with ecological properties. *In* "Soil Organisms as Components of Ecosystems". (U. Lohm and T. Persson, eds), *Ecol. Bull. (Stockholm)* **25**, 56-63.
Perel', T. and Karpachevsky, L. O. (1971). Lumbricidae and leaf litter breakdown in mixed coniferous-broadleaved forest. *In* "IV Colloquium Pedobiologiae" (J. d'Aguilar, ed.), pp. 379-385. Institut National des Recherches Agriculturelles Publ. 71-7, Paris.
Perel', T. S. and Sokolov, D. F. (1964). Quantitative evaluation of the participation of the earthworm *Lumbricus terrestris* Linné (Lumbricidae: Oligochaeta) in the transformation of forest litter. *Zool. Zh.* **53**, 1618-1625.
Perfect, T. J., Cook, A. G., Critchley, B. R., Critchley, U., Moore, R. L., Russell-Smith, A., Smith, M. J. and Yeadon, R. (1977). The effects of DDT on the populations of soil organisms and the processes of decomposition in a cultivated soil in Nigeria. *In* "Soil Organisms as Components of Ecosystems". (U. Lohm and T. Persson, eds). *Ecol. Bull. (Stockholm)* **25**, 565-568.
Persson, T. and Lohm, U. (1977). Energetical significance of the annelids and arthropods in a Swedish grassland soil. *Ecol. Bull. (Stockholm)* **23**, 1-211.
Petal, J., Nowak, E., Jakubczyk, H. and Czerwinski, Z. (1977). Effect of ants and earthworms on soil habitat modification. *In* "Soil Organisms as Components of Ecosystems" (U. Lohm and T. Persson, eds). *Ecol. Bull. (Stockholm)* **25**, 501-503.
Pew, J. C. and Connors, W. J. (1969). New structures from the enzymic dehydrogenation of lignin model p-hydroxy-w-carbinols. *J. Org. Chem.* **34**, 580-584.
Phillipson, J. and Bolton, P. J. (1976). The respiratory metabolism of selected Lumbricidae. *Oecologia (Berl.)* **22**, 135-152.
Phillipson, J. and Bolton, P. J. (1977). Growth and cocoon production by *Allolobophora rosea* (Oligochaeta: Lumbricidae). *Pedobiologia* **17**, 70-82.
Phillipson, J., Abel, R., Steel, J. and Woodell, S. R. J. (1976). Earthworms and the factors governing their distribution in an English beechwood. *Pedobiologia* **16**, 258-285.

References 377

Pianka, E. R. (1970). On r- and K- selection. *Am. Nat.* **104**, 592-597.
Pianka, E. R. (1974). Niche overlap and diffuse competition. *Proc. Natl Acad. Sci. USA* **71**, 2141-2145.
Pickford, G. E. (1926). The Kommetjie Flats. *Blythswood Review* **3**, 57-58. Cape Province, South Africa.
Pickford, G. E. (1937). "A Monograph of the Acanthodriline Earthworms of South Africa". W. Heffer, Cambridge.
Piearce, T. G. (1972a). The calcium relations of selected Lumbricidae. *J. Anim. Ecol.* **41**, 167-188.
Piearce, T. G. (1972b). Acid intolerant and ubiquitous Lumbricidae in selected habitats in north Wales. *J. Anim. Ecol.* **41**, 397-410.
Piearce, T. G. (1978). Gut contents of some lumbricid earthworms. *Pedobiologia* **18**, 153-157.
Piearce, T. G. (1982). Recovery of earthworm populations following salt-water flooding. *Pedobiologia* **24**, 91-100.
Piearce, T. G. and Phillips, M. J. (1980). The fate of ciliates in the earthworm gut: An *in vitro* study. *Microb. Ecol.* **5**, 313-320.
Piearce, T. G. and Piearce, B. (1979). Responses of Lumbricidae to saline inundation. *J. Appl. Ecol.* **16**, 461-473.
Poinar, G. O. (1978). Associations between nematodes (Nematoda) and oligochaetes (Annelida). *Proc. Helminth. Soc. Washington* **45**, 202-210.
Pokarzhevsky, A. D. and Krivolutsky, D. A. (1975). The role of pedobionts in biogeochemical cycles of calcium and strontium-90 in the ecosystems. *In* "Progress in Soil Zoology" (J. Vanek, ed.), pp. 249-254. Proc. 5th Intl Colloq. Soil Zool., Prague, 1973. Junk, The Hague/Academia, Prague.
Pokarzhevsky, A. D. and Titisheva, N. G. (1982). The population dynamics of the earthworm *Eisenia nordenskioldi* in meadow steppe habitats in the USSR. *Pedobiologia* **23**, 266-267.
Ponomareva, S. I. (1948). The rate of formation of calcite in the soil by the earthworms. *Reports Acad. Sci. USSR* **61**, 505-507.
Ponomareva, S. I. (1950). The role of earthworms in the creation of a stable structure in ley rotations. *Pochvovedenie* 1950, 476-486.
Ponomareva, S. I. (1952). The importance of biological factors in increasing the fertility of sod-podzolic soils Z. *Pfl.Ernähr. Düng.* **97**, 205-215.
Ponomareva, S. I. (1953). The influence of the activity of earthworms on the creation of a stable structure in a sod-podzolized soil. *Trud. Pochv. Inst. Dokuchaeva* **41**, 304-378.
Powlson, D. S. (1980). The effects of grinding on microbial and non-microbial organic matter in soil. *J. Soil Sci.* **31**, 77-85.
Prentø, P. (1979). Metals and phosphate in the chloragosomes of *Lumbricus terrestris* and their possible physiological significance. *Cell Tissue Res.* **196**, 123-134.
Prince, A. B., Donovan, J. F. and Bates, J. E. (1981). Vermicomposting of municipal solid wastes and municipal wastewater sludges. *In* "Workshop on the Role of Earthworms in the Stabilization of Organic Residues" (M. Appelhof, ed.), pp. 207-219. Beech Leaf Press, Kalamazoo, Michigan.
Puh, Y. Chiung (1941). Beneficial influence of earthworms on some chemical properties of the soil. *Contrib. Biol. Lab. Sci. Soc. China Zool. Ser.* **15**, 147-155.
Purrini, K. (1983). Soil invertebrates infected by microorganisms. *In* "New Trends in Soil Biology", (Ph. Lebrun, H. M. André, A. de Medts, C. Grégoire-Wibo and G. Wauthy, eds), pp. 167-175. Proc. 8th Intl Colloquium Soil Zool., Louvain-la-Neuve, 1982. Dieu-Brichart, Ottignies-Louvain-la-Neuve.
Puttarudrian, M. and Sastry, K. S. S. (1961). A preliminary study of earthworm damage to crop growth. *Mysore Agric. J.* **36**, 2-11.

Raffy, A. (1930). La respiration des vers de terre dans l'eau. Action de la teneur en oxygène et de la lumière sur l'intensité de la respiration pendant l'immersion. *C. R. Séances Soc. Biol.* **105**, 862-864.

Ralph, C. L. (1957). Persistent rhythms of activity and oxygen consumption in the earthworm. *Physiol. Zool.* **30**, 41-55.

Ranc, D. (1980). Mesure écologique de la température des lombriciens en prairie permanente par la méthode d'inversion du saccharose. D.E.A. Ecologie Microbienne, Univ. Claude Bernard, Lyon I.

Randell, R., Butler, J. D. and Hughes, T. D. (1972). The effect of pesticides on thatch accumulation and earthworm populations in Kentucky bluegrass turf. *Hortscience* **7**, 64-65.

Rao, B. R. C. (1979). Studies on the biological and ecological aspects of certain Indian earthworms. Synopsis Ph.D. Thesis, Mysore University. pp. 1-4. Kasturba Medical College: Manipal 576119.

Rao, K. P. (1966). Biochemical correlates of temperature acclimation. *In* "Molecular Mechanisms of Temperature Adaptation", (C. L. Prosser, ed.), pp. 227-244. American Association for the Advancement of Science Publ. 84, Washington.

Raw, F. (1959). Estimating earthworm populations by using formalin. *Nature (Lond.)* **184**, 1661-1662.

Raw, F. (1960). Earthworm population studies: A comparison of sampling methods. *Nature (Lond.)* **187**, 257.

Raw, F. (1962). Studies of earthworm populations in orchards. I. Leaf burial in apple orchards. *Ann. Appl. Biol.* **50**, 389-404.

Raw, F. (1967). Arthropoda (except Acari and Collembola). *In* "Soil Biology" (A. Burges and F. Raw, eds), pp. 323-362. Academic Press, London.

Reddy, M. V. (1980). Mass migration and mortality of *Amynthas* (= *Pheretima*) *alexandri* (Beddard) (Megascolecidae: Oligochaeta). *Curr. Sci. (Bangalore)* **49**, 606.

Reichle, D. E. (1971a). Systems analysis as applied to ecological processes: A mechanism for synthesis, integration and interpretation of IBP Woodlands Ecosystem research. *In* "Systems Analysis in Northern Coniferous Forests — IBP Workshop" (T. Rosswall, ed.), *Bull. Ecol. Res. Comm. NFR* **14**, 12-28. Swedish National Science Research Council, Stockholm.

Reichle, D. E. (1971b). Energy and nutrient metabolism of soil and litter invertebrates. *In* "Productivity of Forest Ecosystems" (P. Duvigneaud, ed.). Proc. Brussells Symposium, 1969, Ecology and Conservation **4**, 465-477.

Reichle, D. E. (1977). The role of soil invertebrates in nutrient cycling. *In* "Soil Organisms as Components of Ecosystems" (U. Lohm and T. Persson, eds), *Ecol. Bull. (Stockholm)* **25**, 145-156.

Reinecke, A. J. (1974). The upper lethal temperature of *Eisenia rosea* (Oligochaeta). *Wet. Bydraes. P.U.C.H.O. Reeks B Natuurwet.* no. 62.

Reinecke, A. J. (1975). The influence of acclimation and soil moisture on the temperature preference of *Eisenia rosea* (Lumbricidae). *In* "Progress in Soil Zoology" (J. Vanek, ed.), pp. 341-349. Proc. 5th Intl Colloq. Soil Zool., Prague, 1973. Junk, The Hague/Academia, Prague.

Reinecke, A. J. and Kriel, J. R. (1981). Influence of temperature on the reproduction of the earthworm *Eisenia foetida* (Oligochaeta). *S. Afr. J. Zool.* **16**, 96-100.

Reinecke, A. J. and Ljungström, P.-O. (1969). An ecological study of the earthworms from the banks of the Mooi River in Potchefstroom, South Africa. *Pedobiologia* **9**, 106-111.

Reinecke, A. J. and Ryke, P. A. J. (1970a). On the water relations of *Microchaetus modestus* Mich. (Microchaetidae, Oligochaeta). *Wet. Bydraes. P.U.C.H.O. Reeks B Natuurwet.* no. 15.

Reinecke, A. J. and Ryke, P. A. J. (1970b). Casting activity of a South African endemic earthworm, *Microchaetus modestus*. *Wet. Bydraes. P.U.C.H.O. Reeks B Natuurwet.* No. 16.

Reinecke, A. J. and Ryke, P. A. J. (1972). The temperature preferences of *Microchaetus modestus* Mich. and *Eisenia foetida* Sav. *Wet. Bydraes. P.U.C.H.O. Reeks B Natuurwet.* no. 46.

Reinecke, A. J. and Ryke, P. A. J. (1974). The upper lethal temperature of *Microchaetus modestus* (Oligochaeta). *Rev. Ecol. Biol. Sol* **11**, 333–351.

Reinecke, A. J. and Visser, F. A. (1981). Number and size of hatchlings from cocoons of the earthworm species *Eisenia rosea* and *Allolobophora trapezoides* (Oligochaeta). *Rev. Ecol. Biol. Sol* **18**, 473–485.

Reynolds, J. W. (1969). The relationship of earthworm distribution and biomass to soil type in grassland and forest habitats at Oak Ridge National Laboratory. *Reports Oak Ridge National Lab.*, US Atomic Energy Agency, Oak Ridge, Tennessee.

Reynolds, J. W. (1970). The relationship of earthworm (Oligochaeta: Lumbricidae and Medascolecidae) distribution and biomass to soil type in forest and grassland habitats at Oak Ridge National Laboratory. *Assoc. Southeast Biol. Bull.* **17**, 60.

Reynolds, J. W. (1972). The relationship of earthworm (Oligochaeta: Acanthodrilidae and Lumbricidae) distribution and biomass in six heterogeneous woodlot sites in Tippecanoe County, Indiana. *J. Tenn. Acad. Sci.* **47**, 63–67.

Reynolds, J. W. (1973). Earthworm (Annelida: Oligochaeta) ecology and systematics. *In* Proceedings 1st Soil Microcommunities Conf., Syracuse, N.Y., October 1971 (D. L. Dindal, ed.), pp. 95–120. US Atomic Energy Commission, Office of Information Services, Technical Information Centre, Washington, D.C.

Reynolds, J. W. (1974). Are earthworms really hermaphroditic amphimictic organisms? *Biologist* **56**, 90–99.

Reynolds, J. W. (1975). Boiteagan (Oligochaeta: Lumbricidae) Cheap Breatunn. *Megadrilogica* **2**, 1–8.

Reynolds, J. W. (1976). Un aperçu des vers de terre dans les forêts nord-américaines, leurs activités et leurs répartition. *Megadrilogica* **2**, 1–11.

Reynolds, J. W. and Reinecke, A. J. (1976). A preliminary survey of the earthworms of the Kruger National Park, South Africa (Oligochaeta: Glossocolecidae, Megascolecidae and Octochaetidae). *Wet. Bydraes. P.U.C.H.O. Reeks B Natuurwet.* no. 89.

Reynolds, J. W., Krohn, W. B. and Jordan, G. A. (1977). Earthworm populations as related to woodcock habitat usage in central Maine. *Proc. Woodcock Symp.* **6**, 135–146.

Reynoldson, T. B. (1966). The ecology of earthworms with special reference to North Wales habitats. Report North Wales Soils Discussion Group (Rothamsted Experimental Station), no. 1, 25–31.

Rhee, J. A. van (1963). Earthworm activities and the breakdown of organic matter in agricultural soils. *In* "Soil Organisms" (J. Doeksen and J. van der Drift, eds), pp. 55–59. North Holland Publishing Company, Amsterdam.

Rhee, J. A. van (1965). Earthworm activity and plant growth in artificial cultures. *Plant Soil* **22**, 45–48.

Rhee, J. A. van (1967). Development of earthworm populations in orchard soils. *In* "Progress in Soil Biology" (O. Graff and J. E. Satchell, eds), pp. 360–371. North Holland Publishing Company, Amsterdam.

Rhee, J. A. van (1969a). Inoculation of earthworms in a newly drained polder. *Pedobiologia* **9**, 128–132.

Rhee, J. A. van (1969b). Development of earthworm populations in polder soils. *Pedobiologia* **9**, 133–140.

Rhee, J. A. van (1969c). Effects of biocides and their residues on earthworms. *Mededelingen Ryksfacultiet Landbouwweten-schappen Gent* **34**, 682–689.

Rhee, J. A. van (1971). The productivity of orchards in relation to earthworm activities. *In* "IV Colloquium Pedobiologiae" (J. d'Aguilar, ed.), pp. 99–106. Institut National des Recherches Agriculturelles Publ. 71-7, Paris.

Rhee, J. A. van (1975). Copper contamination effects on earthworms by disposal of pig waste in pastures. *In* "Progress in Soil Zoology" (J. Vanek, ed.), pp. 451–457. Proc. 5th Intl Colloq. Soil Zool., Prague, 1973.

Rhee, J. A. van (1977a). A study of the effect of earthworms on orchard productivity. *Pedobiologia* **17**, 107–114.

Rhee, J. A. van (1977b). Effects of soil pollution on earthworms. *Pedobiologia* **17**, 201–208.

Rhee, J. A. van and Nathans, S. (1961). Observations on earthworm populations in orchard soils. *Neth. J. Agric. Sci.* **9**, 94–100.

Richards, K. S. (1978). Epidermis and cuticle. *In* "Physiology of Annelids" (P. J. Mill, ed.), pp. 33–61. Academic Press, London.

Richards, K. S. and Arme, C. (1982). Integumentary uptake of dissolved organic materials by earthworms. *Pedobiologia* **23**, 358–366.

Richards, L. A. (1931). Capillary conduction of liquids through porous medium. *Physics* **1**, 318–333.

Richardson, H. C. (1938). The nitrogen cycle in grassland soils: With special reference to the Rothamsted Park grass experiment. *J. Agric. Sci. Camb.* **28**, 73–121.

Richter, K. O. (1973). Freeze-branding for individually marking the banana slug. *Northwest Sci.* **47**, 109–113.

Ricou, G. (1978). La prairie permanente du nord-ouest France. *In* "Problèmes d'Ecologie: Ecosystèmes Terrestres" (M. Lamotte and F. Bourlière, eds), pp. 17–74. Masson, Paris.

Rigby, B. J. (1967). Relation between the shrinkage of native collagen in acid solution and the melting temperature of the tropocollagen molecule. *Biochem. Biophys. Acta.* **133**, 272–277.

Rigby, B. J. (1968). Temperature relationships of poikilotherms and the melting temperature of molecular collagen. *Biol. Bull. (Woods Hole)* **135**, 223–229.

Robertson, J. D. (1936). The function of the calciferous glands of earthworms. *J. Exp. Biol.* **13**, 279–297.

Robertson, J. D. (1941). The function and metabolism of calcium in the Invertebrata. *Biol. Rev. Camb. Philos. Soc.* **16**, 106–133.

Rogers, C. G. and Lewis, E. M. (1914). The relation of the body temperature of the earthworm to that of its environment. *Biol. Bull. (Woods Hole)* **27**, 267–268.

Roots, B. I. (1955). The water relations of earthworms. I. The activity of the nephridiostome cilia of *Lumbricus terrestris* L. and *Allolobophora chlorotica* (Sav.) in relation to the concentration of the bathing medium. *J. Exp. Biol.* **32**, 765–774.

Rose, C. J. and Wood, A. W. (1980). Some environmental factors affecting earthworm populations and sweet potato production in the Tari Basin, Papua New Guinea Highlands. *Papua New Guinea Agric. J.* **31**, 1–13.

Ross, D. J. and Cairns, A. (1982). Effects of earthworms and ryegrass on respiratory and enzyme activities of soil. *Soil Biol. Biochem.* **14**, 583–587.

Rouelle, J. (1983). Introduction of amoebae and *Rhizobium japonicum* into the gut of *Eisenia fetida* (Sav.) and *Lumbricus terrestris* L. *In* "Earthworm Ecology" (J. E. Satchell, ed.), pp. 375–381. Chapman and Hall, London.

Rovira, A. D. and Greacen, E. L. (1957). The effect of aggregate disruption on the activity of microorganisms in the soil. *Aust. J. Agric. Res.* **8**, 659–673.

Roy, S. K. (1957). Studies on the activities of earthworms. *Proc. Zool. Soc. (Calcutta)* **10**, 81–98.

Rożen, A. (1982). The annual cycle in populations of earthworms (Lumbricidae, Oligochaeta) in three types of oak-hornbeam of the Niepolomicka Forest. I. Species composition, dominance, frequency and associations. *Pedobiologia* **23**, 199–208.

Rudge, M. R. (1968). The food of the common shrew *Sorex araneus* L. *J. Anim. Ecol.* **37**, 565-581.
Rundgren, S. (1975). Vertical distribution of lumbricids in southern Sweden. *Oikos* **26**, 299-306.
Rundgren, S. (1977). Seasonality of emergence in lumbricids in southern Sweden. *Oikos* **28**, 49-55.
Ruppell, R. F. and Laughlin, C. W. (1976). Toxicity of some soil pesticides to earthworms. *J. Kans. Entomol. Soc.* **50**, 113-118.
Ruschmann, G. (1953a). Antibioses and symbioses of soil organisms and their significance in soil fertility. Earthworm symbioses and antibioses. *Z. Acker Pflbau.* **96**, 201-218.
Ruschmann, G. (1953b). Antibioses and symbioses of soil organisms and their significance in soil fertility. 4. Symbiotic and antibiotic earthworm actinomycetes. *Z. Acker. Pflbau* **97**, 101-114.
Ryden, J. C., Syers, J. K. and Harris, R. F. (1973). Phosphorus in runoff and streams. *Adv. Agron.* **25**, 1-45.
Rysavy, B. (1969). Lumbricidae — an important parasitological factor in helminthoses of domestic and wild animals. *Pedobiologia* **9**, 171-174.
Sabine, J. R. (1978). The nutritive value of earthworm meal. *In* "Utilization of Soil Organisms in Sludge Management" (R. Hartenstein, ed.), pp. 122-130. Natl Tech. Inf. Services, PB286932. Springfield, Virginia.
Sabine, J. R. (1983). Earthworms as a source of food and drugs. *In* "Earthworm Ecology" (J. E. Satchell, ed.), pp. 285-296. Chapman and Hall, London.
Salisbury, E. J. (1925). The influence of earthworms on soil reaction and the stratification of undisturbed soils. *J. Linn. Soc. Lond. (Bot.)* **46**, 415-426.
Salt, G. and Hollick, F. S. J. (1944). Studies of wireworm populations. I. A census of wireworms in pasture. *Ann. Appl. Biol.* **31**, 52-64.
Saroja, K. (1961). Seasonal acclimatization of oxygen consumption to temperature in a tropical poikilotherm, the earthworm, *Megascolex mauritii*. *Nature (Lond.)* **190**, 930-931.
Satchell, J. E. (1955a). Some aspects of earthworm ecology. *In* "Soil Zoology" (D. K. McE. Kevan, ed.), pp. 180-202. Butterworth, London.
Satchell, J. E. (1955b). An electrical method of sampling earthworm populations. *In* "Soil Zoology" (D. K. McE. Kevan, ed.), pp. 356-364. Butterworth, London.
Satchell, J. E. (1963). Nitrogen turnover by a woodland population of *Lumbricus terrestris*. *In* "Soil Organisms" (J. Doeksen and J. van der Drift, eds), pp. 60-66. North Holland Publishing Company, Amsterdam.
Satchell, J. E. (1967). Lumbricidae. *In* "Soil Biology" (A. Burges and F. Raw, eds), pp. 259-322. Academic Press, London.
Satchell, J. E. (1969). Methods of sampling earthworm populations. *Pedobiologia* **9**, 20-25.
Satchell, J. E. (1970). Measuring population and energy flow in earth-worms. *In* "Methods of Study in Soil Ecology" (J. Phillipson, ed.), pp. 261-267. Proc. Paris Symp. United Nations Educational Scientific and Cultural Organization, Paris.
Satchell, J. E. (1971). Earthworms. *In* "Methods of Study in Quantitative Soil Ecology: Population, Production and Energy Flow" (J. Phillipson, ed.), pp. 107-127. International Biological Programme Publ 18. Blackwell, London.
Satchell, J. E. (1974). Litter — interface of animate/inanimate matter. *In* "Biology of Plant Litter Decomposition" (C. H. Dickinson and G. J. F. Pugh, eds), Vol. 1, Introduction, pp. xiii-xliv. Academic Press, London.
Satchell, J. E. (1980a). The earthworm populations of experimental *Betula* plots on a *Calluna* podzol. *Pedobiologia* **20**, 151-153.
Satchell, J. E. (1980b). Soil and vegetation changes in experimental birch plots on a *Calluna* podzol. *Soil Biol. Biochem.* **12**, 303-310.

Satchell, J. E. (1980c). Earthworm populations of experimental birch plots on a *Calluna* podzol. *Soil Biol. Biochem.* **12**, 311-316.
Satchell, J. E. (1980d). Potential of the Silpho moor experimental birch plots as a habitat for *Lumbricus terrestris*. *Soil Biol. Biochem.* **12**, 317-323.
Satchell, J. E. (1980e). r worms and K worms: A basis for classifying lumbricid earthworm strategies. *In* "Soil Biology as Related to Land Use Practices" (D. L. Dindal, ed.), pp. 848-854. Proc. 7th Intl Colloq. Soil Zool., Syracuse, 1979. Environmental Protection Agency, Washington D.C.
Satchell, J. E. (Ed.) (1983a). "Earthworm Ecology: From Darwin to Vermiculture". Chapman and Hall, London.
Satchell, J. E. (1983b). Earthworm microbiology *In* "Earthworm Ecology" (J. E. Satchell, ed.), pp. 351-364. Chapman and Hall, London.
Satchell, J. E. (in press). Has vermiculture a future? Implications of current trends in advertising. Proc. International Symposium on Agricultural and Environmental Prospects in Earthworm Farming, Rome, July 1983.
Satchell, J. E. and Lowe, D. G. (1967). Selection of leaf litter in *Lumbricus terrestris*. *In* "Progress in Soil Biology" (O. Graff and J. E. Satchell, eds), pp. 102-119. North Holland Publishing Company, Amsterdam.
Satchell, J. E. and Martin, K. (1981). "A Bibliography of Earthworm Research". Institute of Terrestrial Ecology, Merlewood Research Station, Grange-over-Sands, U.K.
Saussey, M. (1966). Contribution à l'étude des phénomènes de diapause et de régénération caudale chez *Allolobophora icterica* (Savigny) (Oligochète, Lombricien?. *Mém. Soc. Linn. Normandie* n.s. *Sect. Zool.* **3**, 1-158.
Savigny, J. C. (1826). Analyse d'un mémoire sur les lombrics par Cuvier. Mem. Acad. Sci. Inst. Fr. **5**, 176-184.
Scharpenseel, H. W. and Gewehr, H. (1960). Studien zur Wasserbewegung im Boden mit Tritium-Wasser. *Z. Pfl.Ernähr. Düng.* **88**, 35-49.
Schmidt, P. (1918). Anabiosis of the earthworm. *J. Exp. Zool.* **27**, 55-72.
Schread, J. C. (1952). Habits and control of the Oriental earthworm. *Bull. Connecticut Agric. Exp. Stn* **556**, 5-15.
Schultz, W. and Felber, E. (1956). Welche Mikroorganismen spielen im Regenwurmdarm bei der Bildung von Bodenkrumeln eine Rolle? *Z. Acker. Pflbau.* **101**, 471-476.
Schwert, D. P. (1979). Description and significance of a fossil earthworm (Oligochaeta: Lumbricidae) cocoon from postglacial sediments in southern Ontario. *Can. J. Zool.* **57**, 1402-1405.
Schwert, D. P. (1980). Active and passive dispersal of lumbricid earthworms. *In* "Soil Biology as Related to Land Use Practices" (D. L. Dindal, ed.), pp. 182-189. Proc. 7th Intl Colloq. Soil Zool., Syracuse, 1979. Environmental Protection Agency, Washington D.C.
Schwert, D. P. and Dance, K. W. (1979). Earthworm cocoons as a drift component in a southern Ontario stream. *Can. Field-Nat.* **93**, 180-183.
Sears, P. D. (1953). Pasture growth and soil fertility. I. The influence of red and white clovers, superphosphate, lime and sheep grazing, on pasture yields and botanical composition. *N.Z. J. Sci. Tech.* **35A** Suppl. 1, 1-29.
Sears, P. D. and Evans, L. T. (1953). Pasture growth and soil fertility. III. The influence of red and white clovers, superphosphate, lime, and dung and urine on soil composition and on earthworm and grass-grub populations. *N.Z. J. Sci. Tech.* **35A** Suppl. 1, 42-52.
Semenova, L. M. (1966). Relation of the structure of the digestive tract in earthworms (Lumbricidae) to their feeding habits. *Zool. Zh.* **45**, 986-997.
Senapati, B. K. and Dash, M. C. (1981). Effect of grazing on the elements of production in vegetation and oligochaete components of a tropical pasture. *Rev. Ecol. Biol. Sol* **18**, 487-505.

Sergienko, M. I. (1969). The distribution of earthworms in the biocenoses of Chernogora at the ecological profile Vorochta-Goverla. *Pedobiologia* **9**, 112–113.
Seymour, M. (1978). The infinite variety of worms. *New Scientist* **77** (1093), 650–652.
Sharpley, A. N. and Syers, J. K. (1976). Potential role of earthworm casts for the phosphorus enrichment of run-off waters. *Soil Biol. Biochem.* **8**, 341–346.
Sharpley, A. N. and Syers, J. K. (1977). Seasonal variation in casting activity and in the amounts and release to solution of phosphorus forms in earthworm casts. *Soil Biol. Biochem.* **9**, 227–231.
Sharpley, A. N., Syers, J. K. and Springett, J. A. (1979). Effect of surface-casting earthworms on the transport of phosphorus and nitrogen in surface runoff from pasture. *Soil. Biol. Biochem.* **11**, 459–462.
Shields, R. F. (1976). "Earthworm Buyer's Guide". Shields Publications, Elgin, Illinois.
Shilova, S. A., Denisova, A. V., Dmitriev, G. A., Voronova, L. D. and Bardier, M. N. (1971). Effect of some insecticides upon the common mole. *Zool. Zh.* **50**, 886–892.
Shiraishi, K. (1954). On the chemotaxis of the earthworm to carbon dioxide. *Sci. Rep. Tohoku Univ. Fourth Ser. (Biol.)* **20**, 356–361.
Shrikhande, J. C. and Pathak, A. N. (1951). A comparative study of the physicochemical characters of the castings of different insects. *Indian J. Agric. Sci.* **21**, 401–407.
Shutler, R. (1979). A radiocarbon chronology for the New Hebrides. Proc. 9th Int. Congr. Anthropol. Ethnol. Sci., 1968, Tokyo and Kyoto. Vol. 3, pp. 135–137.
Siljak, D. D. (1974). Connective stability of complex systems. *Nature (Lond.)* **249**, 280.
Siljak, D. D. (1975a). When is a complex ecosystem stable? *Math. Biosci.* **25**, 25–50.
Siljak, D. D. (1975b). Connective stability of competitive equilibrium. *Automatica* **11**, 389–400.
Sims, R. W. and Easton, E. G. (1972). A numerical revision of the earthworm genus *Pheretima* auct. (Megascolecidae: Oligochaeta) with the recognition of new genera and an appendix on the earthworms collected by the Royal Society North Borneo Expedition. *Biol. J. Linn. Soc.* **4**, 169–268.
Skoczen, S. (1970). Food storage of some insectivorous mammals (Insectivora). *Przegl. Zool.* **14**, 243–248.
Slater, C. S. and Hopp, H. (1947). Relation of fall protection to earthworm populations and soil physical conditions. *Soil Sci. Soc. Amer. Proc.* **12**, 508–511.
Smirnoff, W. A. and Heimpel, A. M. (1961). Notes on the pathogenicity of *Bacillus thuringiensis* var. *thuringiensis* Bulinu for the earthworm *Lumbricus terrestris* Linnaeus. *J. Insect. Pathol.* **3**, 403–408.
Smith, A. C. (1902). The influence of temperature, odours, light and contact on the movement of the earthworm. *Am. J. Physiol.* **6**, 459–486.
Smith, B. J. and Peterson, J. A. (1982). Studies of the giant Gippsland earthworm *Megascolides australis* McCoy, 1878. *Victorian Nat.* **99**, 164–172.
Smith, W. W. (1893). Further notes on New Zealand earthworms with observations on the known aquatic species. *Trans. N.Z. Inst.* **25**, 111–146.
Solem, A. (1959). Zoogeography of the land and fresh-water Mollusca of the New Hebrides. *Fieldiana Zool.* **43**, 239–343.
Southwood, T. R. E. (1966). "Ecological Methods". Methuen, London.
Southwood, T. R. E. (1977). Habitat, the templet for ecological strategies? *J. Anim. Ecol.* **46**, 337–365.
Spencer, W. B. (1888). On the anatomy of *Megascolides australis*, the giant earth-worm of Gippsland. *Trans. R. Soc. Victoria* **1**, 3–60.
Springett, J. A. and Syers, J. K. (1979). The effect of earthworm casts on ryegrass seedlings. *In* Proc. 2nd Australasian Conf. Grassl. Invert. Ecol. (T. K. Crosby and R. P. Pottinger, eds), pp. 44–47. Government Printer, Wellington.

Standen, V. (1979). Factors affecting the distribution of lumbricids (Oligochaeta) in associations at peat and mineral sites in northern England. *Oecologia (Berl.)* **42**, 359-374.

Stark, N., Pawlowski, P. and Bodmer, S. (1978). Quality of earthworm castings and the use of compost on arid soils. *In* "Utilization of Soil Organisms in Sludge Management" (R. Hartenstein, ed.), pp. 87-102. Natl Tech. Inf. Services, PB286932. Springfield, Virginia.

Stenersen, J. (1979a). Action of pesticides on earthworms. Part I. The toxicity of cholinesterase-inhibiting insecticides to earthworms as evaluated by laboratory tests. *Pestic. Sci.* **10**, 66-74.

Stenersen, J. (1979b). Action of pesticides on earthworms. Part II. Elimination of parathion by the earthworm *Eisenia foetida* (Savigny). *Pestic. Sci.* **10**, 104-112.

Stenersen, J. (1979c). Action of pesticides on earthworms. Part III. Inhibition and reactivation of cholinesterases in *Eisenia foetida* (Savigny) after treatment with cholinesterase-inhibiting insecticides. *Pestic. Sci.* **10**, 113-122.

Stenersen, J., Gilman, A. and Vardanis, A. (1973). Carbofuran: Its toxicity to and metabolism by earthworms (*Lumbricus terrestris*). *J. Agric. Food Chem.* **21**, 166-171.

Stephenson, J. (1930). "The Oligochaeta". Oxford, Clarendon Press.

Stockdill, S. M. J. (1959). Earthworms improve pasture growth. *N.Z. J. Agric.* **98**, 227-233.

Stockdill, S. M. J. (1966). The effect of earthworms on pastures. *Proc. N.Z. Ecol. Soc.* **13**, 68-75.

Stockdill, S. M. J. (1982). Effects of introduced earthworms on the productivity of New Zealand pastures. *Pedobiologia* **24**, 29-35.

Stöckli, A. (1928). Studien über den Einfluss der Regenwürmer auf die Beschaffenheit des Bodens. *Landwirtsch. Jahrb. Schweiz.* **42**, 1.

Stolte, H. A. (1962). Oligochaeta *In* "Klassen und Ordnungen das Tierreiches", Vol. 4, Part 3, pp. 891-982. Geest and Portig, Leipzig, Germany.

Støp-Bowitz, C. (1969). Did lumbricids survive the quaternary glaciations in Norway? *Pedobiologia* **9**, 93-98.

Stout, J. D. and Goh, K. M. (1980). The use of radiocarbon to measure the effects of earthworms on soil development. *Radiocarbon* **22**, 892-896.

St Remy, E. A. de and Daynard, T. B. (1982). Effects of tillage methods on earthworm populations in monoculture corn. *Can. J. Soil Sci.* **62**, 699-703.

Stringer, A. and Lyons, C. H. (1974). The effect of benomyl and thiophanate-methyl on earthworm populations in apple orchards. *Pestic. Sci.* **5**, 189-196.

Stringer, A. and Wright, M. A. (1973). The effect of benomyl and some related compounds on *Lumbricus terrestris* and other earthworms. *Pestic. Sci.* **4**, 165-170.

Sugi, Y. and Tanaka, M. (1978a). Population study of an earthworm, *Pheretima sieboldi*. *In* "Biological Production in a Warm-temperate Evergreen Oak Forest of Japan" (T. Kira, Y. Ono and T. Hosakawa, eds), pp. 163-171. Japanese International Biological Programme Synthesis Vol. 18. University of Tokyo Press, Tokyo.

Sugi, Y. and Tanaka, M. (1978b). Number and biomass of earthworm populations. *In* "Biological Production in a Warm-temperature Evergreen Oak Forest of Japan" (T. Kira, Y. Ono and T. Hosokawa, eds), pp. 171-178. J.I.B.P. Synthesis Vol. 18. University of Tokyo Press, Tokyo.

Summers, L. A. (1980). "The Bipyridinium Herbicides". Academic Press, London.

Svendsen, J. A. (1955). Earthworm population studies: A comparison of sampling methods. *Nature (Lond.)* **175**, 864.

Swaby, R. J. (1950). The influence of earthworms on soil aggregation. *J. Soil Sci.* **1**, 195-197.

Swift, M. J., Heal, O. W. and Anderson, J. M. (1979). "Decomposition in Terrestrial Ecosystems". Blackwell, Oxford.

Syers, J. K., Springett, J. A. and Sharpley, A. N. (1979). The role of earthworms in the cycling of phosphorus in pasture ecosystems. *In* Proc. 2nd Australasian Conf. Grassl. Invert. Ecol. (T. K. Crosby and R. P. Pottinger, eds), pp. 47-49. Government Printer, Wellington.

Takeuchi, N. (1980). Control of coelomic fluid concentration and brain neurosecretion in the littoral earthworm *Pontodrilus matsushimensis* Iizuka. *Comp. Biochem. Physiol. A.* **67**, 357-359.

Tara, B., Wielgus-Serafinska, E., Kawka, E., Dec, R. and Mielzyńska, I. (1979). Localization and accumulation of zinc and lead in some invertebrate tissue from the steel plant "Katowice" area. *Acta Physiol. Pol.* **3**, 196-197.

Taylor, N. H. (1948). Soil map of New Zealand. To accompany: A genetic classification of New Zealand soils. *N.Z. Soil Bureau Bull.* **3**. Government Printer, Wellington.

Taylor, N. H. (1952). Pedology as an aid in animal research. *Aust. Vet. J.* **28**, 183-189.

Taylor, N. H. and Pohlen, I. J. (1962). Soil survey method. *N.Z. Soil Bureau Bull.* **25**. Government Printer, Wellington.

Teotia, S. P., Duley, F. L. and McCalla, T. M. (1950). Effect of stubble mulching on number and activity of earthworms. Nebraska Agricultural Experiment Station Bulletin 165.

Thambi, A. V. and Dash, M. C. (1973). Seasonal changes in numbers and biomass of Enchytraeidae (Oligochaeta) population in tropical grassland soils from India. *Trop. Ecol.* **14**, 228-237.

Thompson, A. R. (1971). Effects of nine insecticides on the numbers and biomass of earthworms in pasture. *Bull. Environ. Contam. Toxicol.* **5**, 577-586.

Thomson, A. J. and Davies, D. M. (1974). Production of surface casts by the earthworm *Eisenia rosea*. *Can. J. Zool.* **52**, 659.

Thornton, M. L. (1970). Transport of soil-dwelling aquatic phycomycetes by earthworms. *Trans. Br. Mycol. Soc.* **55**, 391-397.

Tillinghast, E. K. (1967). Excretory pathways of ammonia and urea in the earthworm *Lumbricus terrestris* L. *J. Exp. Zool.* **166**, 295-300.

Tillinghast, E. K., McInnes, D. C. and Duffill, R. A. (1969). The effect of temperature and water availability on the output of ammonia and urea by the earthworm *Lumbricus terrestris* L. *Comp. Biochem. Physiol.* **29**, 1087-1092.

Tisdall, J. M. (1978). Ecology of earthworms in irrigated orchards. *In* "Modification of Soil Structure" (W. W. Emerson, R. D. Bond and A. R. Dexter, eds), pp. 297-303. Wiley, Chichester.

Tiurin, I. V. (1937). "Organic Matter of the Soil". Selkhozgiz, Moscow.

Tomlin, A. D. (1983). The earthworm bait market in North America. *In* "Earthworm Ecology" (J. E. Satchell, ed.), pp. 331-338. Chapman and Hall, London.

Tréhen, P. and Bouché, M. B. (1983). Place des lombriciens dans les processus de restauration des sols de lande. *In* "New Trends in Soil Biology", (Ph. Lebrun, H. M. André, A. de Medts, C. Grégoire-Wibo and G. Wauthy, eds), pp. 471-486. Proc. 8th Intl Colloquium Soil Zool., Louvain-la-Neuve, 1982. Dieu-Brichart, Ottignies-Louvain-la-Neuve.

Trifonov, D. (1957). Ueber die Bekämpfung der Malwurfsgrille und des Regenwurms mit dem Präparat "Alon Kombi". *Bulgar. Tiûtiûn* **2**, 114-115.

Trueman, E. R. (1978). Locomotion. *In* "Physiology of Annelids" (P. J. Mill, ed.), pp. 243-269. Academic Press, London.

Tsukamoto, J. and Watanabe, H. (1977). Influence of temperature on hatching and growth of *Eisenia foetida* (Oligochaeta, Lumbricidae). *Pedobiologia* **17**, 338-342.

Tsuru, S. (1975). Seasonal variations of number and biomass of earthworms in grassland. *In* "Progress in Soil Zoology" (J. Vanek, ed.), pp. 223-228. Proc. 5th Intl Colloq. Soil Zool., Prague, 1973.

Ustinov, I. P. (1962). Species composition and some problems of the ecology of the earthworms (Lumbricidae) in the Kirov region. *Zool. Zh.* **41**, 133–135.

Usuki, I. (1955). Earthworms and some environmental factors especially oxidation-reduction potentials. *Sci. Rep. Tohoku Univ. Fourth Ser. (Biol.)* **21**, 13–23.

Vail, V. A. (1974). Observations on the hatchlings of *Eisenia foetida* and *Bimastos tumidus* (Oligochaeta: Lumbricidae). *Bull. Tall Timbers Res Stn* **16**, 1–8.

Valiachmedov, V. (1969). Earthworms of the main types of virgin and cultivated soils of Tadjikstan. *Pedobiologia* **9**, 69–75.

Vannier, G. (1973). Originalité des conditions de vie dans le sol due à la présence de l'eau; importance thermodynamique et biologique de la porosphère. *Ann. Soc. R. Zool. Belg.* **103**, 157–167.

Vannier, G. (1983). The importance of ecophysiology for both biotic and abiotic studies of the soil. *In* "New Trends in Soil Biology", (Ph. Lebrun, H. M. André, A. de Medts, C. Grégoire-Wibo and G. Wauthy, eds), pp. 289–314. Proc. 8th Intl Colloquium Soil Zool., Louvain-la-Neuve, 1982. Dieu-Brichart, Ottignies-Louvain-la-Neuve.

Vimmerstedt, J. P. and Finney, J. H. (1973). Impact of earthworm introduction on litter burial and nutrient distribution in Ohio strip-mine spoil banks. *Soil Sci. Soc. Amer. Proc.* **37**, 388–391.

Waksman, S. A. and Martin, J. P. (1939). The conservation of the soil. *Science* **90**, 304–305.

Walsh, A. S. (1936). *Lumbricus* lore. *Turtox News* **14**, 57–60.

Walton, W. R. (1928). Earthworms as pests and otherwise. *U. S. Dep. Agric. Farmers' Bull.* 1569.

Wasawo, D. P. S. and Visser, S. A. (1959). Swampworms and tussock mounds in the swamps of Teso, Uganda. *East Afr. Agric. J.* **25**, 86–90.

Watanabe, H. (1975). On the amount of cast production by the megascolecid earthworm *Pheretima hupeiensis*. *Pedobiologia* **15**, 20–28.

Watanabe, H. and Tsukamoto, J. (1976). Seasonal change in size class and stage structure of lumbricid *Eisenia foetida* population in a field compost and its practical application as the decomposer of organic waste matter. *Rev. Ecol. Biol. Sol* **13**, 141–146.

Waters, R. A. S. (1951). Earthworms and the fertility of pasture. *Proc. 13th Conf. N.Z. Grassl. Assoc., New Plymouth*, pp. 168–175.

Waters, R. A. S. (1955). Numbers and weights of earthworms under a highly productive pasture. *N.Z. J. Sci. Tech.* **A36**, 516–525.

Watkin, B. R. and Wheeler, J. L. (1966). Some factors affecting earthworm populations under pasture. *J. Bro. Grassl. Soc.* **21**, 14–20.

Weber, R. E. (1978a). Respiration. *In* "Physiology of Annelids" (P. J. Mill, ed.), pp. 369–392. Academic Press, London.

Weber, R. E. (1978b). Repiratory pigments. *In* "Physiology of Annelids" (P. J. Mill, ed.), pp. 393–446. Academic Press, London.

Westeringh, W. van de (1972). Deterioration of soil structure in worm free orchards. *Pedobiologia* **12**, 6–15.

Wheatley, G. A. and Hardman, J. A. (1968). Organochlorine insecticide residues in earthworms from arable soils. *J. Sci. Food Agric.* **19**, 219–225.

White, G. (1789). "The Natural History of Selborne". Benjamin White, London.

Whittaker, R. H. (1975). The design and stability of some plant communities. *In* "Unifying Concepts in Ecology" (W. H. van Dobben and R. H. Lowe-McConnell, eds), pp. 169–181. Junk, The Hague.

Wiececk, C. S. and Messenger, A. S. (1972). Calcite contributions by earthworms to forest soils in northern Illinois. *Soil Sci. Soc. Amer. Proc.* **36**, 478–480.

Wiegert, R. G. and Petersen, C. E. (1983). Energy transfer in insects. *Annu. Rev. Entomol.* **28**, 455–486.

Wielgus-Serafinska, E. and Kawka, E. (1976). Accumulation and localization of lead in *Eisenia foetida* (Oligochaeta) tissues. *Folia Histochem. Cytochem.* **14**, 315-320.

Wilcke, D. E. (1952). Zur Kenntnis der Lumbricidenfaunen Deutschlands. *Zool. Anz.* **151**, 104-106.

Wilkinson, G. E. (1975). Effect of grass fallow rotations on the infiltration of water into a savanna zone soil of northern Nigeria. *Trop. Agric. (Trinidad)* **52**, 97-103.

Wittich, W. (1953). Untersuchungen über den Verlauf der Streuzersetzung auf einem Boden mit Regenwurmtätigkeit. *Schr. Reihe Forstl. Fak. Univ. Göttingen* **9**, 1-38.

Wolf, A. V. (1938). Notes on the effect of heat in *Lumbricus terrestris* L. *Ecology* **19**, 346-348.

Wolf, A. V. (1940). Paths of water exchange in the earthworm. *Physiol. Zool.* **13**, 294-308.

Wollny, E. (1890). Untersuchungen über die Beeinflussung der Fruchtbarkeit der Ackerkrume durch die Tatigkeit der Regenwürmer. *Forschn Geb. AgrikPhys. Bodenk.* **13**, 381-395.

Wood, T. G. (1974). The distribution of earthworms (Megascolecidae) in relation to soils, vegetation and altitude on the slopes of Mt Kosciusko, Australia. *J. Anim. Ecol.* **43**, 87-106.

Wright, M. A. (1972). Factors governing ingestion by the earthworm *Lumbricus terrestris* with special reference to apple leaves. *Ann. Appl. Biol.* **70**, 175-188.

Wright, M. A. (1977). Effects of benomyl and some other systemic fungicides on earthworms. *Ann. Appl. Biol.* **87**, 520-524.

Yapp, W. B. (1956). Locomotion of worms. *Nature (Lond.)* **177**, 614-616.

Yeates, G. W. (1976). Earthworm population of a pasture spray-irrigated with dairy shed effluent. *N.Z. J. Agric. Res.* **19**, 387-391.

Yeates, G. W. (1978). Populations of nematode genera in soils under pasture. II. Seasonal dynamics in dryland and effluent-irrigated pastures on a central yellow-grey earth. *N.Z. J. Agric. Res.* **21**, 331-340.

Yeates, G. W. (1981). Soil nematode populations depressed in the presence of earthworms. *Pedobiologia* **22**, 191-195.

Zachariae, G. and Ebert, K. H. (1970). Gefaehrdet chemische Schaedlingsbekaempfung im Forst die Regenwurmer? *Pedobiologia* **10**, 407-433.

Zajonc, I. (1970). Dynamique saisonnière des synusies de lombrics (Lumbricidae) vivant dans les prairies de la Slovaquie méridionale; action des engrais azotés sur la composition de celles-ci. *Pedobiologia* **10**, 286-304.

Zajonc, I. (1971a). Participation des lombrics (Lumbricidae) dans la libération des éléments minéraux des feuilles mortes d'une forêt de hêtres et de chênes. *In* "IV Colloquium Pedobiologiae" (J. d'Aguilar, ed.), pp. 387-395. Institut National des Recherches Agriculturelles Publ. 71-7, Paris.

Zajonc, I. (1971b). Synusia analysis of earthworms (Lumbricidae Oligochaeta) in the oak-hornbeam forest in south-west Slovakia. *In* "Productivity of Forest Ecosystems" (P. Duvigneaud, ed.), Proc. Brussels Symposium, 1969, Ecology and Conservation 4, 443-452. United Nations Educational, Scientific and Cultural Organization, Paris.

Zajonc, I. (1971c). La distribution quantitative des lombrics (Lumbricidae Oligochaeta) dans les grands types mondiaux d'écosystèmes forestiers. *In* "Productivity of Forest Ecosystems". (P. Duvigneaud, ed.). Proc. Brussels Symposium, 1969, Ecology and Conservation 4, 453-462. United Nations Educational, Scientific and Cultural Organization, Paris.

Zajonc, I. (1975). Variations in meadow associations of earthworms caused by the influence of nitrogen fertilizers and liquid-manure irrigation. *In* "Progress in Soil Zoology" (J. Vanek, ed.), pp. 497-503. proc. 5th Intl Colloq. Soil Zool., Prague, 1973.

Zicsi, A. (1958a). Einfluss der Trockenheit und der Bodenbearbeitung auf das Leben der Regenwürmer in Ackerboden. *Acta Agron. Acad. Sci. Hung.* **8**, 67-75.

Zicsi, A. (1958b). Determination of number and size of sampling unit for estimating lumbricid populations of arable soils. *In* "Progress in Soil Zoology" (P. W. Murphy, ed.), pp. 68-71. Butterworth, London.

Zicsi, A. (1958c). Freilandsuntersuchungen zur Kenntnis der Empfindlichkeit einiger Lumbriciden-Arten gegen Trockenperioden. *Acta Zool. Acad. Sci. Hung.* **3**, 369-383.

Zicsi, A. (1967). Die Auswirkung von Bodenbearbeitungsverhafen auf Zustand und Besatzdichte von einheimischen Regenwürmen. *In* "Progress in Soil Zoology" (O. Graff and J. E. Satchell, eds), pp. 290-298. North Holland Publishing Company, Amsterdam.

Zicsi, A. (1969). Über die Auswirking der Nachfrucht und Bodenbearbeitung auf die Akitivität der Regenwürmer. *Pedobiologia* **9**, 141-145.

Zicsi, A. (1983). Earthworm ecology in deciduous forests in central and southeast Europe. *In* "Earthworm Ecology" (J. E. Satchell, ed.), pp. 171-177. Chapman and Hall, London.

Zietek, B. and Pytasz, M. (1979). Lead and zinc accumulation in the soil and tissues of the earth-worm (*Lumbricus terrestris*) along the highroads in the province of Katowice. *Acta Physiol. Pol.* **30**, 223.

Zimmermann, P. (1971). Die zentralnervöse Kontrolle der Dehydration bei *Lumbricus terrestris* L. *Z. Zellforsch. Mikrosk. Anat. Abt. Histochem.* **112**, 551-571.

Zimmermann, P. (1973) Zur neuronalen Regulation des Wasserhaushalts bei *Lumbricus terrestris* L. *Z. Zellforsch. Mikrosk. Anat. Abt. Histochem.* **145**, 103-118.

Zrazhevskii, A. I. (1957). Dozdevye cervi kak faktor plodorodija lesnych pocv. Akad. Nauk Ukr. SSR, Kiev.

Zrazhevskii, A. I. (1958). Earthworms and the interrelations of tree and shrub species. *Priroda (Mosc.)* **47**, 96-98.

Author Index

A

Abdel-Fattah, R. F., 231
Abel, R., 38, 57, 63, 92, 95, 96, 99–101, 112, 126, 128–129, 130, 131
Abrahamsen, G., 335, 343, 349
Abrams, B. I., 321
Agarwal, G. S., 199
Aichberger, R. von., 24
Aldag, R., 210, 212, 214, 268
Altwegg, A., 151, 303, 305
Alvarado, R. H., 60, 61
Andersen, C., 230, 231, 236, 238, 239
Anderson, J. M., 18, 31
Andrews, R., 140
Arlidge, E. Z., 303
Arme, C., 29
Arthur, D. R., 178, 226
Ash, C. P. J., 238, 239
Aspöck, H., 303
Atlavinyte, O., 25, 155, 222, 267, 269, 273, 282, 303
Ausubel, S., 165
Avel, M., 74, 86
Axelsson, B., 336

B

Babb, M. R., 315, 328
Baermann, G., 342
Bagdanaviciene, Z., 267, 273
Bagnoli, G., 221
Bahl, K. N., 214, 215, 217, 222
Baker, G. H., 97, 340
Baltzer, R., 38
Banage, W. B., 341
Bano, K., 23, 82, 318
Bardier, M. N., 314
Barker, R. J., 314
Barley, K. P., 23, 90, 91, 182, 183, 188, 204, 207, 212, 214, 219, 223, 250, 257, 261, 282
Barnes, B. T., 284, 341
Bassalik, K., 26, 176

Bates, J. E., 322
Bauer, K., 303, 304
Baweja, K. D., 275
Beadle, L. C., 50–52
Beaugé, C., 183
Belousova, N. K., 282
Bengtson, S.-A., 97, 146, 161, 279
Benham, W. B., 65, 161
Benz, G., 151, 303, 305
Beugnot, M., 338, 347
Bevelander, G., 226
Beven, K., 194
Beyer, W. N., 310, 314
Beynon, K. I., 303, 310
Bhandari, G. S., 211
Bhat, J. V., 24, 275
Bhatnagar, T., 208, 209
Bird, S. J. G., 322
Bishop, S. H., 214
Black, W. M., 303
Blancke, E., 176
Block, W., 341
Bocock, K. L., 221
Bodenheimer, F. S., 41
Bodmer, S., 321
Bogatyrev, I. O., 242
Bolton, P. J., 18, 49, 50, 69, 73, 74, 76, 136, 137, 139, 144, 175, 179
Bornebusch, C. H., 96
Bosse, I., 286
Bottner, P., 198, 203, 260
Bouché, M. B., 9, 14, 18, 20, 39, 52, 57, 58, 59, 60, 63, 67, 68, 70, 72, 85, 86, 87, 90, 96, 107–112, 115, 119, 121, 122, 123, 125, 129, 134, 142–145, 152, 153, 164, 183, 185, 187, 189, 195, 198, 201–202, 208, 219, 220, 231, 248, 278–279, 304, 318, 320, 328, 334, 338, 339, 341, 342, 345, 346, 347, 348, 349
Bradbury, J. F., 221
Braunig, G., 54
Brauns, A., 92

Bremeyer, A., 142
Bretscher, K., 334
Breznak, J. A., 221
Brill, W. J., 221
Brinkhurst, R. O., 64
Brookfield, H. C., 160
Brown, A. W. A., 303, 313
Brown, B. A., 153
Brown, S. M., 303
Brüsewitz, G., 24
Budviciene, L., 267, 273
Bull, K. R., 239
Buntley, G. J., 246, 261
Burnett, A. J., 164
Butler, J. D., 303
Byzova, J. B., 44, 48, 88, 115

Coppel, H. C., 221
Correll, R., 130
Cosgrove, W. B., 48
Cotton, D. C. F., 90, 324, 325, 326
Coughtrey, P. J., 235
Couperus, H., 349
Couture, G., 242
Critchley, B. R., 225, 285, 303, 310, 312
Critchley, U., 225, 285, 303, 310, 312
Crossley, D. A., 242
Cuendet, G., 148
Cuénot, L., 231
Curry, J. P., 90, 324, 325, 326
Czerwinski, Z., 180, 206, 208, 210, 211, 222, 225
Czochanska, Z., 26

C

Cairns, A., 208
Calamante, R., 60
Calow, P., 119
Campbell, C. J., 303
Campbell, J. W., 214
Carley, W. W., 34–35
Carter, A., 189, 235
Carter, G. S., 50
Caseley, J. C., 303
Cernosvitov, L., 64
Chandler, R. F., 339
Chapron, C., 226, 227
Charles, J. C., 205, 261
Chen, C. M., 276
Christensen, M. J., 24
Christensen, R. E., 235
Citernesi, U., 221
Clark, R. B., 67, 85
Cleland, W., 30
Clements, R. O., 191, 303
Cole, C. V., 140, 144
Coleman, D. C., 140, 141, 144
Coles, G. C., 52
Collett, N., 314
Collier, J., 321, 322
Collins, N. M., 93, 167, 170
Combault, A., 226
Connors, W. J., 30, 31
Contreras, E., 27
Cook, A. G., 225, 285, 303, 310, 312
Cook, M. E., 303, 311
Cooke, A., 24, 25

D

Daciulyte, J., 273, 303
Dales, R. P., 151
Dance, K. W., 155
Darwin, C. R., 32, 60, 103, 155, 174, 180, 181, 182, 183, 226, 244, 251, 256, 263, 276, 282
Dash, M. C., 29, 82, 87, 88, 91, 98, 141, 142, 143, 180, 213, 219, 275
Davies, D. M., 179
Davis, B. N. K., 310
Davis, J. B., 50
Day, G. M., 24
Daynard, T. B., 341
Dec, R., 230
Denisova, A. V., 314
Derby, J. M., 325
Dexter, A. R., 197
Dhennin, Léone, 153
Dhennin, Louis, 153
Diaz Cosin, D. J., 97
Dietz, S., 144, 198, 203, 260
Dietz, T. H., 60, 61
Dindal, D. L., 327
Dmitriev, G. A., 314
Doane, C. C., 303
Dobrovol'skii, G. V., 177, 179
Dobzansky, T., 121
Doeksen, J., 85, 349
Dommergues, Y., 207
Donovan, J. F., 322
Dorsett, D. A., 7, 53
Douglas, J. T., 180

Drift, J. van der, 303
Drummond, H., 252
Dubash, P. J., 180, 272
Duff, H. A., 266
Duffill, R. A., 214, 217
Duley, F. L., 176, 186, 189, 206, 283
Dzangaliev, A. D., 282

E

Easton, E. G., 157, 159, 213
Eaton, T. H., 339
Ebert, K. H., 303
Edwards, C. A., 32, 59, 88, 151, 152, 176, 179, 186, 221, 224, 242, 243, 268, 276, 282, 284, 287, 290, 291, 303, 304, 306, 310, 334, 335, 337, 339, 348
Edwards, P. J., 303
Edwards, W. M., 192-194
Ehlers, W., 187, 189, 192-194
Eijsackers, H. J. P., 162, 190, 249, 258
Eitminaviciute, I., 273
Ellenby, C., 275
Ellis, F. B., 284
Ellis, J. E., 341
Emerson, W. W., 178
Emiliani, F., 38, 59, 90, 92
Eno, C. F., 242, 303
Enwezor, W. P., 222
Evans, A. C., 68, 69, 71, 73, 74, 102, 176, 282, 339
Evans, L. T., 90, 265, 291, 324, 344

F

Farrell, D. A., 197
Felber, E., 206, 207
Ferrière, G., 20, 129
Filippi, C., 221
Finck, A., 282
Finlayson, D. G., 303
Finney, J. H., 203, 261
Fleming, T. P., 229
Florkin, M., 214, 217
Forrest, J. A., 197
Fosgate, O. T., 315, 328
Franz, H., 202
French, J. R. J., 221
Friedman, A. A., 315

G

Galluzzi, R., 221
Galvelis, A., 303

Ganong, W. F., 349
Gansen-Semal, P. van, 226
Ganti, S. S., 180, 272
Gardefors, D., 336
Gardner, C. R., 7
Gardner, R. H., 334
Gates, G. E., 14, 73, 87, 149, 155, 156, 159, 160, 175, 185, 246
Gavrilov, K., 164, 270, 273
Gerard, B. M., 68, 71, 83, 288, 291, 338, 346
Germann, P., 194
Gewehr, H., 189
Ghabbour, S. I., 34, 36, 40, 41, 43, 54, 56, 61, 62, 91, 97, 152, 214, 215, 217, 218, 290, 303
Ghilarov, M. S., 44, 289
Giesecke, F., 176
Gilman, A., 303
Gish, C. D., 235, 310, 314
Gissel-Nielsen, G., 238, 240
Gissel-Nielsen, M., 238, 240
Glaessner, M. F., 3
Goh, B. S., 95
Goh, K. M., 203
Goodman, G. T., 239
Gordon, W. T., 257, 261
Goss, M. J., 180
Gotwald, W. W., 150
Graff, O., 8, 41, 42, 46, 71, 102, 182, 183, 210, 211, 212, 214, 222, 268, 272, 276, 282, 286, 316, 319, 330
Grant, J. D., 273-274
Grant, W. C., 36-37, 41, 42, 44, 72
Greacen, E. L., 197, 205, 212
Greenslade, P., 129
Greenslade, P. J. M., 120-123, 124-125, 126, 127, 129
Gregg, P. H., 223
Gregg, R. E. H., 223
Griffiths, D. C., 303
Griffiths, M., 149, 150
Grimaldi, D. A., 165
Grime, J. P., 121
Gruia, L. P., 91, 169
Gruner, B., 50
Guardabassi, A., 226
Guerrero, R. D., 319
Guild, W. J. McL., 68, 69, 71, 73, 74, 96, 102, 189, 204, 282, 339, 346
Gupta, M. L., 222

Guthrie, T. F., 235
Gutiérrez, T., 60

H
Haantjens, H. A., 254–255
Hale, I. M., 322
Hall, F. G., 36
Handley, W. R. C., 32
Handreck, K. A., 4, 5
Hanschko, D., 291
Hansen, R. P., 26
Harding, D. J. L., 221
Hardman, J. A., 310
Harker, W., 254
Harmon, N. P., 346
Harris, R. F., 223
Harrison, D. L., 314
Hart, D., 160
Hartenstein, R., 30, 31, 32, 42, 43, 64, 70, 73, 75, 82, 221, 316, 317, 318, 320, 321, 322
Hay, R. K. M., 291
Heal, O. W., 18, 31
Heath, G. W., 31, 32, 203, 337
Heath, R. G., 313
Heim de Balsac, H., 153
Heimpel, A. M., 150, 305
Heinonen, J., 189
Helmke, P. A., 240, 326
Herrera, R., 97, 112
Hess, W. N., 11, 51
Heungens, A., 203, 276, 303
Hill, D., 180
Hill, E. F., 313
Hoeksema, K. J., 185, 187, 286
Hoffmann, J. A., 275
Hogben, L., 40, 41
Hole, F. E., 179, 203, 245
Hollick, F. S. J., 338
Holter, P., 204
Hoogerkamp, M., 162, 190, 249, 258
Hopkins, A. R., 303
Hopkins, H. T., 282
Hopp, H., 42, 46, 179, 185, 187, 189, 253, 264, 266, 267, 282, 286
Hornor, S. G., 321
Houssiau, M., 325
Howell, C. D., 51
Huang Foo Chen, 322
Hughes, M. K., 228, 236, 238

Hughes, T. D., 303
Huhta, V., 92, 290, 291
Hurwitz, S. H., 58
Hutchinson, S. A., 24, 274
Hyche, L. L., 303

I
Imam, M., 303
Inoue, T., 276
Inskip, M. J., 239
Ireland, M. P., 231, 232, 234–235, 236, 237, 240

J
Jackson, R. B., 305
Jacobson, G. M., 222, 225
Jaenike, J., 165
Jakubczyk, H., 180, 206, 208, 210, 211, 222, 225
Jamieson, B. G. M., 3, 64, 224, 226, 227, 261
Jeanson, C., 177, 197, 198, 208
Jefferies, D. J., 148
Jefferson, P., 179
Jennings, A. C., 207, 212
Jesus, B. J., 97
Jeuniaux, C., 30
Joest, E., 349
Johansen, K., 9, 10, 50
Johnson, M. L., 49
Jones, M. P., 238
Jongerius, A., 185, 187, 286
Jordan, G. A., 148, 164
Josens, G., 141
Joshi, N. V., 176
Joyner, J. W., 346

K
Kale, R. D., 23, 58, 82, 227, 318
Kamel, M., 24, 274
Kanyonyo, J., 28
Kaplan, D. L., 42, 43, 64, 70, 73, 75, 82, 221, 316, 317, 318, 321, 322
Karavyanskii, N. S., 242
Karimullah, 180
Karpachevsky, L. O., 92, 203
Karpinnen, E., 92, 290
Kawka, E., 229, 230
Kelkar, B. V., 176
Kelsey, J. M., 303

Author Index

Kenney, E. A., 235
Keogh, R. G., 24, 220, 303, 311
Khalaf El-Duweini, A., 34, 40, 41, 43, 54, 56, 61, 62, 91, 97, 214, 215, 217, 218, 290
Khambata, S. R., 24, 275
Khan, A. W., 242, 268
Khan, N., 180
King, H. G. C., 31, 32
Kirita, H., 204
Kirk, R. L., 40, 41
Kirk, V. M., 303
Kitazawa, Y., 89, 92
Kleinig, C. R., 257, 261
Kollmannsperger, F., 183, 206
Kondo, K., 276
Korotev, R. L., 240, 326
Kozlovskaya, L. S., 222
Kretzschmar, A., 19, 134, 195-196, 208
Kriel, J. R., 71
Krishnamoorthy, R. V., 23, 58, 82, 227, 318
Krivolutsky, D. A., 241, 243
Krohn, W. B., 148
Kubiena, W. L., 179
Kudrjaseva, I. V., 203
Kuginyte, Z., 155
Kühle, J. C., 288

L

Ladd, J. N., 144
Lakhani, K. H., 74, 76, 83, 132, 141, 220, 341, 348
Lal, R., 190, 222, 285
Lan, H. an der, 303
Langdon, F. E., 11
Langmaid, K. K., 248
Laughlin, C. W., 303, 306
Laursen, J., 230, 231
Lautour, M.-L. de, 90
Lavelle, P., 15, 20, 21, 22, 27, 28, 29, 38, 70, 76, 77-81, 83-84, 87, 88, 91, 93, 94, 95, 97, 108, 112, 113, 116-119, 121, 122, 123, 124-125, 129, 130, 132-133, 136, 137, 139, 141, 150, 179, 180, 183, 203, 210, 336, 341, 348, 349
Laverack, M. S., 30, 51, 59, 63, 213, 214, 226
Lawrence, R. D., 219, 328
Lebrun, Ph., 325

Lee, D. L., 238, 239
Lee, K. E., 8, 19, 54, 57, 59, 60, 65, 66, 72, 91, 95, 98, 101, 102, 103-107, 108, 109, 110, 111, 112, 116, 119, 122, 123, 129, 130, 144, 145, 146, 156, 160, 161, 162, 166, 167, 169, 174, 175, 179, 180, 185, 210, 211, 219, 221, 245, 250, 254, 261, 280, 281, 289
Legg, D. C., 303
Leitenberger, L., 202
Lemasson-Florenville, M., 325
Lepidi, A. A., 221
Lepp, N. W., 228, 236, 238
Lesser, E. J., 30
Lewis, E. M., 49
Liebmann, E., 231
Liepenis, A., 273
Linder, P. J., 42, 46, 286
Littleton, I., 329
Liu, C. L., 276
Livingstone, D., 322
Ljungström, P.-O., 9, 38, 60, 67, 83, 87, 90, 91, 93, 98, 159, 164, 180, 183, 253, 281, 335
Loehr, R. C., 233
Lofty, J. R., 59, 88, 151, 152, 176, 186, 224, 243, 268, 276, 282, 284, 287, 288, 289, 303, 304, 306, 310, 334, 339
Lohm, U., 92, 110, 112, 141, 334, 336, 348
Loquet, M., 206, 208
Low, A. J., 282
Lowe, D. G., 31, 32
Lugauskas, A., 273, 303
Lunt, H. A., 222, 225
Luxton, M. S., 24
Lyons, C. H., 303

M

Ma, Wei-Chun, 237
MacArthur, R. H., 94, 119-120
McCalla, T. M., 176, 186, 189, 206, 283
McColl, H. P., 15, 90
Macdonald, D. W., 147, 149
Macfadyen, A., 244, 315, 349
McInnes, D. C., 214, 217
McIntyre, A. D., 229
Mackay, A. D., 223
McRill, M., 277
Madge, D. S., 37, 40, 42, 44, 54, 59, 81, 87, 93, 98, 149, 174, 179, 180, 183, 204, 341, 348

Makeschin, F., 272
Maldague, M., 49, 93, 203, 242, 263
Malecki, M. R., 42, 64, 233, 316, 317, 318
Malone, C. E., 203
Mamajev, B. M., 289
Mangum, C. P., 45
Marialigeti, K., 26
Marshall, V. G., 270
Martin, A. W., 9, 10, 50
Martin, J. P., 178
Martin, K., 000
Martin, L. W., 303
Martin, M. H., 235
Martin, N. A., 14, 86, 205, 261, 303, 338, 341
Maskina, M. S., 211
May, R. M., 95, 99
Mazanceva, G. P., 86, 348
Mazaud, D., 39, 346
Mba, C. C., 288
M'Dowall, J., 226
Meijer, J., 156
Meinhardt, U., 345
Merker, E., 54
Mertins, J. W., 221
Messenger, A. S., 226
Meyer, J. A., 132-133, 139
Meyer, L., 178
Michaelsen, W., 226
Michel, C., 30
Michon, J., 74, 85
Mielzyńska, I., 230
Mikhaltsova, Z., 243
Miles, H. B., 25, 41, 42
Mill, P. J., 11, 51
Millar, H. R., 219, 328
Miller, R. B., 161, 280
Mills, C. F., 229
Mishra, C. C., 29, 275
Mitchell, M. J., 133-134, 153, 320, 321
Miyasaka, M., 174
Moeed, A., 148
Mohammad, Ashraf, 180
Mohr, G., 329
Moore, A. W., 222
Moore, R. L., 303, 312
Moreau, J.-P., 327
Moreno, A. G., 97
Mountford, M. D., 99
Müller, P. E., 18, 245
Murchie, W. R., 75

Muskett, C. J., 238
Musso, H., 30

N
Nakahara, H., 226
Nakamura, Y., 23, 86, 98, 327
Nakatsugawa, T., 309
Nathans, S., 91, 177
Neal, E. G., 148
Needham, A. E., 198, 202, 214, 215, 216, 217, 219, 220, 226
Neely, D., 303
Negi, L. S., 199
Neglia, R., 221
Nelson, J. M., 336
Nelson, P. A., 309
Németh, A., 94, 97, 112
Neuhauser, E. F., 30, 31, 42, 43, 64, 70, 73, 75, 82, 233, 316, 317, 318, 322
Newell, P. F., 221
Nielsen, C. O., 30, 264-265
Nielsen, G. A., 179, 203, 245
Nielson, R. L., 257, 270
Nilsson, A., 97, 146, 161, 279
Nishioka, M., 204
Noble, J. C., 257, 261
Nomura, S., 63, 64
Nordström, S., 39, 41, 46, 55, 86, 90, 96, 97, 99, 101, 112, 146, 161, 185, 279, 341, 348
Nowak, E., 69, 70, 72, 74, 96, 141, 180, 206, 208, 210, 211, 220, 222, 225, 279
Nurminen, M., 92, 290
Nuti, M. P., 221
Nye, P. H., 174, 175, 180, 183, 208, 225

O
Oglesby, L. C., 33, 60, 86
Olive, P. J. W., 67, 85
Oliver, J. H., 151
Omodeo, P., 158, 164, 165
Orellana, J. A. de, 38
Otanes, F. G., 276

P
Papendick, R. J., 246, 261
Parle, J. N., 24, 26, 27, 178, 206, 207, 211, 214
Patel, H. K., 276
Patel, R. M., 276
Pathak, A. N., 176

Patra, U. C., 88, 91, 141, 142, 180, 213, 219
Pawlowski, P., 321
Payarskaite, A. I., 155
Peredel'skii, A. A., 242
Perel', T. S., 17, 114–115, 129, 203
Perfect, T. J., 225, 285, 303, 310, 312
Persson, T., 92, 110, 112, 141, 334, 336, 348
Petal, J., 180, 222
Petersen, C. E., 136
Peterson, J. A., 194
Pew, J. C., 30
Phillips, M. J., 25, 29, 275
Phillipson, J., 18, 38, 49, 50, 57, 63, 69, 73, 76, 92, 95, 96, 99–101, 112, 126, 128–129, 130, 131, 144, 175, 178
Phipps, D. A., 228, 236, 238
Pianka, E. R., 119, 121, 130, 131
Pickford, G. E., 253
Piearce, B., 63
Piearce, T. G., 19, 25, 29, 63, 206, 227, 275, 348
Pileckis, S., 155
Ploeg, R. R. van der, 192–194
Pociene, C., 25
Pohlen, I. J., 11, 12
Poinar, G. O., 151–152
Pokarzhevsky, A. D., 110, 241
Ponomareva, S. I., 177, 178, 179, 183, 226, 267, 282
Poryadkova, N. A., 242
Powlson, D. S., 204
Preiss, W. V., 3
Prentø, P., 226, 227
Priano, L., 38, 60
Prince, A. B., 322
Puh, Y. Chiung, 222
Purdy, L. H., 275
Purrini, K., 151
Puttarudrian, M., 276
Pytasz, M., 238

R

Raffy, A., 49
Ralph, C. L., 88
Ranc, D., 39
Randell, R., 303
Randhawa, N. S., 211
Rangel, P., 28
Rao, B. R. C., 275
Rao, K. P., 45
Rao, K. S. K., 199
Raw, F., 203, 221, 246, 303, 335, 336, 337, 338, 340
Reddy, M. V., 155
Reichle, D. E., 131–132, 145, 203, 242, 263
Reid, C. C. P., 140, 144
Reinecke, A. J., 9, 37, 38, 40, 41, 42, 45, 46, 71, 73, 83, 86, 90, 98, 164, 180, 183, 253, 281, 335
Reynolds, J. W., 73, 91, 92, 97, 148, 159, 164, 336, 341, 344, 348
Reynoldson, T. B., 90, 91, 96
Rhee, J. A. van, 83, 91, 162, 177, 188, 202, 232, 239, 249, 258, 261, 267, 269, 303, 326
Richards, K. S., 4, 29, 35, 61, 229, 231, 232, 236
Richards, L. A., 192, 198
Richardson, H. C., 290
Richter, K. O., 346
Ricou, G., 143
Rigby, B. J., 40
Robarge, W. P., 240, 326
Roberts, H. A., 303
Roberts, R. D., 239
Robertson, J. D., 226
Rodionova, L. Z., 242
Rogaar, H., 162, 190, 249, 258
Rogers, C. G., 49
Roots, B. I., 65
Rose, C. J., 149
Ross, D. J., 208
Rouelle, J., 26
Rovira, A. D., 205
Roy, S. K., 180, 183
Rożen, A., 101
Rudge, M. R., 148
Rundgren, S., 39, 46, 53, 90, 96, 97, 99, 101, 110, 112, 146, 161, 185, 279, 335, 341, 348
Ruppell, R. F., 303, 306
Ruschmann, G., 178, 206
Russell-Smith, A., 303, 312
Ryden, J. C., 223
Ryke, P. A. J., 37, 38, 42, 180, 183
Rysavy, B., 152

S

Sabine, J. R., 219, 316, 328, 329
Sagar, G. R., 277

Sakal, R., 222
Salisbury, E. J., 224
Salt, G., 338
Saroja, K., 50
Sasson, A., 140, 141
Sastry, K. S. S., 276
Satchell, J. E., 18, 26, 27, 31, 32, 41, 48, 50, 56, 59, 68, 71, 74, 76, 83, 87, 102, 121, 122, 132, 141, 143, 146, 185, 202, 219, 220, 221, 244, 245, 246, 291, 319, 321, 322, 323, 325, 329, 330, 336, 340, 341, 342, 343, 345
Saussey, M., 85
Savigny, J. C., 000
Schaefer, R., 20, 21, 22, 27, 130, 131
Scharpenseel, H. W., 189
Schmidt, P., 36, 42
Schomberg, P. J., 240, 326
Schread, J. C., 303
Schultz, W., 206, 207
Schwartz, J. B., 48
Schwert, D. P., 155, 156, 158, 327
Sears, P. D., 90, 265, 291, 324
Semenova, L. M., 114
Senapati, B. K., 29, 82, 87, 98, 275
Sergienko, M. I., 169
Seritti, A., 221
Seymour, M., 4, 7, 8, 184
Shain, S. S., 242
Shakir, S. H., 56
Sharpley, A. N., 175, 179, 183, 209, 210, 211, 222, 223, 252
Shields, R. F., 317
Shilova, S. A., 314
Shiraishi, K., 50
Shrikhande, J. C., 176
Shutler, R., 160
Siljak, D. D., 95
Sims, R. W., 68, 157, 213
Singh, J. S., 140
Sison, P. L., 276
Skoczen, S., 148
Slater, C. S., 189, 264, 266, 267, 282, 286
Slater, W. K., 50
Slepetine, J., 273
Smirnoff, W. A., 150, 305
Smith, A. C., 42
Smith, B. J., 194
Smith, M. J., 303, 312
Smith, W. W., 157

Sokolov, D. F., 203
Solem, A., 160
Southwood, T. R. E., 95, 122, 123, 124–125, 126, 127
Sow, B., 20, 21, 22, 130, 131
Spann, J. W., 313
Spencer, W. B., 8
Springett, J. A., 209, 210, 211, 223, 252, 266, 271
Standen, V., 96
Stark, N., 321
Steel, J., 38, 57, 63, 92, 95, 96, 99–101, 112, 126, 128–129, 130, 131
Stenersen, J., 303, 307, 308–309
Stephenson, J., 6, 49, 50, 72, 73, 230, 270, 275, 276
Stockdill, S. M. J., 163, 188, 189, 190, 248, 250, 257, 261, 262, 265, 312
Stöckli, A., 27, 186
Stolte, H. A., 51, 114, 115
Støp-Bowitz, C., 158
Stout, D. J., 161, 203, 280
Strazdiene, V., 273
St Remy, E. A., de, 341
Stringer, A., 303
Studdard, R. A., 221
Sugi, Y., 81, 87, 204, 345, 348
Sukackiene, I., 273
Summers, L. A., 303
Svendsen, J. A., 90, 335, 339
Swaby, R. J., 177, 211, 269
Swait, A. A. J., 303, 311
Swift, M. J., 18, 31
Syers, J. K., 175, 179, 183, 209, 210, 211, 222, 223, 252, 266, 271

T

Takeuchi, N., 66
Tanaka, M., 81, 87, 204, 345, 348
Tara, B., 230
Taschenberg, E. W., 30
Taylor, N. H., 11, 12, 103, 106
Tenow, O., 336
Teotia, S. P., 176, 186, 189, 206, 283
Thambi, A. V., 141
Theoret, L., 327
Thompson, A., 303, 310
Thomson, A. J., 179
Thornton, M. L., 274
Tichomirova, A. L., 243

Tillinghast, E. K., 213, 214, 217
Timmenga, H., 235
Tisdall, J. M., 91, 98, 190, 286, 289, 303
Titisheva, N. G., 110
Titkova, N. F., 177, 179
Tiurin, I. V., 203
Tomlin, A. D., 329
Tréhen, P., 248
Trifonov, D., 276
Trueman, E. R., 4, 8
Tsukamoto, J., 72, 73, 75, 316
Tsuru, S., 98
Turcaninova, V. A., 243
Turner, G. L., 221

U

Ustinov, I. P., 92
Usuki, I., 63, 64

V

Vail, V. A., 73, 316
Valiachmedov, V., 292
Valpas, A., 92, 290
Vanagas, J., 222, 269
Vannier, G., 13
Vardanis, A., 303
Verge, J., 159
Villez, E. J. de, 30
Vimmerstedt, J. P., 203, 261
Visser, F. A., 73
Visser, S. A., 52, 210, 225, 254
Voronova, L. D., 314
Vries, J. de, 189

W

Waksman, S. A., 178
Walsh, A. S., 73
Walter, H. R., 3
Walton, W. R., 151, 303
Wasawo, D. P. S., 52, 210, 225, 254
Watanabe, H., 72, 73, 75, 81, 87, 175, 180, 183, 210, 225, 316
Waters, R. A. S., 90, 203, 256, 265
Watkin, B. R., 266, 281

Wauthy, G., 325
Weber, R. E., 49, 50, 52
Westeringh, W. van de, 179, 187, 188, 249, 269, 303
Wheatley, G. A., 310
Wheeler, J. L., 266, 281
White, G., 276
Whitehead, P. H., 303, 311
Whiting, A. E., 337
Whittaker, R. H., 120, 126
Wiececk, C. S., 226
Wiegert, R. G., 136
Wielgus-Serafinska, E., 229, 230
Wiggans, S. C., 303
Wilcke, D. E., 75
Wilkinson, G. E., 190, 191, 285
Williams, J. D., 313
Wilson, E. O., 119–120
Wittich, W., 247
Wolf, A. V., 33, 41, 214
Wollny, E., 185, 266
Wood, A. W., 18, 149
Wood, T. G., 93, 98, 166, 168, 175, 180, 281
Woodell, S. R. J., 38, 57, 63, 92, 95, 96, 99–101, 112, 126, 128–129, 130, 131
Wooton, R. J., 237
Wright, M. A., 25, 32, 303, 307, 311, 312

Y

Yapp, W. B., 7
Yeadon, R., 225, 303, 310, 312
Yeates, G. W., 29, 275, 323, 324

Z

Zachariae, G., 303
Zaidi, Z., 27
Zajonc, I., 92, 96, 203, 222, 291
Zebe, E., 52
Zhdannikova, E. N., 222
Zicsi, A., 22, 38, 282, 283, 285, 335
Zietek, B., 238
Zimmerman, P., 35
Zrazhevskii, A. I., 270

Subject Index

A

Abundance
  altitudinal effects, 166-169
  effects of cultivation, 282-286
  effects of irrigation, 288-289, 323-325
  effects of mulching, 286-288
  introduced species, 161-163, 188-189, 257, 258
  methods of estimation, 333-347
  relation to pedogenesis, 246-247
  relation to soil variability, 101
  in various habitats, 89-93
Acanthomoeba triangularis, 26
Acaricides
  effects on earthworms, 304-305
  see also Biocides
Aeration of soils, 185, 188, 195-196
Aestivation, *see* Seasonal rhythms of activity
Agastrodrilus dominicae, 29, 150
Agastrodrilus multivesiculatus, 29, 150
Agastrodrilus opisthogynus, 29, 76, 114, 116-119, 123, 126, 150
Agastrodrilus spp., 336
Agathis australis, 101, 245
Age structure of populations, 79, 81, 83-84
Agrostis tenuis, 264, 277
Algae, as food for earthworms, 23
Allez-les-Vers model, 132-133
Allolobophora bartolii, 115
Allolobophora chlorotica, 19, 30, 36, 41, 63, 65, 68, 69, 71, 83, 86, 100, 115, 126-129, 151, 162, 164, 236, 238, 239, 243, 249, 260, 266, 268, 284, 287, 289, 290, 307, 309, 310, 324, 336
Allolobophora eiseni, 59, 227, 247, 248, 342
Allolobophora parva, 115
Allolobophora sturanyi, 115
Alma emini, 50-53, 254
Alma nilotica, 43
Alma stuhlmanni, 36, 52, 61-62, 254, 290
Alternaria solani, 25
Altitudinal zonation, 166-169
Ambient temperature, 39, 40
Amynthas agrestis, 53
Amynthas alexandri, 155
Amynthas asiaticus, 319
Amynthas communissimus, 63-64
Amynthas corticus, 149
Amynthas diffringens, 344
Amynthas hawayanus, 164, 318, 323
Amynthas hilgendorfi, 23, 164
Amynthas hupeiensis, 30, 36-37, 42, 44, 46, 81, 87
Amynthas megascolidioides, 40
Amynthas rodericensis, 318, 323
Amynthas sieboldi, 81-82, 87, 345
Anas platyrhynchos, 314
Anatomy of earthworms, 4-6
Andropogon gayanus, 190
Anéciques
  absence in tropical soils, 112
  casts, 201
  characteristics, 107-109, 201
  C:N ratios, optimum, 57
  diapause, 87
  effects of land clearance on, 278-279
  food preferences, 20
  soil moisture preferences, 39
  see also Topsoil species
Aphanascus oryzivorus, 276
Aphodius rufipes, 204
Aporrectodea caliginosa, 14, 15, 19, 20, 23, 26, 27, 28, 34, 35, 36, 37, 38, 40, 41, 47, 48, 53, 54, 55, 57, 58, 61-62, 63, 68, 70, 72, 73, 74, 83, 86-87, 88, 95, 99, 100, 115, 126-129, 141, 147, 161, 162, 163, 164-165, 176, 188, 190, 197, 203, 204, 205, 206, 207, 208-209, 212, 214, 215, 217, 219, 222, 227, 232, 236, 237, 238, 240, 248, 249, 250, 252, 257,

400 Subject Index

Aporrectodea caliginosa—continued
258, 260, 261, 264, 265, 266, 268, 269, 270, 271, 272, 273, 279, 280, 281, 284, 286, 287, 288, 289, 290, 291, 307, 309, 310, 313, 323, 326, 346, 348
Aporrectodea giardi, 248
Aporrectodea hortensis, 242
Aporrectodea icterica, 197, 219
Aporrectodea longa, 19, 23, 27, 41, 42, 55, 58–59, 68, 72, 75–76, 115, 126–129, 143, 162, 164, 176, 177, 190, 204, 206–207, 219, 230, 238, 258, 267, 271, 273, 274, 275, 281, 284, 287, 288, 290, 310, 323, 328
Aporrectodea rosea, 18, 19, 30, 38, 40, 44–45, 46, 49, 50, 55, 57, 58, 63, 69, 70, 73, 74, 83, 86–87, 95, 99, 100, 126–129, 136–140, 142–143, 165, 174, 190, 219, 236, 238, 249, 250, 257, 260, 261, 264, 273, 281, 284, 287, 289, 310, 326, 340, 342, 343, 346, 348
Aporrectodea velox, 61, 231
Aquatic habitats, 33, 64–66
Arable lands, species associations in, 94
Arboreal habitats, 107, 169
Ascaridia galli, 152
Ascaris suum, 152
A-Selection, *see r-, K-, A*-Selection strategies
Asplenium spp., 107
Assimilation
 definition, 136
 efficiency, 138–139
 rates, 138, 142–145
 in temperate ecosystems, 141
 termites compared with earthworms, 141
 in tropical ecosystems, 141–142
 use in demographic profiles, 116–119
Astelia spp., 107

B

Bacillus cereus, 24, 27
Bacillus thuringiensis, 150–151, 305
Bacterial gums in casts, 177, 211
Bark of trees, as earthworm habitat, 70, 103
Bimastos antiquus michalisi, 67
Bimastos samerigera, 41, 46
Bimastos tumidus, 73
Bimastos zeteki, 75
Biocides, effects on earthworms

 accumulation, 310
 application methods, 306–307
 behaviour, 310–312
 contamination of earthworm protein, 330
 dispersal, 312–313
 interspecific variability, 307–309
 mechanisms of action, 307–310
 method of assessment, 304
 toxicity, 292–307
 transfer along food chains, 313–314
Biomass
 altitudinal effects on, 166–169
 energy content, 137, 142, 143
 methods of estimation, 347–349
 nitrogen from, 218–219
 production, 142, 316–319
 use in demographic profiles, 116–119
 in various habitats, 89–93, 135–146 *passim*
Birds
 dispersing earthworms, 156
 as predators, 147–148
Blood
 changes in response to low temperatures, 44, 45
 circulation, 9–10
 pH regulation, 214, 226
 respiratory pigment, 47
 see also Erythrocruorin
Body temperature, 39, 40
Body wall
 modification for caudal respiration, 50–52
 role in respiration, 47
 structure, 4–6
Brachypodium phoenicoides, 203, 260
Brown bodies, 230
Bulk density, effects of earthworms on, 188, 191
Burrows
 as air-filled soil pores, 195–196
 classification, 14, 182
 construction, 7, 184, 196–199
 diameter, 185
 ecological significance, 13–15
 energy use in construction, 139, 198
 frequency, 187, 189
 geometry, 173, 184, 185, 188, 189
 hydraulic significance, 188–194, 195
 length related to food supply, 14–15
 linings, 196–199, 208–209

Subject Index **401**

pedogenetic significance, 245–249
as proportion of total soil pores, 185–187
relation with ecological classification, 14
significance for plant growth, 184
volume, 185–186

C

Calciferous glands
  description, 224
  function, 50, 60–61, 139, 226–228
  heavy metal excretion, 236
Calluna vulgaris, 56–57, 246–247, 248
Calorific value
  of food, 21
  of plant litter, 21
  of soil, 21
Carbon, 56–58
  in casts, 209–211
  flow in ecosystems, 131–132, 142–145
  isotopes, use as tracers, 143–145, 198, 203, 260–261
  limiting earthworm populations, 56, 166–167
  as sole source of energy, 143
Carbonates in soils
  derived from calciferous glands, 226–227
  effects on earthworms, 63
Carbon dioxide
  effect of concentration on respiration, 50
  role of calciferous glands in transport, 50, 139
  transport system, 50
Carbon:nitrogen (C:N) ratio, 56–58
  of casts, 209–211
  as indication of food quality, 56–57, 145–146
  maximum and minimum limits, 57–58
  optimum for earthworms, 57–58
  relationship with fulvic acid:humic acid ratio, 58
Carica papaya, 275
Casts
  burying stones, 180–181
  chemical composition, 205, 209–211, 212–213, 222–223, 224
  damaging plants, 179, 304
  dimensions, 174–175
  erosion of, 251–253
  microbial decomposition in, 206–208
  microrelief associated with, 253–255
  morphology, 174–175

  nitrogen in, 212–213
  as nuisance, 179, 304
  organic matter in, 205–206
  pedogenetic effects, 246, 251–253
  periodicity of deposition, 88, 181–182
  phosphatase activity in, 28
  physical composition, 173–178
  physical stability, 177–179
  as plant-potting medium, 315–316, 320, 321
  quantities and rates of deposition, 179–182, 253–254
  soil horizons of origin, 175, 181
  subsurface deposition, 173
  surface deposition, 173, 179–182
Cellulose, digestion of, 30
Cernosvitova biserialis, 115
Chaetomium globosum, 25
Chemical repellents, 339–342
Chemoreceptors, 11
Chloragogen cells, 230–232
Cholinesterase, 307–309
Chuniodrilus zielae, 76, 112, 116–119, 123, 126, 130–131
Circadian rhythms, 88
Cladosporium cladosporioides, 25
Clostridium beijerinckii, 221
Clostridium butyricum, 221
Clostridium paraputrificum, 221
"Cocoonization", 72
Cocoons
  hatching rates, 72–73, 79, 82, 83
  juvenile weight related to cocoon weight, 73
  numbers of ova per cocoon, 72–73
  production rates, 68–71, 78, 79, 80, 83, 116, 316–319
  seasonality of production, 69, 71, 76, 77, 81–82, 84
  size, dependence on adult size, 70, 73
  see also Hatching
Cole's C_{AB} index, 95
Colinus virginianus, 313
Collagen
  of cuticle, 35
  interspecific differences, 35, 40, 44
Colpydium campylum, 25–26
Comminution of soil materials, 173, 176
Communities, earthworm, *see* Species associations
Compaction, influence of earthworms on, 8, 182, 196–197

Compost
  effect on earthworm abundance, 248
  production, 315-323
Consumption
  definition, 135
  rates, 136-138
  in temperate ecosystems, 141
  in tropical ecosystems, 141-142
Corticoles, 107
Costelytra zealandica, 262, 312
Coturnix coturnix, 313
Criodrilus lacuum, 52, 164
Criodrilus sp., 276
Cropping cycles, effects on earthworms, 190
Crops
  mineral uptake, related to earthworms, 269
  production increase due to earthworms, 266-268
  protein content, increase due to earthworms, 268-269
  tillage practices, effects on earthworms, 186-187, 190
  see also Tillage practices
Cultivation, 282-286
  see also Tillage practices
Cuticle
  composition, 35
  control of water loss by, 37
  modifications for respiration, 47
  relation to locomotion, 7-8
  structure, 4, 35
  see also Collagen
Cyperus haspan, 254

D

Dactylis glomerata, 274, 277
Danthonia sp., 264
Dead tissue
  heavy metals in, 234-235
  nitrogen in, 211, 217-219
Decachaetus violaceus, 65
"Decomposer industry", 244-255
Deinodrilus agilis, 281
Demographic indices, 116-119
Demographic profiles, 81, 116-119
Dendrobaena attemsi, 115
Dendrobaena hortensis, 115
Dendrobaena octaedra, 30, 48-49, 57, 58, 59, 94, 115, 164, 247, 248, 279, 340

Dendrobaena veneta, 28, 236
Dendrodrilus rubidus, 19, 42, 45, 49, 59, 63, 68, 73, 86, 94, 99-100, 115, 126-129, 139, 158, 161, 165, 227, 234-235, 236, 237, 240, 271, 280, 324
Deschampsia flexuosa, 342
Desiccation, see Water loss by earthworms
Detritivorous species
  casts, 206
  demographic profiles, 118-119
  food, 18-22
  quantities of litter ingested, 118-119
Diapause, see Seasonal rhythms of activity
Dichogaster agilis, 21, 76, 112, 116-119, 123, 126, 130-131
Dichogaster bolaui, 157, 160, 164
Dichogaster saliens, 157
Dichogaster terrae-nigrae, 20-21, 22, 76, 114, 116-119, 123, 126, 130-131
Digaster longmani, 40
Digestion
  of cellulose, 30
  digestive capabilities, 30-31, 209-211
  enzymes, 30-31
  of lignin, 30-31
  of phenolic compounds, 30-32
"Digging muscles", 108
Dioctophyme sp., 152
Diplocardia floridana, 344
Diplocardia mississipiensis, 344
Diplocardia sp., 266
Diporochaeta aquatica, 65
Diporochaeta chathamensis, 65
Diporochaeta minima, 94
Diporochaeta obtusa, 281
Diporochaeta punctata, 94
Dispersal
  active, 154-155
  passive, 155-156
  of peregrine species, 157-160
  rates, 161-163
Diversity, 94-102
Dorylus (Anomma) gerstaeckeri, 150
Dorylus (Anomma) nigricans, 150
Drawida bahamensis, 157
Drawida calebi, 82-83
Drawida willsi, 82, 87
Drilosphere, 208-209, 213
Drought, effect on earthworms, 250-251
Dung
  as fertilizer, 291-292

as food for earthworms, 23, 288
nutritional value, 23
rates of incorporation, 204-205
toxicity, 23
use of earthworms for disposal, 23
Dung beetles, compared with earthworms, 204-205

E

Echinochloa pyramidalis, 254
Ecological energetics, 135-146
　of *Apporectodea rosea*, 136-140
　definition, 135-136
　of a megascolecid-eudrilid species association, 136-140
　use in demographic profiles, 116-119
　see also Assimilation; Biomass; Carbon; Consumption; Egestion; Energy flow; Production; Respiration
Ecological plasticity, 130-131
Ecological strategies, 102-131
　comparison of classifications, 109-119
　relation to evolution, 110-112, 114
Egestion
　definition, 136
　rates, 136-139
Eisenia fetida, 23, 25-26, 27-28, 29, 30, 31, 32, 35, 42, 45, 48, 50, 58, 63-64, 70-71, 72, 73, 75, 76, 82, 133-134, 153, 161, 164, 198, 214, 215, 217, 219, 221, 233, 242, 268-269, 270-271, 272, 304, 309, 316, 317, 318, 319, 320, 321, 322, 323, 326, 328, 329, 330
Eisenia japonica, 23
Eisenia lucens, 26-27, 115
Eisenia nordenskioldi, 44, 115
Eiseniella tetraedra, 48, 49, 64, 73, 82, 151, 164, 236, 248, 336
Electrical sampling method, 343-344
Electrolyte concentrations
　regulation, 60-61
　tolerance of, in soils, 61-63
　see also Osmotic pressure of body fluids; pH; Salinity
Enchytraeus albidus, 29
Endogées
　casts, 201
　characteristics, 107-109, 201
　dispersal, 154
　effects of land clearance on, 279
　food preferences, 20

longevity, 81
mortality, 81
optimum C:N ratios, 57
soil moisture preferences, 39
see also Subsoil species
Energy flow
　contribution of earthworms in ecosystems, 135-146 *passim*
　pathways, 142-145
Environmental requirements, classification, 15-18
Enzymes, 27-28, 30-31, 208-209, 214
Eodrilus pallidus, 161, 280
Eodrilus paludosus, 65
Eophila savignyi, 54
Epidermis, 4, 5
　mucus gland cells of, 35
　role in respiration, 47
Epiendogées, 108
Epigées
　characteristics, 107-109, 201
　cocoonization, 72
　dispersal, 154
　effects of land clearance on, 278-279
　faeces, 201
　food preferences, 20
　longevity, 81
　mortality, 81
　optimum C:N ratios, 57
　see also Litter species
Erica cinerea, 248
Erosion, effects of earthworms, 251-253, 253-255
Erythrocruorin, 47-48
　as energy source, 88
　see also Respiration
Escherichia coli, 24
Eudrilus eugeniae, 37, 40, 43, 44, 54-55, 59, 82, 157, 174, 242, 285, 288, 310, 312, 315, 323, 330
Eukerria halophila, 60
Eukerria sp., 164
Euryhaline species
　dispersal, 65-66
　habitats, 65
　of oceanic islands, 60
　osmoregulation, 66
Eutyphoeus incommodus, 242
Eutyphoeus waltoni, 211
Exchangeable cations, 224

F

Fertility, *see* Soil fertility
Fertilizers
  effects on earthworm populations, 289–292
  incorporation by earthworms, 261
  organic wastes, 315–327
  vermicompost, 320
Fire, effect on earthworms, 161–162
Flotation, as sampling method, 338–339
Food
  calorific value, 21, 136
  chemical composition, 20–21
  C:N ratio, 56–57
  detritivorous species, 18–22
  geophagous species, 18–22, 136
  integumentary uptake, 29
  mor, 17–18
  mull, 17–18
  palatability, 31–32
  quality
    effect on cocoon production, 70, 82
    effect on growth rate, 75
  quantities consumed, 20–22, 136
  relation to ecological strategies, 102–103 *passim*
  selection from soil, 18–20, 212
  toxicity, 31–32
  types, 19–29, 275
Fumigants, effects on earthworms, 305
  see also Biocides
Fungi
  as food for earthworms, 19, 23–26, 31
  palatability, 25
  stabilizing casts, 178
Fungicides, effects on earthworms, 187, 305, 307, 311–312
  see also Biocides
Fusarium oxysporum, 25
Fusarium sp., 275

G

Geophagous species
  casts, 173, 206
  chemical analyses of food, 21
  demographic profiles, 118–119
  ecological energetics, 136–140
  ecological strategies, 112
  food, 20–22
  quantity of soil ingested, 118–119

Glaciation, survival of earthworms, 158–159
Glossoscolex giganteus, 8–9, 10, 50
Glycogen
  as energy source when inactive, 88
  storage, 231
  use in anaerobic respiration, 52–53
Glyphidrilus sp., 52, 254
Growth
  effect of food quality, 75
  effect of temperature, 75
  of endogées, 81
  of épigées, 81
  life stages, 74, 78, 79, 81
  rates, 74–75, 78, 79, 116
  seasonal variation, 81

H

Habitat templet, 123–129
  definition, 123–124
  relation to r-, K-, A-selection, 126–127
  relation to species associations, 126–129
Hand sorting methods, 334–337
Hatching
  effect of cocoon size, 73
  effect of temperature, 73, 82–83
  numbers of juveniles per cocoon, 72–73
Heat extraction, as sampling method, 342–343
Heath lands
  soil modification, 246–247
  species associations in, 94
Heavy metals
  absorption, 228
  concentration in earthworms, 234–239
  contamination of earthworm protein, 330
  contamination of vermicompost, 322
  definition, 228
  earthworm analysis to monitor, 240–241
  excretion, 229–231
  immobilization, 230–232
  from leaded fuels, 238
  toxicity, 232–233, 325–326
  transfer in food chains, 240
  in waste water, 325–326
Helodrilus antipae, 251, 288
Herbicides, effects on earthworms, 306
  see also Biocides
Heterodera rostochiensis, 275
Hibernation, *see* Seasonal rhythms of activity

Hippopera nigeriae, 174
Histiostoma murchiei, 151
Histomonas sp., 153
Hoplochaetina spp., 109
Humic materials
  in casts, 206
  as food for earthworms, 19
  formation by earthworms, 31
  fulvic acid:humic acid ratio, 21, 58
Humus feeders, 114-115
  see also Detritivorous species;
  Geophagous species
Humus formers, 114-115
  see also Detritivorous species;
  Geophagous species
Hyperiodrilus africanus, 37, 40, 43, 44, 45, 59, 81, 87, 174, 204, 285
Hyperiodrilus spp., 310, 312
Hypoendogées, 108

I

Incubation time
  delayed hatching, 72
  range in Lumbricidae, 71-72
Infiltration, *see* Water infiltration
Insecticides
  effects on earthworms, 191, 304-305, 312
  incorporation by earthworms, 261-262, 312-313
  see also Biocides
Intestinal transit time, 109
Intestine, form related to morpho-ecological groups, 114-115
Introduction of earthworms
  effects on plant production, 264-266
  pedogenetic effects, 247-250
  peregrine species, 158-160
  rates of spread, 161-163, 257, 258, 265-266
  root mats, incorporation, 257-261
  selection of species, 250-251, 258
Ipomoea batatis, 149
Iron-rich materials, lining burrows, 198-199
Irrigation, 288-289, 323-327

K

Kritodrilus calarensis, 115
K-Selection, *see* r-, K-, A-Selection strategies

L

Lampito mauritii, 29, 50, 58, 82, 87, 88, 213, 219, 275
Land clearance
  association with peregrine species, 161-163
  effect on earthworms, 250, 278-282
  successions of species, 280-281
Larix sp., 325
Leptospermum spp., 280
Light
  cause of death after heavy rain, 54
  limits of tolerance, 53
  reaction to, 53
Lignin, digestion of, 30-32
Litter, *see* Plant litter
Litter species, 103-131 *passim*, 161
  characteristics, 103-105, 108
  relation to morpho-ecological groups, 106-107
  see also Epigées
Locomotion
  coordination, 6-7
  hydraulic pressures, 8
  mechanics, 6-8, 108-109
Logs, as earthworm habitat, 103, 107
Lolium perenne, 203
Longevity, 75-76, 78, 79, 81-82, 83, 116
Lumbricus castaneus, 19, 23, 49, 57, 58, 59, 63, 68, 69, 72, 83, 95, 99, 100, 126-129, 139, 164, 205, 269, 279, 324, 336
Lumbricus festivus, 82, 205, 248, 324
Lumbricus polyphemus, 22
Lumbricus rubellus, 14-15, 19, 20, 23, 24, 26, 28, 29, 41, 45, 57, 58, 59, 63, 68, 86, 101, 115, 147, 161, 162, 164, 188, 190, 202, 204, 205, 206, 209, 211, 222, 223, 227, 231, 232, 235, 236, 238, 248, 250, 252, 258, 261, 264, 265, 266, 270-271, 273, 279, 280, 281, 287, 289, 307, 309, 323, 324, 326, 348
Lumbricus terrestris, 8, 10, 11, 14, 19-20, 23, 24, 25, 26, 27, 29, 30, 31-32, 33, 34-35, 39, 41, 48, 49, 50, 53, 55, 56-57, 58-59, 60-61, 63, 74, 75-76, 83, 88, 95, 99, 115, 126-129, 130, 141, 143, 148, 149, 152, 153, 163, 175, 182, 194, 196, 197, 198, 202, 203, 205, 212, 214, 215, 217, 219, 220, 221, 226, 230, 231, 236, 237, 238, 239, 240, 242,

Lumbricus terrestris—continued
  245–246, 247, 248, 249, 258, 260, 261, 267, 268, 269, 270, 271, 272, 273, 274, 275, 277, 279, 280, 284, 287, 289, 290, 307, 310, 311, 315, 328, 340, 341, 342, 343, 346, 348

M

Macroporosity, significance of earthworms, 184–196
Madhuca spp., 306
Malabaria paludicola, 276
Maoridrilus montanus, 94
Maoridrilus ruber, 161, 280
Mark-recapture methods, 345–346
Mass migration, 155
Megascolex exiguus, 257
Megascolex insignis, 275
Megascolex macleayi, 257
Megascolides australis, 8, 195
Metapheretima elongata, 276
Metapheretima jocchana, 174, 255
Metaphire californica, 34, 36, 43, 61, 164, 215, 217, 251, 290
Metaphire posthuma, 214, 217, 242, 268, 272
Metastrongylus sp., 152
Microchaetus sp., 9, 253
Microorganisms
  associated with burrows, 208–209
  in casts, 177–178, 206–208
  as food for earthworms, 23–26, 31, 220
  nitrogen fixers in gut, 220–221
  pathogenic to earthworms, 150, 153, 305
  pathogens carried by earthworms, 150–153
  resident in gut, 23, 26–29
  survival after ingestion, 24
  symbiosis with earthworms, 26–29
  synergistic relationships, 26–29
Microrelief associated with earthworms, 253–255
Microscolex aucklandicus, 65
Microscolex campbellianus, 65
Microscolex dubius, 38, 157, 164, 257, 287
Microscolex kerguelarum, 60
Microscolex modestus, 37, 38, 40, 43
Microscolex phosphoreus, 157, 164, 342
Millsonia anomala, 15, 20–21, 27, 28, 38–39, 76, 83–84, 112, 116–119, 123, 126, 130–131, 132–133, 139

Millsonia ghanensis, 20, 21, 22, 76, 114, 116–119, 123, 126, 130–131
Millsonia lamtoiana, 21, 76, 78, 81, 112, 116–119, 123, 126, 130–131
Millsonia spp., 336
Models
  of earthworm effects on infiltration, 192–194
  of earthworm populations, 131–134
Monocystis spp., 151, 230
Mor, 17–18, 245
Morpho-ecological groups of earthworms
  in European species, 107–115 *passim*
  in non-European species, 108, 112–114
  relation to litter/topsoil/subsoil groups, 107–108, 109–112, 115
  see also Anéciques, Endogées, Epigées
Mortality rates, 73, 78, 79, 83–84, 217–218, 242–243
Mucor hiemalis, 24, 25
Mucus
  composition, 35
  energy use in production, 142
  epidermal, 35, 243
  heavy metal absorption by, 229
  intestinal, 28, 35
  lining burrows, 184, 198–199
  as lubricant, 28, 35
  nitrogen in, 211, 213–217
  quantity produced, 28
  regulating ionic balance, 61
  as respiratory film, 35
Mulching
  effects on earthworm abundance, 190–191, 286–288
  protection of earthworms from freezing, 264
Mull, 17–18, 245
Murchieona minuscula, 100–101, 126–129
Muscle layers, 4, 5, 6

N

Nematicides, effects on earthworms, 304–305
  see also Biocides
Nematodes
  as food for earthworms, 29, 275
  as parasites in earthworms, 151–152
  viability after ingestion, 275
Neodrilus polycystis, 7

Nephridia
  function, 213-217
  types related to ecological strategies, 109
Nereis diversicolor, 66
Niche partitioning, as ecological strategy, 129-131
Nitrogen
  in casts, 207, 209-211, 212-213
  in dead tissue, 211, 217-219
  fixation, 220-221
  flow through earthworms, 142-145, 214-217, 219-220
  limiting earthworm populations, 5, 145-146, 247
  mineralization, 207
  in mucoproteins, 211, 213-217
  ^{15}N, use as tracer, 143-145
  transformations, 211-221
  in urine, 211, 213-217
  in vermicompost, 322
Notoscolex montikosciuskoi, 281
Nutrient cycling
  fast and slow cycling, 144
  nitrogen, 143-145, 211-220
  phosphorus, 222-223
  relationship to energy used, 263

O

Oceanic islands, earthworms of, 60, 65, 66, 156, 160, 161, 169
Ocnerodrilus occidentalis, 157, 160, 164
Octochaetona surensis, 82-83, 87
Octochaetus multiporus, 14, 95, 109, 161, 162, 176, 198, 265, 280
Octolasion cyaneum, 42, 49, 82, 100, 126-129, 139, 162, 174, 261, 279, 280, 281, 284, 289, 310
Octolasion lacteum, 48, 49, 70, 86, 101, 242, 279
Octolasion tyrtaeum, 30, 221
Ojk indices, 129-131
Orchards, earthworms of, 190-191, 269-270
Ordination, Mountford's method, 99
Organic matter, *see* Soil organic matter
Organic-mineral particle complexes
  in casts, 173
  rupturing by earthworms, 177
Osmotic pressure of body fluids
  effects of water loss, 35
  interspecific differences, 36
  modification for extreme cold, 44
  regulatory mechanisms, 60-61, 66, 225-227
Oxidation-reduction potential
  definition, 64
  effects on earthworms, 63-64
  of sewage sludge, 64
  of soils, 64
Oxygen consumption, 45, 48-50
Oxygen exchange
  mechanisms, 47-52
  role of erythrocruorin, 48
  see also Respiration

P

Palatability of food, 31-32
Para-diapause, *see* Seasonal rhythms of activity
Paramecium sp., 273
Parasites of earthworms, 150-153, 329-330
Paratylenchus sp., 275
Parthenogenesis, 68, 164-165
Particle size, selection by earthworms, 175-177
Pathogens
  of earthworms, 150-153
  earthworms as carriers, 151-153, 274-275
  viability after ingestion, 275
Pastures
  abundance of earthworms, 264-265, 280
  botanical composition, changes, 264-265, 266
  production increase due to earthworms, 264-266
Peat lands, species associations in, 94
Pedogenesis
  casts, 173-182
  plant litter incorporation, 244-245
  role of earthworms in, 103, 244-255
Penicillium digitatum, 25
Penicillium sp., 24
Peregrine earthworms
  association with humans, 156, 165
  dispersal, 157-163
  parthenogenesis, 164-165
  population growth, 161-163
  r-, K-, A-selection, 122
  species, 157
  of temperate regions, 157-159
  of tropical regions, 157, 159-160
Perelia diplotetratheca, 115

Perelia kaznekovi, 289
Peromyscus sp., 240
Perionyx excavatus, 23, 58, 82
Perionyx helophilus, 65
Perionyx sp., 155, 276
pF, relation to earthworm activity, 188
pH
 effects on earthworm abundance, 58, 169, 247
 of fertilizers, effect on earthworms, 290–291
 physiological effects on earthworms, 58–60
 relation to electrolytes, 59
 soil, effect of earthworms, 224, 325
Phasianus colchicus, 314
Phenolic compounds, digestion of, 30–32
Pheretima s.l., 159–160
Pholéophiles, 107
Phosphatase
 in casts, 28, 208–209, 223
 in earthworm gut, 27–28
 microbial origin, 28
 pH optimum, 28
Phosphorus
 availability to plants, 222–223, 291
 in casts, 222–223
Photoreceptors, 11
Pithomyces chartarum, 24, 311
Plant growth
 promotion by earthworms, 256–273, 322
 reduction due to earthworms, 179, 276
 significance of burrows, 184, 268
 stimulants produced by earthworms, 270–272
Plant introduction, association with peregrine earthworms, 157, 159, 160
Plant litter
 calorific value, 21
 as food for earthworms, 19
 incorporation into soil, 244–245, 248–289, 256–261, 269, 270
 nitrogen content, 219–220
 quantities consumed, 21, 245
Plant pests, earthworms as, 276–277
Plant roots
 as food for earthworms, 19
 stimulation by earthworms, 268, 269, 270–272
Plutellus attenuatus, 257
Pluvialis apricaria, 147

Poa pratensis, 274
Poa trivialis, 274
Pollenia rudis, 151
Polyploidy, 164–165
Pontodrilus lacustris, 65
Pontodrilus matsushimensis, 66
Pontoscolex corethrurus, 28, 58, 149, 157, 160, 164, 227, 272
Poronia piliformis, 25
Porosity of soils, effect of earthworms on, 184–196
Porosphere, 13
Predators of earthworms, 147–150
 biocide accumulation, 313–314
 relation to ecological strategies, 102–111 *passim*
Production, of earthworms
 definition, 135
 efficiency, 138–139, 144–145
 rates, 138–139, 142, 316–319
 use in demographic profiles, 116–119
Production, primary
 index of utilization, 142
 proportion used by earthworms, 140–142 *passim*
Proprioceptors, 11
Protein
 content of earthworms, 219, 316
 economics of production, 329
 nitrogen content, 219
 production, 68–71, 73, 328–330
Protozoa, as food for earthworms, 19, 25–26, 31
Pseudomonas aeruginosa, 24
Pseudomonas fluorescens, 24
Pythium aphanidermatum, 275

Q

Q_{10}, 40, 138–139
 definition, 135
Quiescence, *see* Seasonal rhythms of activity

R

Radioactive elements in environment
 earthworm analysis to monitor, 243
 mortality, 242–243
 plant uptake, influence of earthworms, 242
 uptake, 241–242

Rain, as dispersal agent, 155
REAL model, 134
Redox potential, see Oxidation-reduction potential
Reproduction
  cocoon production, 68-71
  incubation time, 71-72, 78
  reproductive strategies, 67-68
  uniparental parthenogenesis, 68
Resource limitation of populations, nitrogen, 145-146
Resource utilization, 129-131
Respiration
  aerobic, 47-52
  anaerobic, 52-53
  caudal, 50-52, 254
  definition, in energetic terms, 135
  effect of carbon dioxide content of air, 50
  interspecific and intraspecific variation, 48-52
  physiology of, 47-53
  Q_{10}, 40
  rates, 138-139, 142
  relation to soil water, 33-34
  relation to temperature, 40, 44
  role of calciferous glands, 50, 226, 227-228
  R.Q., 135
  use in demographic profiles, 116-119
Respiratory quotient (R.Q.), 135
Respiratory system, 47-53
Rhododrilus aquaticus, 65
Rhododrilus cockaynei, 65
Rhododrilus leptomerus, 65
Rhododrilus similis, 162, 280, 281
r-, K-, A-Selection categories, 119-129 *passim*
  definitions, 120-121
  relation to feeding habits, 121
  relation to litter/topsoil/subsoil groups, 121-122, 123
  relation to morpho-ecological groups, 121-122
Root mat, incorporation by earthworms, 248-249, 256-261
Roots, see Plant roots
r-Selection, see r-, K-, A- Selection strategies

S
Saccamoeba stagnicola, 26
Saccharomyces cerevisiae, 273
Salinity
  effect on hatching of cocoons, 72
  euryhaline species, 65-66
  limiting earthworms, 60-61, 317
  tolerance, 61-63
  of vermicompost, 321
Salmo gairdneri, 35
Salmonella enteridis, 153
Saprorhizophages, 108
Satchellius mammalis, 19, 57, 58, 63, 95, 99, 100, 340
Scherotheca (Opothedrilus) occidentalis thibauti, 9
Seasonal rhythms of activity, 69, 71, 76, 77, 81-82, 85-88, 109
Seeds
  burial by earthworms, 273
  as food for earthworms, 19, 274, 277
  viability after ingestion, 274, 277
Sense organs, 10-11
Septa, 6-7
Serratia marcescens, 24
Setae, 6-7
Sewage sludge
  oxidation-reduction potential, 64
  processing by earthworms, 315-327
  toxicity, 321
Similarity coefficients, 95
Size of earthworms
  meridional trend, 94-95
  range, 8-9
  relation to cocoon size, 70, 73
  relation to demographic profiles, 116
  relation to longevity, 81
  relation to metabolic rate, 139
  relation to species diversity in ecosystems, 94
Soil aggregates formed by earthworms
  casts, 173-182, 209-211, 212-213
  microbial activity in, 177-178
  pedological significance, 248, 249
  related to tillage practices, 283
  stability, 177-179, 211, 269
Soil fertility, effects of earthworms, 263-270
  biological fertility, 263
  inherent mineralogical fertility, 263

Soil moisture
  effects on biomass, 166–167
  effects of earthworms on, 188–195
  effect on oxidation–reduction potential, 64
  effect on seasonal rhythms, 85
  effect on temperature preferences, 46
  optima, 37, 38–39
  related to earthworm activity, 13, 34
  see also Water content of earthworms
Soil organic matter
  carbon:nitrogen ratio, 56–58, 209–211
  decomposition in casts, 206–208
  decomposition by earthworms, 205–206, 209–211
  lining earthworm burrows, 198–199
  and oxidation–reduction potentials, 63–64
  physical nature, in casts, 205
  rates of incorporation, 202–204
  see also Food; Mor; Mull
Soil profile
  description, 19
  effects of earthworms on, 244–251
Soil temperature
  effects on assimilation rates, 136–140 passim
  effects on earthworm behaviour, 44, 85, 86
  limits for earthworms, 39
  optimum for earthworms, 41–43
  spatio-temporal variability, 39
Soil texture
  effect on abundance, 54–55
  effect on biomass, 166–167
Spatial relationships in earthworm communities
  vertical stratification, 102–114
Stability of communities, relation to complexity, 95
Species associations
  altitudinal effects on, 166–169
  comparison with other soil animals, 94
  connective stability, 95
  ecological fullness, 101
  meridional effects, 94
  methodology, 95, 99–101
  numbers of species, 94–102 passim
  relation to ecological strategies, 102–131
  relation to habitat variability, 126–129, 130–131, 266, 279–281
  relation to mulching, 288
  relation to vegetation types, 94, 99–101
  significance, 95, 99–101
  successions of, 280–281
Stonehenge, burial of stones, 180–181
Stones
  burial, 180–181
  earthworms living under, 103
Straminicoles, 107
Streams, as dispersal agent, 155
Streptomyces lipmanii, 26–27
Stuhlmannia porifera, 29, 130–131
Subsoil species, 103–131 *passim*, 161
  altitudinal effects on, 166–167
  characteristics, 103–105, 109
  relation to morpho-ecological groups, 107–108
  relation to soil properties, 166
  see also Endogées
Surface runoff, effects of earthworms, 192–194, 253
Surface:volume ratio, 9, 348–349
Symbiotic organisms, 26–29

T

Tactile sense organs, 11
Temperature, 39–47
  influence on cocoon production, 70–71
  see also Ambient temperature; Body temperature; Soil temperature; Temperature limits; Temperature preferences
Temperature limits
  effects of acclimatization, 44–45
  lower lethal limits, 44
  LT_{50}, definition, 40
  upper lethal limits, 40, 44
  for various species, 41–43
  see also Temperature preferences
Temperature preferences
  acclimatization, 44–45
  intraspecific variability, 45–46
  of Lumbricidae, 45–46
  methods of estimation, 45
  relation to differences in cuticle, 35
  relation to soil moisture, 46
  temperate zone species, 45–46
  tropical zone species, 46
  see also Temperature limits

Subject Index **411**

Termites, compared with earthworms, assimilation rates, 141
Thecamoeba sp., 26
Tillage practices
 effect on earthworm populations, 187, 190, 283-286
 effect on numbers of burrows, 186-187
Tilletia controversa, 275
Topsoil species, 103-131 *passim*, 161
 altitudinal effects on, 166-167
 characteristics, 103-105, 108
 relation to morpho-ecological groups, 106-107
 relation to soil properties, 166
 see also Anéciques
Toxicity
 of aromatic plant constituents, 32
 of biocides, 292-307
 of heavy metals, 228, 232-233, 235, 239
 initiating quiescence, 85
 of organic wastes, 23, 316-317, 321
Trapping methods, 154, 344-345
Trichoderma viride, 25
Turdus merula, 314
Turdus migratorius, 313
Turdus philomelos, 314

U

Ulex europaeus, 248, 280
Urine
 ammonia-urea balance in, 214-217
 composition, 213-217
 discharge into gut, 213, 215, 251
 nitrogen in, 211, 213-217
 rate of production, 214-217
 see also Water loss by earthworms

V

Vascular system, 9-10
Vermicides, effects on earthworms, 306
 see also Biocides
Vermicompost, 315-327
Vermisols, 246
Vertical stratification of earthworm communities, 102-114

 relation to soil profile morphology, 103-107
Vibrio sp., 27
Vigna unguiculata, 285
Vitamin production associated with earthworms, 272-273

W

Washing/sieving methods, 337-338
Waste disposal by earthworms
 disease hazards, 329-330
 economics, 321-322
 heavy metal contamination, 322, 330
 methods, 315-316, 320-327
 suitable species, 71, 73, 316-319
Water content of earthworms, 33
 regulatory mechanisms, 60, 214-217
 see also Water loss by earthworms; Water uptake by earthworms; Soil moisture
Water-holding capacity of soils, effects of earthworms, 188, 249
Water infiltration, effect of earthworms, 186, 188-195, 249, 252-253, 285, 287
Water loss by earthworms
 effect on osmotic pressures, 35
 physiological control of rate, 35, 37
 rates, 33, 34-35, 214-217, 251
 regulation, 37
 survival limits, 34-35, 36-37
 survival in resting stages, 36
Water uptake by earthworms
 absorption from air, 34-35
 ingestion of free water, 34-35
 rate related to temperature, 40
 resorption from gut, 60
"Withdrawal reflex", 53, 108
Worm.For Model, 133-134

X

Xenopus laevis, 240

Z

Zaglossus bruijnii, 149-150